Endocrine FGFs and Klothos

ADVANCES IN EXPERIMENTAL MEDICINE AND BIOLOGY

Editorial Board:

NATHAN BACK, *State University of New York at Buffalo*

IRUN R. COHEN, *The Weizmann Institute of Science*

ABEL LAJTHA, *N.S. Kline Institute for Psychiatric Research*

JOHN D. LAMBRIS, *University of Pennsylvania*

RODOLFO PAOLETTI, *University of Milan*

Recent Volumes in this Series

Volume 720
HUMAN CELL TRANSFORMATION
Johng S. Rhim and Richard Kremer

Volume 721
SPHINGOLIPIDS AND METABOLIC DISEASE
L. Ashley Cowart

Volume 722
RNA INFRASTRUCTURE AND NETWORKS
Lesley J. Collins

Volume 723
RETINAL DEGENERATIVE DISEASES
Matthew M. LaVail, Joe G. Hollyfield, Robert E. Anderson, Christian Grimm
and John D. Ash

Volume 724
NEURODEGENERATIVE DISEASES
Shamim I. Ahmad

Volume 725
FUZZINESS: STRUCTURAL DISORDER IN PROTEIN COMPLEXES
Monika Fuxreiter and Peter Tompa

Volume 726
VIRAL MOLECULAR MACHINES
Michael G. Rossman

Volume 727
NOTCH SIGNALING IN EMBRYOLOGY AND CANCER
Jörg Reichrath and Sandra Reichrath

Volume 728
ENDOCRINE FGFS AND KLOTHOS
Makoto Kuro-o

A Continuation Order Plan is available for this series. A continuation order will bring delivery of each new volume immediately upon publication. Volumes are billed only upon actual shipment. For further information please contact the publisher.

Endocrine FGFs and Klothos

Edited by

Makoto Kuro-o, MD, PhD
*Department of Pathology, University of Texas Southwestern Medical Center at Dallas,
Dallas, Texas, USA*

Springer Science+Business Media, LLC

Landes Bioscience

Springer Science+Business Media, LLC
Landes Bioscience

Copyright ©2012 Landes Bioscience and Springer Science+Business Media, LLC

All rights reserved.
No part of this book may be reproduced or transmitted in any form or by any means, electronic or mechanical, including photocopy, recording, or any information storage and retrieval system, without permission in writing from the publisher, with the exception of any material supplied specifically for the purpose of being entered and executed on a computer system; for exclusive use by the Purchaser of the work.

Printed in the USA.

Springer Science+Business Media, LLC, 233 Spring Street, New York, New York 10013, USA
http://www.springer.com

Please address all inquiries to the publishers:
Landes Bioscience, 1806 Rio Grande, Austin, Texas 78701, USA
Phone: 512/ 637 6050; FAX: 512/ 637 6079
http://www.landesbioscience.com

The chapters in this book are available in the Madame Curie Bioscience Database.
http://www.landesbioscience.com/curie

Endocrine FGFs and Klothos, edited by Makoto Kuro-o. Landes Bioscience / Springer Science+Business Media, LLC dual imprint / Springer series: Advances in Experimental Medicine and Biology.

ISBN: 978-1-4614-0886-4

While the authors, editors and publisher believe that drug selection and dosage and the specifications and usage of equipment and devices, as set forth in this book, are in accord with current recommendations and practice at the time of publication, they make no warranty, expressed or implied, with respect to material described in this book. In view of the ongoing research, equipment development, changes in governmental regulations and the rapid accumulation of information relating to the biomedical sciences, the reader is urged to carefully review and evaluate the information provided herein.

Library of Congress Cataloging-in-Publication Data

Endocrine FGFs and klothos / edited by Makoto Kuro-o.
    p. ; cm. -- (Advances in experimental medicine and biology ; 728)
 Includes bibliographical references and index.
 ISBN 978-1-4614-0886-4 (hardback)
    I. Kuro-o, Makoto, 1960- II. Series: Advances in experimental medicine and biology ; 728. 0065-2598
 [DNLM: 1. Fibroblast Growth Factors--metabolism. 2. Endocrine System--metabolism. 3. Glucuronidase--metabolism. 4. Receptors, Fibroblast Growth Factor. W1 AD559 v.728 2011 / QU 107]
 LC classification not assigned
 573.4--dc23
                              2011026182

PREFACE

Fibroblast growth factors (FGFs) have been recognized primarily as autocrine/paracrine factors that regulate embryonic development and organogenesis. However, recent studies have revealed that some FGFs function as endocrine factors and regulate various metabolic processes in adulthood. Such FGFs, collectively called endocrine FGFs, are comprised of three members (FGF15/19, FGF21, and FGF23: FGF15 is the mouse ortholog of human FGF19). These endocrine FGFs share a common structural feature that enables the endocrine mode of action at the expense of the affinity to FGF receptors (Chapter 1). To restore the affinity to FGF receptors in their target organs, the endocrine FGFs have designated the Klotho family of transmembrane proteins as obligate co-receptors (Chapter 2). By expressing Klothos in a tissue-specific manner, this unique co-receptor system also enables the endocrine FGFs to specify their target organs among many tissues that express FGF receptors.

The first documented endocrine axis mediated by endocrine FGFs and Klothos was the FGF23-Klotho system. FGF23 was identified as the gene mutated in patients with a hereditary phosphate-wasting syndrome (Chapter 3). Functions of FGF23 as a phosphaturic hormone and a counter-regulatory hormone for vitamin D and parathyroid hormone are comprehensively discussed in Chapters 4 and 6. On the other hand, Klotho was originally identified as the gene mutated in mice that inherited a syndrome resembling human aging associated with abnormal mineral metabolism (Chapter 2). The close relation between these two seemingly unrelated proteins (FGF23 and Klotho) became evident when mice lacking FGF23 developed not only abnormal mineral metabolism but also complex aging-like phenotypes, which are indistinguishable from those seen in mice lacking Klotho (Chapter 5). In addition to functioning as FGF23 receptor, Klotho has multiple FGF23-independent activities, which are discussed in Chapters 7 and 9. An exponential increase in research interests in the FGF23-Klotho endocrine axis has been observed since their defects turned out to be universal in patients with chronic kidney disease, which affects more than 10% of total population in developed countries. Chapters 8, 9 and 10 discuss potential of FGF23 and Klotho as novel diagnostic markers and therapeutic targets for chronic kidney disease.

Chapters 11, 12, 13, and 14 overview our current knowledge on FGF19 and FGF21. Unlike FGF23, FGF19 and FGF21 require βKlotho, another member of the Klotho family proteins, as their obligate co-receptor. Although their involvement in human diseases remains to be clarified, FGF19 and FGF21 are hormones indispensable for metabolic adaptation to feeding and fasting, respectively. In addition, FGF19 is essential for maintaining bile acid homeostasis and mitogenic. Thus, FGF19 and FGF21 likely participate in pathophysiology of obesity, diabetes, hypercholesterolemia, gallstone and cancer and possess potential for novel diagnostic markers and therapeutic targets for these highly prevalent diseases.

It was a pleasant surprise to know that such major endocrine axes had remained unrevealed until quite recently. *Endocrine FGFs and Klothos* will add a new massive chapter to endocrinology textbooks. I hope this volume help stimulate research interests in various fields of biomedical science and promote better understanding of these unique endocrine systems.

Makoto Kuro-o, MD, PhD
Department of Pathology
University of Texas Southwestern Medical Center at Dallas
Dallas, Texas, USA

ABOUT THE EDITOR...

MAKOTO KURO-O is an Associate Professor in the Pathology Department at the University of Texas Southwestern Medical Center at Dallas (UT Southwestern). He also has an appointment to the Pak Center for Mineral Metabolism and the Simmons Comprehensive Cancer Center in UT Southwestern. He received his MD and PhD from the University of Tokyo, Japan. His major research interests include molecular mechanisms of aging and age-related diseases with special reference to chronic kidney disease. He is a member of the American Society of Nephrology (ASN).

PARTICIPANTS

Andrew Beenken
Department of Pharmacology
New York University School of Medicine
New York, New York
USA

Clemens Bergwitz
Endocrine Unit
Massachusetts General Hospital
and Harvard Medical School
Boston, Massachusetts
USA

Luc R. Desnoyers
Oncology Biomarker Development/
Pharmacodynamic Biomarkers
Genentech, Inc.
South San Francisco, California
USA

James D. Dunbar
Eli Lilly and Company
Lilly Corporate Center
Indianapolis, Indiana
USA

Seiji Fukumoto
Japan Division of Nephrology
and Endocrinology
Department of Medicine
University of Tokyo Hospital
Tokyo
Japan

Ming Chang Hu
Charles and Jane Pak Center for Mineral
Metabolism and Clinical Research
Departments of Internal Medicine,
Physiology, Pediatrics and Pathology
University of Texas Southwestern
Medical Center
Dallas, Texas
USA

Chou-Long Huang
Department of Medicine
UT Southwestern Medical Center
Dallas, Texas
USA

Stacey A. Jones
Drug Discovery
Research and Development
GlaxoSmithKline
Research Triangle Park, North Carolina
USA

Harald Jüppner
Endocrine Unit
and
Pediatric Nephrology Unit
Massachusetts General Hospital
and Harvard Medical School
Boston, Massachusetts
USA

Alexei Kharitonenkov
Eli Lilly and Company
Lilly Corporate Center
Indianapolis, Indiana
USA

Makoto Kuro-o
Department of Pathology
University of Texas Southwestern
  Medical Center at Dallas
Dallas, Texas
USA

Yang Li
Amgen, Inc
South San Francisco, California
USA

Benjamin C. Lin
Oncology Biomarker Development/
  Pharmacodynamic Biomarkers
Genentech, Inc.
South San Francisco, California
USA

Aline Martin
University of Tennessee Health
  Science Center
Department of Medicine
Memphis, Tennessee
USA

Orson W. Moe
Charles and Jane Pak Center for Mineral
  Metabolism and Clinical Research
Departments of Internal Medicine,
  Physiology, Pediatrics and Pathology
University of Texas Southwestern
  Medical Center
Dallas, Texas
USA

Moosa Mohammadi
Department of Pharmacology
New York University School of Medicine
New York, New York
USA

Tally Naveh-Many
Minerva Center for Calcium
  and Bone Metabolism
Nephrology Services
Hadassah Hebrew University
  Medical Center
Jerusalem
Israel

L. Darryl Quarles
University of Tennessee Health
  Science Center
Department of Medicine
Memphis, Tennessee
USA

M. Shawkat Razzaque
Department of Oral Medicine, Infection
  and Immunity
Harvard School of Dental Medicine
Boston, Massachusetts
USA
and
Department of Pathology
Nagasaki University School
  of Biomedical Science
Nagasaki
Japan

Takashi Shimada
Innovative Drug Research Laboratories
Kyowa Hakko Kirin Co., Ltd,
Tokyo
Japan

Justin Silver
Minerva Center for Calcium
  and Bone Metabolism
Nephrology Services
Hadassah Hebrew University
  Medical Center
Jerusalem
Israel

Patricia Wahl
Division of Nephrology and Hypertension
University of Miami Miller School
  of Medicine
Miami, Florida
USA

PARTICIPANTS

Myles Wolf
Division of Nephrology and Hypertension
University of Miami Miller School
    of Medicine
Miami, Florida
USA

Xinle Wu
Amgen, Inc
South San Francisco, California
USA

Yang Zhao
Eli Lilly and Company
Lilly Corporate Center
Indianapolis, Indiana
USA

CONTENTS

1. THE STRUCTURAL BIOLOGY OF THE FGF19 SUBFAMILY 1

Andrew Beenken and Moosa Mohammadi

Abstract.. 1
Introduction: An Overview of FGF-FGFR Signaling.. 2
HS Modulates FGF Signaling through Multiple Mechanisms........................... 3
Paracrine FGFs Mediate a Mesenchymal-Epithelial Signaling Loop
    within Tissues 5
Endocrine FGFs Regulate Key Metabolic Processes
    in a Klotho-Dependent Fashion 7
The Molecular Basis for the FGF19 Subfamily's Klotho Coreceptor Requirement............ 14
Structural Basis for the Specificity and Affinity of Endocrine FGFs' Binding
    to FGFR 16
Characterization of the Binding of FGF23 To αKlotho 17
Pharmacological Impact of Structural Insights into FGF19 Subfamily............ 18
Conclusion 18

2. KLOTHO AND βKLOTHO 25

Makoto Kuro-o

Abstract.. 25
Introduction... 25
Phenotypes of *KLOTHO*$^{-/-}$ mice 26
Klotho Protein Function.. 29
Phosphate Homeostasis 30
A Potential Link between Phosphate and Aging 31
Secreted Klotho 32
βKlotho.. 33
KLPH 35
Nuclear Receptor Regulation of Endocrine FGFs 35
Conclusion 36

3. FGF23 AND SYNDROMES OF ABNORMAL RENAL PHOSPHATE HANDLING 41

Clemens Bergwitz and Harald Jüppner

Abstract.. 41
Introduction... 42
Autosomal Dominant Hypophosphatemic Rickets (ADHR, OMIM 193100)...................... 42
Other Disorders of Renal Phosphate Excretion 48
Inherited Hypophosphatemic Disorders other than ADHR 49
Hyperphosphatemic Disorders 54
Conclusion 57

4. EVIDENCE FOR FGF23 INVOLVEMENT IN A BONE-KIDNEY AXIS REGULATING BONE MINERALIZATION AND SYSTEMIC PHOSPHATE AND VITAMIN D HOMEOSTASIS..............65

Aline Martin and L. Darryl Quarles

Abstract.. 65
Introduction... 65
Evolving Insights into Skeletal Functions... 67
Genetic Disorders Define FGF23 Function... 70
Physiological Functions of FGF23 72
Conclusion 78

5. FGF23, KLOTHO AND VITAMIN D INTERACTIONS: WHAT HAVE WE LEARNED FROM IN VIVO MOUSE GENETICS STUDIES?..84

M. Shawkat Razzaque

Abstract.. 84
Introduction... 84
FGF23.. 85
FGF23 and CKD 85
Klotho... 86
FGF23 Signaling.. 86
In Vivo Importance of Klotho in FGF23 Function 87
Vitamin D and FGF23-Klotho Axis 87
Conclusion 88

6. FGF23 AND THE PARATHYROID 92

Justin Silver and Tally Naveh-Many

Abstract.. 92
Introduction... 92
Effect of FGF23 on the Parathyroid... 93
Expression of Klotho in the Parathyroid 93

CONTENTS

Functional Effect of FGF23 on PTH Expression 94
FGF23 Activates MAPK in the Parathyroid 94
FGF23 Acts on Bovine Parathyroid Cells to Decrease PTH and Increase
    25(OH) Vitamin D 1α Hydroxylase Expression 95
FGF23 Decreases Serum PTH Levels in Transgenic Mice Expressing
    the Human PTH Gene 95
FGF23 in Chronic Kidney Disease (CKD).. 95
The Mechanism of the Resistance to FGF23 in the Parathyroid............................... 95
The Mechanism of the Increased FGF23 Levels in CKD.. 96
Conclusion 97

7. REGULATION OF ION CHANNELS BY SECRETED KLOTHO 100

Chou-Long Huang

Abstract.. 100
Introduction.. 100
Homology with the Family 1 Glycosidases and the Glucuronidase Activity
    of Klotho 101
Sialidase Activity and Regulation of Ion Channels by Secreted Klotho 101
Control of the Resident Time of Membrane Glycoproteins
    via Galectin-N-Glycan Interaction 103
Regulation of Ion Channels by Secreted Klotho via other Mechanisms............... 104
Role of Regulation of TRPV5 by Secreted Klotho in the Overall Anti-Aging
    Function of Klotho 104
Conclusion and Future Perspectives 105

8. FGF23 IN CHRONIC KIDNEY DISEASE 107

Patricia Wahl and Myles Wolf

Abstract.. 107
Introduction: Burden of Chronic Kidney Disease 107
Overview of Normal Mineral Metabolism and Its Disorders in CKD 110
Disordered Phosphorus Metabolism and Adverse Clinical Outcomes 116
Conclusion 121

9. SECRETED KLOTHO AND CHRONIC KIDNEY DISEASE 126

Ming Chang Hu, Makoto Kuro-o and Orson W. Moe

Abstract.. 126
Introduction.. 126
Kidney: The Major Site for Klotho Synthesis 127
Chronic Kidney Disease: A State of Klotho Deficiency 129
Role of Klotho Deficiency in Progression and Complications of CKD 136
Mechanisms of How Secreted Klotho Preserves Kidney Function
    and Ameliorates Complications in CKD.. 141
Value of Measuring Klotho in CKD 143
Potential Strategies to Increase Klotho Protein 145
Conclusion and Perspectives.. 149

10. FGF23 AS A NOVEL THERAPEUTIC TARGET 158

Takashi Shimada and Seiji Fukumoto

Abstract... 158
Introduction... 158
Bioactivities of Recombinant Protein of FGF23... 159
Hypophosphatemic Rickets/Osteomalacia Due to Excess Action of FGF23......................... 161
Inhibition of Bioactivity of Circulating FGF23... 162
The Pharmacological Effect of Neutralizing Antibodies to FGF23
    in Hypophosphatemic Mice... 165
The Possible Hormone Replacement Therapy of FGF23 in Tumoral Calcinosis 166
Conclusion 167

11. PHYSIOLOGY OF FGF15/19 171

Stacey A. Jones

Abstract... 171
Introduction... 171
Discovery of Mouse *Fgf15* and Human FGF19... 172
Signaling Mechanisms 173
*Fgf*19 and *Fgf15* are Orthologs.. 173
Regulation of Bile Acid Biosynthesis 174
Regulation of Gallbladder Filling.. 176
Chronic Diarrhea and Defective Ileal FGF19 Release 177
Metabolic Effects... 178
Conclusion and Future Directions.. 179

12. FGF19 AND CANCER 183

Benjamin C. Lin and Luc R. Desnoyers

Abstract... 183
Introduction... 183
FGF19/FGF15 185
FGF19 Binding Specificity: FGFR4 and the βKlotho Coreceptor 185
FGF19 and Cancer.. 186
Conclusion 192

CONTENTS xvii

13. UNDERSTANDING THE STRUCTURE-FUNCTION RELATIONSHIP BETWEEN FGF19 AND ITS MITOGENIC AND METABOLIC ACTIVITIES 195

Xinle Wu and Yang Li

Abstract..195
Introduction..195
FGF19 and Cancer...196
Structural Basis for the Interactions between FGF19 Subfamily Members
 and Their Receptors..198
Role of FGFR4 in Glucose Metabolism and Mitogenesis..............................207
Conclusion 210

14. FGF21 AS A THERAPEUTIC REAGENT 214

Yang Zhao, James D Dunbar and Alexei Kharitonenkov

Abstract..214
Introduction..214
FGF21 In Vitro Findings 217
FGF21 Genetic Models 220
FGF21 Chronic and Acute Administration In Vivo..221
FGF21 in Humans..223
Conclusion 224

INDEX..229

ACKNOWLEDGEMENTS

I thank Eva Riedmann at Landes Bioscience, who invited me to contribute a book on endocrine FGFs and Klothos. I thank Cynthia Conomos, Erin O'Brien and Daniel Olasky at Landes Bioscience, whose patience and editorial efforts made it possible to publish this book. I thank my wife Kumiko Kuro-o for her support and encouragement to keep my Klotho research going.

CHAPTER 1

THE STRUCTURAL BIOLOGY
OF THE FGF19 SUBFAMILY

Andrew Beenken and Moosa Mohammadi*

Department of Pharmacology, New York University School of Medicine, New York, New York, USA.
*Corresponding Author: Moosa Mohammadi—Email: Moosa.Mohammadi@nyumc.org

Abstract:	The ability of the Fibroblast Growth Factor (FGF) 19 subfamily to signal in an endocrine fashion sets this subfamily apart from the remaining five FGF subfamilies known for their paracrine functions during embryonic development. Compared to the members of paracrine FGF subfamiles, the three members of the FGF19 subfamily, namely FGF19, FGF21 and FGF23, have poor affinity for heparan sulfate (HS) and therefore can diffuse freely in the HS-rich extracellular matrix to enter into the bloodstream. In further contrast to paracrine FGFs, FGF19 subfamily members have unusually poor affinity for their cognate FGF receptors (FGFRs) and therefore cannot bind and activate them in a solely HS-dependent fashion. As a result, the FGF19 subfamily requires α/βklotho coreceptor proteins in order to bind, dimerize and activate their cognate FGFRs. This klotho-dependency also determines the tissue specificity of endocrine FGFs. Recent structural and biochemical studies have begun to shed light onto the molecular basis for the klotho-dependent endocrine mode of action of the FGF19 subfamily. Crystal structures of FGF19 and FGF23 show that the topology of the HS binding site (HBS) of FGF19 subfamily members deviates drastically from the common topology adopted by the paracrine FGFs. The distinct topologies of the HBS of FGF19 and FGF23 prevent HS from direct hydrogen bonding with the backbone atoms of the HBS of these ligands and accordingly decrease the HS binding affinity of this subfamily. Recent biochemical data reveal that the αklotho ectodomain binds avidly to the ectodomain of FGFR1c, the main cognate FGFR of FGF23, creating a de novo high affinity binding site for the C-terminal tail of FGF23. The isolated FGF23 C-terminus can be used to effectively inhibit the formation of the FGF23-FGFR1c-αklotho complex and alleviate hypophosphatemia in renal phosphate disorders due to elevated levels of FGF23.

Endocrine FGFs and Klothos, edited by Makoto Kuro-o.
©2012 Landes Bioscience and Springer Science+Business Media.

INTRODUCTION: AN OVERVIEW OF FGF-FGFR SIGNALING

The human fibroblast growth factors (FGFs) compose a family of secreted polypeptides that are encoded by 18 distinct genes (FGF1-FGF10 and FGF16-FGF23). FGFs play pleiotropic roles in human development and metabolism by binding and activating FGF receptor tyrosine kinases (FGFRs) that are encoded by four genes in humans (FGFR1-4).[1-3] Based on sequence homology and phylogeny, the eighteen mammalian FGFs are grouped into five paracrine subfamilies and one endocrine subfamily.[4-6] The paracrine subfamilies include the *FGF1* subfamily comprising FGF1, 2; the *FGF7* subfamily comprising FGF3, 7, 10, 22; the *FGF4* subfamily comprising FGF4, 5, 6; the *FGF8* subfamily comprising FGF8, 17, 18; and the *FGF9* subfamily comprising FGF9, 16, 20. The endocrine-acting *FGF19* subfamily comprises FGF19, 21, 23. The paracrine-acting FGF1, FGF4, FGF7, FGF8 and FGF9 subfamilies play essential roles in spermatogenesis,[7-9] mesoderm induction,[10,11] somitogenesis,[12-16] organogenesis,[17-20] and pattern formation,[21] whereas the FGF19 subfamily acts in an endocrine fashion to regulate major metabolic processes including glucose,[22] lipid, cholesterol and bile acid metabolism,[23-25] and serum phosphate/vitamin D homeostasis.[26] Based on sequence homology, four additional genes, namely FGF11-FGF14, have also been considered to be members of the FGF family. Functionally speaking however, FGF11-FGF14 are not bona fide FGFs as they remain intracellular and lack key residues necessary for binding to FGFR.[27-29]

FGFs share a core homology region of about 120 amino acids consisting of twelve antiparallel β-strands arranged into three sets of four-stranded β-sheets (Fig. 1A).[30-32] This globular core domain is flanked by N- and C-terminal regions that are highly divergent with respect to both length and sequence among FGFs, particularly across subfamilies.[3] Moreover, even within some subfamilies the sequence identity at the N-terminus can be rather limited. The sequence identity of the N-terminal regions of FGF4 and FGF6 is only 36% compared to 69% for their core regions and that of FGF9 and FGF20 is only 38% compared to 86% for their core regions. Comparison of the crystal structures of several paracrine FGFs bound to their cognate FGFRs has shown that the FGFR binding specificity/promiscuity profile of a given FGF is principally dictated by the primary sequence of its N-terminal region.[33-35] The structural data have begun to pinpoint the common primary and secondary structural elements within the N-termini of members of a given subfamily that explain their overlapping FGFR binding specificity/promiscuity profile.[3]

The prototypical FGFR is composed of three extracellular immunoglobulin domains (D1-D3), a transmembrane domain and an intracellular tyrosine kinase domain.[36] (Fig. 1B). Structural studies have shown that ligand binding requires both D2 and D3 domains.[34,37,38] The D1 domain and D1-D2 linker, that harbors the acid box (AB), are dispensable for ligand binding and in fact suppress FGF and HS binding affinity of the D2-D3 region.[39,40] The specificity of FGFR1-3 for ligand binding is modulated by alternative splicing of mutually exclusive 'b' and 'c' exons in the second half of the D3 domain.[41,42] This D3 alternative splicing event is tissue specific, with the b and c exons being preferentially used in epithelial and mesenchymal tissues respectively.[41,43] Importantly, the D3 alternative splicing event elaborates the number of principal FGFRs from four to seven: FGFR1b, FGFR1c, FGFR2b, FGFR2c, FGFR3b, FGFR3c, FGFR4. Structural studies have shown that D3 alternative splicing modulates the FGF binding specificity/promiscuity of FGFRs by switching the primary sequence of key ligand binding epitopes in D3.[35]

THE STRUCTURAL BIOLOGY OF THE FGF19 SUBFAMILY

HS MODULATES FGF SIGNALING THROUGH MULTIPLE MECHANISMS

A wealth of genetic studies in mice and flies as well as cell-based studies have established that FGF signaling requires HS.[44-48] HS is a highly sulfated linear polymer of alternating glucuronate (GlcA) and N-acetylglucosamine (GlcNAc) monosaccharides that undergoes heterogenous deacetylation, N-sulfation on GlcNAc, O-sulfation on both GlcA and GlcNAc, and epimerization on GlcA.[49] HS impinges on FGF signaling through multiple mechanisms, including coordination/stabilization of FGF-FGFR binding and dimerization,[50] control of FGF gradients in the extraceullular matrix (ECM),[51] and protection of FGFs against thermal instability and proteolytic degradation.[52,53]

All FGFs interact with HS, albeit with differing affinities.[54] The HBS of all FGFs is located within the core region and is composed of residues from the β1–β2 loop and the segment spanning the β10 to β12 strand. FGFRs also interact with HS via residues from the gA helix, the gA-βA' loop, the βA'-βB loop, and the βB strand. Structural studies have shown that HS promotes formation of a 2:2:2 FGF-FGFR-HS cell surface signaling unit in which each ligand binds both receptors in the complex and the two receptors additionally make contact with one another.[50,55] (Fig. 2A,B) Two HS molecules bind in a symmetric fashion to a positively-charged HS-binding cleft formed from the union of the HBS of the two FGFs and two FGFRs at the membrane-distal end of the dimer. By simultaneously engaging the HBS of both FGF and FGFR, HS stabilizes protein-protein contacts both within the 1:1 FGF-FGFR complex and between the two FGF-FGFR complexes in the dimer. In addition to promoting FGF-FGFR binding and dimerization, emerging data show that HS also controls the diffusion and morphogenetic gradients of paracrine FGFs in the extracellular matrix,[51] and that the HS affinity of a ligand ultimately determines whether that FGF acts in a paracrine or endocrine fashion.[56]

Dimerization of the extracellular domains of FGFRs juxtaposes the intracellular kinase domains, affording them with sufficient opportunity to trans-phosphorylate each other on the A-loop tyrosines. A-loop tyrosine phosphorylation increases the intrinsic kinase activity of FGFR kinase by stabilizing the active conformation of the kinase.[57] Activated kinases then further trans-phosphorylate each other on tyrosines within the C-tail, kinase insert and juxtamembrane regions.[58-60] The phosphorylated tyrosines in the C-tail and juxtamembrane regions of activated FGFR serve as recruitment sites for SH2 domains of PLCγ[61,62] and CRKL,[63] respectively. In the case of PLCγ, this recruitment serves two purposes: (i) it facilitates phosphorylation of PLCγ to increase its enzymatic activity, (ii) it brings PLCγ to the vicinity of its substrate PIP2 in the plasma membrane. Hydrolysis of PIP2 generates two second messengers: IP3 and DAG that stimulate Ca^{2+} release from intracellular stores and PKC activation respectively.[64-66] In contrast, CRKL is an adaptor protein that lacks intrinsic enzymatic activity. Recruitment of CRKL to the phosphorylated tyrosine in the juxtamembrane region of FGFR1 and FGFR2 leads to translocation of associated Rac1/Cdc42 to the plasma membrane which culminates in cytoskeletal reorganization and cell motility.[63] Lastly, activated FGFR phosphorylates FRS2α,[67] another adaptor protein that, unlike PLCγ and CRKL, associates constitutively (independently of receptor phosphorylation) with the juxtamembrane region of FGFR.[68-70] Phosphorylation of FRS2α by the activated FGFR generates docking sites for the SH2 domains of the adaptor protein GRB2[67] and the phosphatase Shp2,[71] leading to activation of the Ras-MAPK and PI3K-Akt pathways (Fig. 1C).[72]

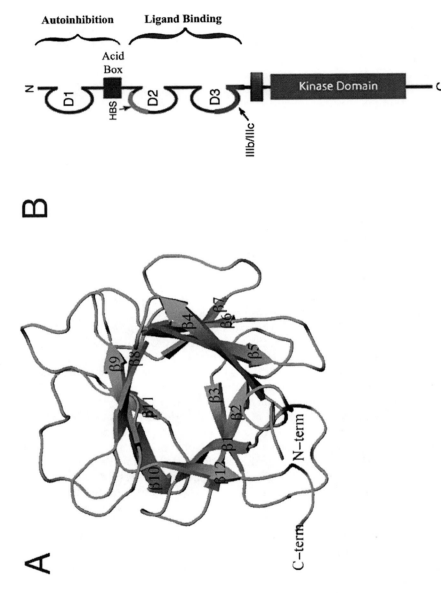

Figure 1. See figure legend on following page.

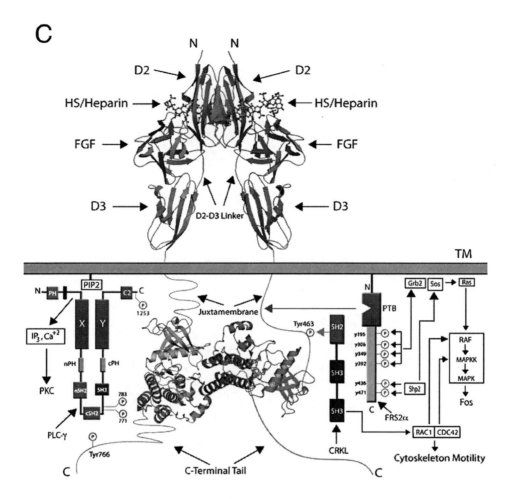

Figure 1. An overview of FGF and FGFR structural biology. A, viewed on previous page) FGF1 is represented as a cartoon. N- and C-termini are labeled and the twelve β-sheets that compose the β-trefoil core are indicated. B, viewed on previous page) A schematic of FGFR shows its three Ig-like domains. D1 and the acid box (AB) are involved in autoinhibition, the heparan sulfate binding site (HBS) is located on D2 and alternative splicing takes place in the latter half of D3. D2 and D3 are necessary and sufficient for ligand binding. An intracellular kinase domain mediates downstream signaling. C) The formation of a 2:2:2 FGF:FGFR:HS dimer on the cell surface leads to intracellular transphosphorylation of the FGFR kinase domains and downstream signaling through PLCγ, FRS2α, and CRKL.

PARACRINE FGFs MEDIATE A MESENCHYMAL-EPITHELIAL SIGNALING LOOP WITHIN TISSUES

Historically, FGFs have been viewed as paracrine factors known for their wide ranging roles in tissue patterning and organogenesis during embryonic development and the FGF1, FGF4, FGF7, FGF8 and FGF9 subfamilies fall under this category. Members of these five FGF subfamilies have significant affinity for HS, limting their diffusion in the HS-rich ECM and accounting for their paracrine mode of action. Superimposition

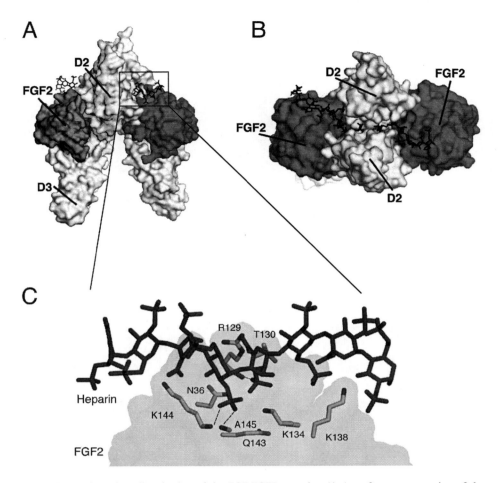

Figure 2. HS-dependent dimerization of the FGF-FGFR complex A) A surface representation of the FGF2-FGFR1c-heparin ternary complex, PDB ID: 1FQ9.[55] FGF2 is in dark grey, the D2 and D3 domains of FGFR1c are in light grey and heparin is represented as sticks in black. B) The complex has been rotated ninety degrees around an axis parallel to the plane of the page to reveal heparin binding the HBS in the membrane distal cleft of the complex. C) The boxed region from Figure 2A is expanded to show the FGF2 residues involved in hydrogen bonding to HS. There a total of 16 hydrogen bonds between FGF2 and HS. For the sake of clarity, only those three hydrogen bonds mediated by backbone atoms are shown.

of crystal structures of paracrine FGFs shows that the HBS of paracrine FGFs all adopt nearly identical topologies. In these FGFs, the two HS binding regions, namely the β1–β2 loop and the segment spanning β10 to β12, are juxtaposed and form a continuous positively-charged flat surface.[32-35,37,73-76]

Structure-based sequence alignment of paracrine FGFs shows that both the length and the primary sequence of the β1–β2 HS-binding loop differ across paracrine FGF subfamilies. In contrast, with the single exception of FGF5, the length of the HS binding segment spanning β10 to β12 is constant among all paracrine FGFs. Notably, all paracrine FGFs possess a prominent GXXXXGXX(T/S) motif in the HS binding β10–β12 segment.

THE STRUCTURAL BIOLOGY OF THE FGF19 SUBFAMILY

This motif, sometimes known as the glycine box,[77] plays a key role in imparting the common conformation of the region between the $\beta 10$ and $\beta 12$ strands in paracrine FGFs (Figs. 3A, 5B,D). The first glycine of this motif makes hydrogen bonds with a conserved glycine in $\beta 3$ and the second glycine hydrogen bonds with a fully conserved glycine in $\beta 7$. These hydrogen bonds are essential for the formation of the $\beta 11$ strand. Lastly, the side chain hydoxyl group of the threonine/serine hydrogen bonds with the backbone of the second glycine in the GXXXXGXX(T/S) motif. Crystal structures of FGF1 and FGF2 in complex with heparin oligosaccharides and SOS have demonstrated that both backbone and side chain atoms of the ligand's HBS region partake in HS binding.[55,78-80] Likewise, both sulfate groups and sugar backbone atoms of HS are engaged in FGF binding (Fig. 2C). The hydrogen bonds between backbone atoms of the HBS with HS are considered to be a key provider of binding energy, since they should incur less entropic loss upon HS binding than those hydrogen bonds involving the exposed flexible side chains.

Generally, paracrine FGF subfamilies exhibit tissue-specific expression patterns and are expressed in either epithelial or mesenchymal compartments within organs. The epithelially-expressed FGFs typically show specificity for FGFRc isoforms expressed in the mesenchyme and vice versa, resulting in the establishment of an epithelial-mesenchymal signaling loop (Fig. 4).[81-85] It is well documented that FGF7 and FGF10, which are expressed exclusively in the mesenchyme, specifically activate FGFR2b to mediate the mesenchymal-to-epithelial signaling required for the development of multiple organs and glands including lung, thyroid, pituitary, lacrimal and salivary glands.[84,86-90] In contrast, the members of the FGF4, FGF8 and FGF9 subfamilies are expressed in the epithelium and activate the mesenchymal FGFRc isoforms to govern patterning and morphogenesis of multiple tissues and organs, including brain, lung, heart, kidney, eye, limb and ear.[7,20,91,92]

ENDOCRINE FGFs REGULATE KEY METABOLIC PROCESSES IN A KLOTHO-DEPENDENT FASHION

The perception of FGFs as paracrine factors acting mainly during embryonic development has been overturned by the discovery that members of the FGF19 subfamily are humoral factors that regulate key metabolic processes. The involvement of FGF23 in phosphate homeostasis was discovered simultaneously by clinical studies of autosomal dominant hypophosphatemic rickets (ADHR) and tumor-induced osteomalacia (TIO), two human phosphate wasting disorders. The ADHR consortium identified mutations in FGF23 affecting either of two arginines at a RXXR motif that lies at the boundary between its core region and its 72 residue-long C-terminal tail (Fig. 3A).[93] Later studies showed that the ADHR mutations interfered with the natural process of proteolytic inactivation of FGF23, leading to an increase in the serum concentration of FGF23 that in turn induces phosphate wasting.[56,94,95] Shimada and colleagues showed that FGF23 secreted from the tumors of TIO patients was capable of causing hypophosphatemia.[96] FGF21's role in metabolism was originally discovered by Kharitonenkov and colleagues through experiments on *FGF21* transgenic mice and murine models of diabetes,[22] and FGF19 was first identified as an important regulator of energy metabolism through studies of *FGF19* transgenic rats by Tomlinson and colleagues.[25]

The endocrine FGFs require α/βklotho proteins as coreceptors in order to exert their metabolic actions. Klotho proteins are single-pass transmembrane proteins with an extracellular domain consisting of two tandem glycosidase-like domains, termed

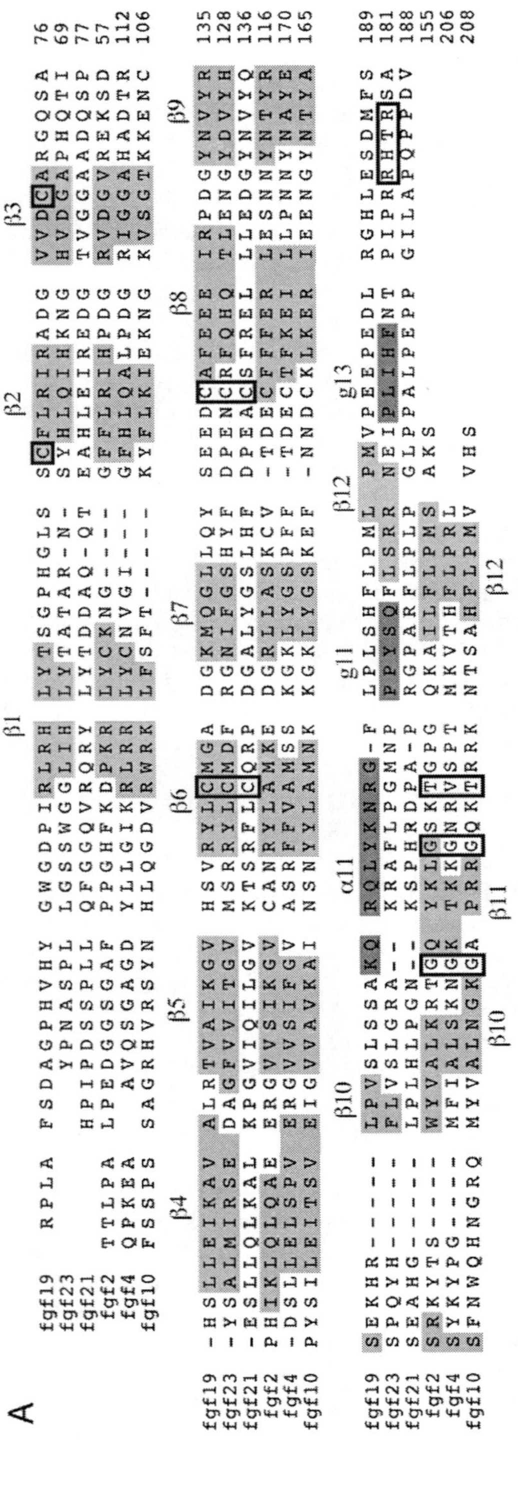

Figure 3. See figure legend on following page.

THE STRUCTURAL BIOLOGY OF THE FGF19 SUBFAMILY

B

β10   α11   β12

```
     humanFGF19  L P V S L S S A K Q R Q L Y K N R G F L P L S H F L P M  168
    monkeyFGF19  . . . . . . . . . . . . . . . . . . . . . . . . . . . .  168
    bovineFGF19  . . . . . . . S R . . . . F . S . . . . . . . . . . . .  306
    canineFGF19  . . I . . . . . . . . . . . . G . . . . . . . . . . . I  147
      ratFGF15  . H I I F I K . . P . E Q L Q G Q - - - K P . N . I . I  172
     mouseFGF15  . H I I F I Q . . P . E Q L Q D Q - - - K P . N . I . V  172
     chickFGF19  I S . . . . . . . . . Q F . G K D . . . . . . . . . .  173
      frogFGF19  V A . . . . K E . . K . Q . . G K . Y . . . . . . . V  169
 zebrafishFGF19  A L . T . . G . . N K - - L H S N D G T S A . Q . . . .  156
```

```
     humanFGF21  L P L H L P G N K S P H R D P A P R G P A R F L P L P  168
chimpanzeeFGF21  . . . . . . . . . . . . . . . . . . . . . . . . . . .  168
       pigFGF21  . . . R . . P H R . S N . . L . . . . . . . . . . . .  167
    bovineFGF21  . . . R . . P Q R . S N . . . . . . . . . . . . . . .  168
    canineFGF21  . . . R . R P H N . A Y . . L . . . . . . . . . . . .  168
      ratFGF21  . . . R . . Q K D . - - Q . . . T . . . V . . . . .  167
     mouseFGF21  . . . R . . Q K D . . N Q . A T S W . . V . . . . M .  169
 zebrafishFGF21  . . V S . L S K R Q K . G N . L - - - - S . . . . V S  149
```

β10   g11   β12

```
     humanFGF23  F L V S L G R A K R A F L P G M N P P P Y S Q F L S R R N  162
chimpanzeeFGF23  . . . . . . . . . . . . . . S . . . . . . . . . . . .  162
    bovineFGF23  . . . . . . . . . . . . T . . . . . A . . . . . . .  157
    canineFGF23  . . . . . . Q . . . . . . T . . . . . . . . . . . .  224
     chickFGF23  . . . . . . . T . Q V . F . . . . . . . . . . . . . .  166
      ratFGF23  Y . . . . . . S . . I . Q . . T . . . . F . . . . A . . .  162
     mouseFGF23  Y . . . . . . . . . I . Q . . T . . . F . . . . A . . .  162
 zebrafishFGF23  I . L N . D G . . Q . . I A . Q . L . Q S . L . . . E K .  172
```

β10   ∨   ∨   ∨   β12

```
      humanFGF2  W Y V A L K R T G Q Y K L G S K T G P G Q K A I L F L P M S  152
 chimpanzeeFGF2  . . . . . . . . . . . . . . . . . . . . . . . . . . . . .  152
     monkeyFGF2  . . . . . . . . . . . . . . . . . . . . . . . . . . . . .  152
     bovineFGF2  . . . . . . . . . . . . . . . . . . P . . . . . . . . . .  152
      sheepFGF2  . . . . . . . . . . . . . . . . . . P . . . . . . . . . .  152
    opossumFGF2  . . . . . . . . . . . . . . . . . . . . . . . . . . . . .  153
      chickFGF2  . . . . . . . . . . . . . . . P . P . . . . . . . . . . .  155
        ratFGF2  . . . . . . . . . . . . . . . . . . . . . . . . . . . . .  151
      mouseFGF2  . . . . . . . . . . . . . . . . . . . . . . . . . . . . .  151
       frogFGF2  . . . . . . . . . . . . . N . . S . . . . . . . . . . . .  152
 zebrafishFGF2  . . . . . . . . . . . . . S . . . . S . . . . . . . . . .  151
```

Figure 3. Sequence alignments of FGFs. A, viewed on previous page) Sequences of human FGF19, FGF23, FGF21, FGF2, FGF4 and FGF10 are aligned. N- and C-terminal regions of some ligands are truncated for the sake of presentation. β-sheets are highlighted with grey in the alignment and the helical secondary structures in FGF19 and FGF23 are highlighted with dark grey. Numerous important residues are enclosed in boxes to emphasize their importance: the Cys-58 and Cys-70 residues that support FGF19's unusual β10–β12 segment structure are boxed, along with other cysteines that form disulfide bridges; the GXXXXGXX(T/S) motif in the paracrine FGFs is indicated with boxes; the RXXR motif in FGF23 is also boxed.. B) Orthologs of FGF19, FGF21, FGF23 and FGF2 are aligned and the β10–β12 segment is shown for the solved structures. FGF23 exhibits greater conservation in its HBS than do either FGF19 or FGF21. The location of the GXXXXGXX(T/S) motif is indicated with arrowheads above the alignment of FGF2 orthologs.

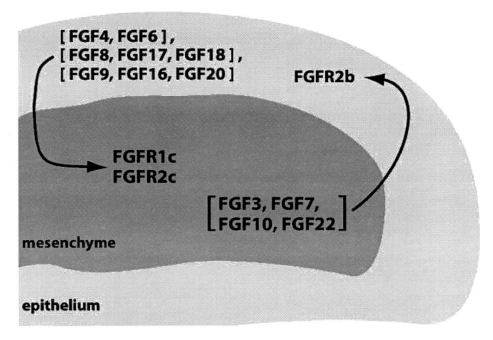

Figure 4. Paracrine FGFs mediate an epithelial-mesenchymal signaling loop. Ligands expressed in the epithelium signal through receptors expressed in the mesenchyme and vice versa.

KL1 and KL2.[97] FGF19 and FGF21 require βklotho,[98-103] and FGF23 requires αklotho in order to bind and activate their cognate FGFRs.[104,105] Klotho-dependency restricts signaling of FGF19 subfamily members to those tissues where klotho proteins are expressed. αklotho was originally discovered as an aging suppressor gene in mice and its requirement for FGF23 signaling was inferred from the phenotypic similarity between FGF23 and αklotho knockout mice.[106,107] Likewise, mice deficient for FGF15 (the mouse ortholog of human FGF19), βKlotho-knockout mice, and FGFR4 knockout mice all have overlapping defects in bile acid metabolism, a correlation that facilitated the identification of βklotho as a coreceptor for FGF19.[108-110]

Each of the FGF19 subfamily members participates in an endocrine signaling axis that is critical for maintaining homeostasis. The bone-kidney axis mediated by FGF23, FGFR1c (its main cognate FGFR)[111] and αklotho plays a vital role in serum phosphate regulation. In response to rising serum phosphate, FGF23 is secreted from bone and activates FGFR1c in the kidney in an αklotho-dependent fashion,[104,105,111] thus promoting phosphate excretion and suppressing vitamin D biosynthesis.[112-118]

FGF19, FGFR4 (its main cognate FGFR) and βklotho are essential components of an intestine-liver axis that form a postprandial negative feedback loop to regulate bile acid synthesis and release. FGF19 is secreted from intestinal epithelium in response to bile acid release into the intestinal lumen upon food intake.[108,119] FGF19 reaches the liver via the hepatic portal vein where it activates hepatic FGFR4 in a βklotho-dependent fashion,[99,100,103] thereby suppressing expression of the CYP7A1 gene that encodes the

THE STRUCTURAL BIOLOGY OF THE FGF19 SUBFAMILY

rate-limiting enzyme for bile acid synthesis.[24] Additionally, FGF19 acts to promote gallbladder filling through FGFRs other than FGFR4.[120]

The liver-fat axis mediated by FGF21, FGFR1c (its main cognate FGFR), and βklotho is critical for inducing metabolic adaptation in response to fasting. FGF21 is secreted from the liver[121] upon fasting,[122-125] and it activates FGFR1c in adipocytes in a βklotho-dependent fashion[98,99,101,102] to stimulate gluconeogenesis, ketogenesis and fatty acid oxidation.[22,126-128]

Interest in the FGF19 subfamily of ligands has been stimulated not only by their fascinating biology but also by their potential for the treatment of a variety of human diseases that have placed a major burden on health care. FGF21 agonists hold promise for the treatment of Type 2 diabetes and obesity.[129] FGF19 antagonists hold promise for the treatment of hypercholesterolemia, gallstone disorders, hepatocellular carcinoma and colon cancer. FGF23 has been implicated in the pathogenesis of many congenital diseases, including FGF23 gain-of-function disorders such as ADHR,[93] TIO,[130,131] FGF23 loss-of-function disorders such as familial tumoral calcinosis (FTC),[132-135] and X-linked hypophosphatemia (XLH),[130,131] a disorder involving increased FGF23 levels that results from loss of function mutations in the metalloproteinase PHEX that is thought to cleave FGF23 at its RXXR motif. More recently, elevations in FGF23 serum concentration have been correlated with the progression of chronic kidney disease (CKD).[136] FGF23 antagonists could be used to alleviate hypophosphatemia in inherited and tumor-induced phosphate wasting disorders as well as hypophosphatemia associated with other conditions such as organ transplantation and parenteral iron therapy. Conversely, FGF23 agonists could be used to correct hyperphosphatemia in FTC patients.

The recent crystal structures and biochemical studies of FGF19 and FGF23 have begun to elucidate the molecular basis for these ligands' endocrine behavior and the mechanism of action of their klotho coreceptors. Advances in the structural biology of FGF23 are already being translated towards the discovery of drugs for renal phosphate wasting disorders.

Structure-Function Relationships of Endocrine FGFs

Relative to the five paracrine-acting FGF subfamilies, the FGF19 subfamily exhibits the least sequence identity amongst its members. The pairwise sequence identity between the core regions of members of FGF19 subfamily ranges between 33% for FGF21 and FGF23 and 38% for FGF19 and FGF21 (Fig. 3A). In comparison, the identity between the core regions of members of paracrine FGF subfamilies is significantly higher and ranges between 88% for FGF9 and FGF16 to 54% for FGF7 and FGF10. Most of the sequence divergence between FGF19 subfamily members stems from the HBS regions, namely the β1-β2 loop and the segment between the β10 and β12 strands of these ligands. The identity between the HBS of FGF19 subfamily members is at best 13%. Exclusion of the HBS from the alignment improves the identity between FGF19 subfamily members to above 40%. In paracrine FGFs, however, the degree of sequence divergence at the HBS region is much less and is comparable with the degree of divergence in other regions of the trefoil core.

The HBS is the site of greatest divergence between the core regions of FGF19 subfamily ligands and those of other paracrine FGFs. The HS binding segment between β10 and β12 in endocrine FGFs is shortened and lacks the critical GXXXXGXX(T/S) motif, suggesting that this region cannot adopt the same conformation as in paracrine FGFs. Moreover, the β1–β2 loop of all three members of the FGF19 subfamily is longer than those of paracrine FGF subfamilies. Consistent with these major primary sequence divergences between the FGF19 subfamily and other FGF subfamilies, the crystal structures of FGF19 alone and of FGF23 bound to the heparin analogue sucrose octasulfate (SOS)[56] revealed that the HBS of FGF19 and FGF23 take on completely different conformations than that seen in paracrine FGFs. Notably, the conformations of the HBS seen in these two endocrine FGFs are incompatible with the hydrogen bonding of HS to backbone atoms in the HBS regions of these endocrine FGFs. This finding provides a molecular basis for the weak binding of FGF19 subfamily members for HS and explains the subfamily's ability to act in an endocrine fashion.

FGF19 Structural Findings

The first crystal structure of FGF19 was reported by the Blundell laboratory.[137] In this structure, the HS binding segment between the β10 and β12 strands is disordered, leading the authors to suggest that the HBS of FGF19 is inherently flexible. It was proposed that the HBS region of FGF19 assumes an ordered conformation upon HS binding and that the entropic penalty associated with HS binding was what caused the ligand's reduced HS binding affinity.[137] In the second FGF19 crystal structure published 3 years later, however, both FGF19 copies in the asymmetric unit of the crystal displayed a well ordered β10–β12 region (Fig. 5A). Since the β10–β12 region of these two FGF19 molecules are in different crystal packing environments, it can be argued that crystal packing did not bias the conformation of the HBS region in the FGF19 structure.[56]

As anticipated based on the lack of sequence homology between FGF19 and paracrine FGFs at the HBS region between β10 to β12, the conformation of this region deviates completely from the common conformation that is seen in paracrine FGFs. The Cα trace of FGF19 in the region of the HBS begins to diverges from that of paracrine FGFs at Leu-145 and converges again at Leu-162 (Fig. 5B). Residues 149 to 158 in the β10–β12 region of FGF19 form an α-helix (α11) that bulges out from the β-trefoil core. The atypical conformation of the β10–β12 region of FGF19 is accompanied by other structural differences between FGF19 and paracrine FGFs in the core. In FGF19, the conserved β3 glycine of paracrine FGFs is replaced by a cysteine (Cys-70). In the FGF19 structure, Cys-70 forms a disulfide bond with Cys-58 (also unique to FGF19) in β2 that packs against Leu-145 and Leu-162 at the divergent and convergent ends of the β10–β12 region. This hydrophobic interaction helps to partially shield the disulfide bridge from solvent (Fig. 5C).

The β1–β2 loop of FGF19 is the longest among the FGFs and extends out from the β-trefoil core in the same direction as the α11 helix. In contrast to paracrine ligands where the β1–β2 loop and β10–β12 segment are juxtaposed and form a contiguous HBS, in FGF19 there is a spatial separation between the two regions as they do not engage in any intramolecular interactions (Fig. 5A, 5D).

FGF23 Structural Findings

Attempts to crystallize the full length FGF23 were unsuccessful, likely because of the flexibility of its (73-amino acid long) C-terminus. The core domain was thus

THE STRUCTURAL BIOLOGY OF THE FGF19 SUBFAMILY

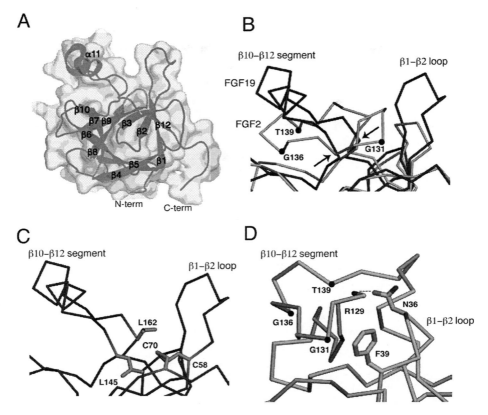

Figure 5. The FGF19 structure A) The FGF19 structure, PDB ID: 2P23,[56] is shown as a surface and as a cartoon. The β–sheets and α–helix are labeled along with the N- and C-termini of the protein. B) The HBS of FGF19 is superimposed onto that of FGF2. Both ligands are represented as ribbons, with FGF19 in dark grey and FGF2 in light grey. The atoms of the GXXXXGXX(T/S) motif are marked with dots in the FGF2 structure and identified. Arrows indicate the positions of the Leu-145 and Leu-162 where the FGF19 Cα trace diverges from the FGF2 Cαtrace. C) The FGF19 HBS is shown and the C58-C70 disulfide bridge residues are represented as sticks along with the Leu-162 and Leu-145 residues that protect them from solvent. D) The FGF2 HBS is shown, the GXXXXGXX(T/S) motif is identified, and intramolecular contacts between the β1–β2 loop and β10–β12 segment are shown. Asn-36 of the β1–β2 loop interacts with the backbone atom of Arg-129 in the β10–β12 segment and Phe-39 of the β1–β2 region engages in Van der Waals interactions with the backbone of the β10–β12 region.

crystallized in complex with SOS (Figs. 6A,B). Consistent with the major primary sequence divergence at the β10–β12 segment between FGF23 and paracrine FGFs, the Cα trace of FGF23 diverges from that of paracrine FGFs at Leu-138 and converges again at Pro-153. (Fig. 6C). Predictably, the conformation of this region is also completely different from that of FGF19 (Fig. 6D) since the two ligands share no homology in their β10–β12 regions. Like FGF19, FGF23 does not exhibit the β11 strand of paracrine FGFs and instead PROCHECK assigns a g-helix to the C-terminal region of the β10–β12 region in FGF23. Again as for FGF19, the altered conformation of the β10–β12 region in FGF23 is accompanied by changes in other regions of the core. Compared to paracrine FGFs, FGF23 has a very short β9–β10 loop which is needed

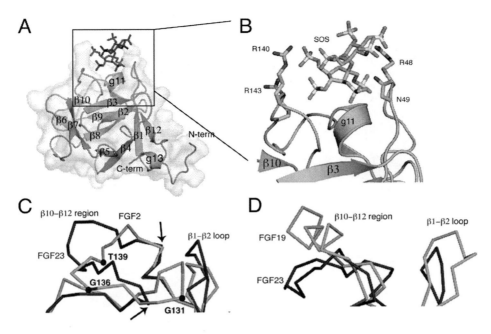

Figure 6. The FGF23 structure A) The FGF23 structure, PDB ID: 2P39,[56] is shown as a surface and as a cartoon. The β-sheets and α-helix are labeled along with the N- and C-termini of the protein. The SOS molecule that is bound to FGF23 is represented as sticks in dark grey. B) An expansion of the box in Figure 6A showing a close-up of the SOS molecule and the residues in the HBS of FGF23 with which it binds. C) An overlay of the HBS of FGF23 with that of FGF2 showing their divergence. FGF2 is in light grey and FGF23 is in dark grey. The residues of the GXXXXGXX(T/S) motif are labeled and the Leu-138 and Pro-153 residues where the FGF23 Cα trace diverges from that of FGF2 are indicated with arrows. D) The FGF19 and FGF23 HBS regions are superimposed, with FGF19 shown in light grey and FGF23 in dark grey.

to accommodate the novel HBS topology of FGF23. Structural analysis shows that even a one residue insertion in the β9–β10 loop of FGF23 would interfere with the β10–β12 conformation seen in FGF23. The HBS of FGF23 contains a cleft similar to that in FGF19 between the β1–β2 loop and β10–12 region (Fig. 6D). The SOS molecule binds with its sulfated fructose ring facing down into the cleft in the FGF23 HBS and it engages in hydrogen bonds with Arg-140 and Arg-143 in β10–β12 and Arg-48 and Asn-49 in β1–β2 (Fig. 6B). Hence, based on the crystal structure these four residues of FGF23 mediate binding of FGF23 to HS.

THE MOLECULAR BASIS FOR THE FGF19 SUBFAMILY'S KLOTHO CORECEPTOR REQUIREMENT

Reduced Affinity of FGF19 Subfamily Members for HS

Superimposition of the FGF19 and FGF23 crystal structures onto FGF2 in the FGF2-FGFR1c-heparin ternary complex structure (PDB ID; 1FQ9)[55] illuminates the impact of these ligands' altered HBS topology on their interaction with HS. This superimposition shows that HS would clash with the HBS of FGF19 and FGF23 if it attempted to bind to the ligand as it does in paracrine FGFs (Figs. 7A,B). These clashes could be avoided by translating HS away from the ligands' core domains, but such a translation would be detrimental for the HS binding affinity since the backbone atoms of the two ligands would no longer be able to make direct hydrogen bonds with the N-sulfate group from 4GlcN(S)6O(S) or a 2-O-sulfate group from 5IdoA (Fig. 2C). Thus, the altered topology of the β10–β12 region in FGF19 and FGF23 should impart a major loss in HS binding. This lowered HS binding affinity is a prerequisite for the endocrine behavior of this subfamily of ligands since it allows them to diffuse unhindered through the HS-rich pericellular space and enter the bloodstream.

The orientation of SOS bound to FGF23 is perpendicular to the orientation of SOS bound to FGF1,[80] and of heparin oligosaccharides bound to FGF1 and FGF2.[55,78,79] This structural observation further corroborates the notion that HS cannot bind to FGF19 subfamily ligands as it classically binds to paracrine FGFs.

Biochemical experiments have confirmed the conclusions drawn from structural analysis. Surface plasmon resonance (SPR) experiments showed that FGF19, FGF21 and FGF23 have poor binding to heparin relative to paracrine ligands.[56] Moreover, by mutating FGF19 residues Lys-149 and Arg-157 that based on the FGF19-FGFR-heparin model are predicted to mediate the residual HS binding, the affinity of FGF19 for heparin was reduced even further.[56] Likewise, by mutating residues in FGF23 that were involved in binding to SOS, the affinity of FGF23 for heparin was further decreased.[56] It remains to be tested if the residual HS binding affinity of the endocrine FGFs still plays a role in their signaling.

The HS binding segment between β10 and β12 in FGF19 and FGF23 displays two unique conformations (Fig. 6D), both of which sterically hinder HS from hydrogen bonding with the backbone atoms of HBS in these ligands (Fig. 7). Structure-based sequence alignment of FGF21 with FGF19 and FGF23 shows that the β10–β12 region of FGF21 would adopt yet a third unique topology. There is no sequence identity between the β10–β12 segment of FGF21 and FGF19 and only 11% sequence identity between

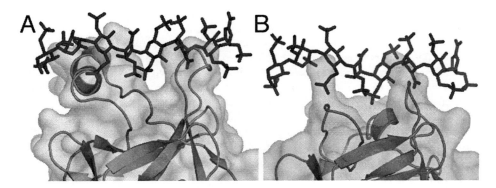

Figure 7. Superimposition of heparin onto the FGF19 and FGF23 structures. A) FGF19 is superimposed onto the FGF2-FGFR1c-heparin structure (PDB ID: 1FQ9)[55] and the heparin from that structure is grafted onto FGF19 to show the clashes that would occur were heparin to attempt to bind FGF19 in the same fashion as it binds FGF2. B) Similarly for FGF23, heparin is superimposed to reveal the clashes that would occur were heparin to attempt to bind in a FGF2-like fashion.

FGF21 and FGF23 at this region (Fig. 3A). These observations raise the question whether the specific conformation of the HBS of FGF19 subfamily members is important or whether their HBS needs simply to provide a steric obstacle in any way possible against HS bonding. If the former were the case, then one would expect an evolutionary pressure to preserve the sequence of the HBS in the endocrine FGFs.

To address this question, we aligned the orthologs of FGF19, FGF21, FGF23 and FGF2, a prototypical paracrine FGF (Fig. 3B). The alignments show gaps in the β10-β12 region of rat and zebrafish FGF19 and FGF21. FGF19 shows an additional gap in the β10-β12 of mouse (Fig. 3B). These differences between the sequences of the orthologs in FGF19 and FGF21 indicate that the specific conformation of their β10–β12 region is not as critical as it is for FGF23, where the sequence of the protein is more conserved across the different orthologs and no gaps in the sequence are seen. In contrast to the endocrine FGFs, the sequence alignment of the FGF2 orthologs shows almost total identity in the β10–β12 region (Fig. 3B), underscoring how important the primary and secondary structure elements in the HBS are for HS binding to FGF2.

STRUCTURAL BASIS FOR THE SPECIFICITY
AND AFFINITY OF ENDOCRINE FGFs' BINDING TO FGFR

The principal cognate receptors for the FGF19 subfamily ligands have been well characterized. FGF23 signaling in the kidney is mediated by FGFR1c, since administration of FGF23 to mice with a conditional knockdown of FGFR1 in the metanephric mesenchyme did not induce hypophosphatemia.[111] FGF19 signaling is mediated by FGFR4, since FGF15 (the mouse ortholog of human FGF19) could not suppress CYP7A1 activity in FGFR4 knockout mice,[108] FGF19-induced hepatocyte proliferation is driven through FGFR4,[138] and FGFR4- and FGF15-deficient mice both have elevated bile acid pools.[108,110]

The crystal structures of FGF19 and FGF23 offer a molecular basis for the receptor binding specificity of these two ligands. Comparison of the receptor-bound crystal structures of several paracrine FGFs, including FGF1, FGF2, FGF3, FGF8 and FGF10,[33-35,37-39] have shown that the extent to which β1 strand pairs with β4 plays a decisive role in the discrimination of FGFR binding specificity. Compared to β1 of FGF2, the β1 strands of the FGF7 subfamily are extended two residues N-terminally due to additional strand pairing with β4. This extension is essential for the ability of the N-termini of FGF7 subfamily members to engage in specific contacts with the alternatively spliced βC'-βE loop of FGFR2b. Conversely, the N-terminally extended β1 strand of the FGF7 subfamily clashes with the alternatively spliced loop of FGFR2c, thus actively discouraging binding of this subfamily to FGFR2c. Interestingly, akin to the FGF7 subfamily, the β1 strand of FGF19 is extended two residues longer than the β1 strand of the FGF1 subfamily (Fig. 3A) and the length of the βC'-βE loop of FGFR4, FGF19's cognate receptor, is the same length as in FGFR2b—two residues shorter than the corresponding loop in FGFRc isoforms. Although FGF19 does not bind to FGFR2b, we predict that the lengthened β1 strand of FGF19 confers its specificity for FGFR4. Conversely, the β1 strand of FGF23 is the same length as in FGF2, which is consistent with FGF23's preference for FGFR1c.

The crystal structures of FGF19 and FGF23 also afford a potential molecular basis for the poor affinity of these ligands for their cognate FGFRs. SPR, size exclusion and co-immunoprecipitation data have shown that FGF19 subfamily members have unusually low affinity for their cognate FGFRs.[3,98-101,103-105] Superimposition of FGF23 onto FGF2

THE STRUCTURAL BIOLOGY OF THE FGF19 SUBFAMILY

in the FGF2-FGFR1c-heparin structure (PDB ID: 1FQ9)[55] shows that substitution of His-117 in FGF23 for the highly conserved glutamate in the β8 strand of all other FGFs would lead to a major loss in FGFR binding affinity. In the crystal structure of FGF2 and other paracrine FGFs complexed with FGFR, this glutamate (Glu-105 in FGF2) makes direct hydrogen bonds with D3. Indeed, mutation of this glutamate to alanine in FGF2 has been shown to dramatically reduce the binding of the ligand to FGFR.[139]

In summary, the structural and biochemical data along with sequence analysis reveal that FGF19 subfamily members have poor HS and FGFR binding affinity. Consequently, these FGFs are incapable of binding, dimerizing and activating their cognate FGFRs in a solely HS-dependent fashion. To overcome these deficiencies the endocrine FGFs must rely on klotho proteins as coreceptors. Klotho coreceptors act by constitutively associating with the cognate FGFRs of endocrine FGFs, thereby enhancing the affinity of the FGFR for endocrine FGFs to levels sufficient for FGFR dimerization and activation.

CHARACTERIZATION OF THE BINDING OF FGF23 TO αKLOTHO

Determining why FGF19 subfamily members depend on klotho coreceptors for signaling was an important step towards understanding the biology of this interesting FGF subfamily. The next question to be studied is how the FGF19 subfamily members, their cognate FGFRs and klotho coreceptor interact to form ternary complexes. Through a combination of SPR, size exclusion and co-immunoprecipitation experiments, the nature of FGF23's binding to FGFR1c and αklotho has been recently investigated.[94] First, it was shown that the soluble ectodomains of FGFR1c and αklotho are sufficient to form a ternary complex with FGF23. The ability to reconstitute the ternary complex in vitro enabled a thorough characterization of the bimolecular interactions among the three components of the ternary complex. αklotho ectodomain was shown to bind to the ectodomain of FGFR1c with high affinity (72 nM). Interestingly, although FGF23 bound poorly to αklotho ectodomain, it nonetheless bound avidly to a preformed 1:1 complex of FGFR1c and αklotho ectodomains. Since FGF23 binds poorly to both FGFR1c and αklotho, these data indicate that in the context of the FGFR1c-αklotho complex a de novo site for FGF23 is generated. Importantly, truncated FGF23 lacking its C-terminal tail past the RXXR motif is unable to bind a 1:1 binary complex of FGFR1c-αklotho in the absence of its own C-terminal tail, revealing that the C-terminal tail of FGF23 mediates binding of FGF23 to this de novo site created at the interface of FGFR1c and αklotho.

Consistent with this finding, a SPR competition assay showed that the C-terminal tail of FGF23 was able to compete away binding of full length FGF23 from the FGFR1c –αklotho binary complex. The binding region within the FGF23 C-tail for this de novo site in the αklotho-FGFR1c binary complex was further narrowed down to the region between Ser-180 and Thr-200. Accordingly, this fragment was also able to compete away binding of full length FGF23 from the FGFR1c-αklotho binary complex, albeit with less potency than the full C-tail. Consistent with this region serving as a minimal epitope, a FGF23 ligand truncated at Thr-200 was able to elicit similar levels of FRS2α phosphorylation as the full length ligand.

PHARMACOLOGICAL IMPACT OF STRUCTURAL INSIGHTS INTO FGF19 SUBFAMILY

To date, the primary treatment for disorders of phosphate wasting has been intravenous phosphate therapy along with 1,25-dihydroxyvitamin D_3 to increase phosphate absorption in the intestine. This relatively primitive therapy sometimes results in side-effects of over-shoot hyperphosphatemia, and it is also simply ineffective in cases of chronic phosphate wasting such as occur in XLH.

Thus, the finding that the FGF23 C-terminus effectively inhibits the FGF23 interaction with FGFR1c-αklotho holds great promise for the development of therapeutics. Developing pharmacologically active C-terminal peptides, peptidomimetics or organomimetics thereof would introduce an important new tool for the treatment of phosphate wasting disorders. This method of inhibition has an advantage over neutralizing FGF23 by antibodies, since the C-terminal peptide only inhibits the αklotho-specific activity of FGF23, leaving any potential klotho-independent activity of FGF23 intact. C-terminal peptides thus may also serve as a useful tool to dissect the αklotho dependent and αklotho independent functions of FGF23.

These studies of the inhibition of FGF23 activity by a FGF23 fragment produced through proteolysis are also fascinating at the level of general biological principle. Here, we observed a hormone being inhibited by the product of its own degradation. Accordingly, in pathological hyperphosphatemic states such as FTC where serum levels of C-terminal peptides of FGF23 are high, this C-terminal fragment may be aggravating the disease by inhibiting any residual function of the full length mutant FGF23 ligand found in FTC.

CONCLUSION

Undoubtedly, solving the crystal structures of the FGF19-FGFR4-βklotho, FGF21-FGFR1c-βklotho, and FGF23-FGFR1c-αklotho complexes will be the immediate priority for this fast moving and exciting field. The structure of the FGF23-FGFR1c-αklotho complex should unveil the details of how a de novo site for FGF23's C-tail is created when αklotho and FGFR1c ectodomain interact. It would be fascinating to see whether βklotho also utilizes the same mechanism as αklotho to promote the signaling of FGF19 and FGF21 through their respective cognate receptors, FGFR4 and FGFR1c. The structures of these ternary complexes will also inform us of the determinants of the FGFR binding specificity/promiscuity of klotho proteins. Biochemically, it will be important to find out if the residual HS binding affinity of the FGF19 subfamily is still required for signaling. If this were the case then we could envision the model shown in Figure 8 for how the cell surface signaling unit of an endocrine FGF might look like. The signaling unit will still have a 2:2:2 FGF-FGFR-HS dimer at its core that is further stabilized by interactions of klotho proteins with FGFR and endocrine FGFs.

The discovery that a major difference in HS binding was responsible for the endocrine mode of action of the FGF19 subfamily suggested the corollary that subtle differences in the HS binding affinity of paracrine FGFs may also play a role in distinguishing their distinct biological activity. Indeed, a recent study has shown that differences in the HS affinity of two members of the FGF7 subfamily, FGF7 and FGF10, was responsible for differences in their branching morphogenetic potentials.[51] Using a branching morphogenesis

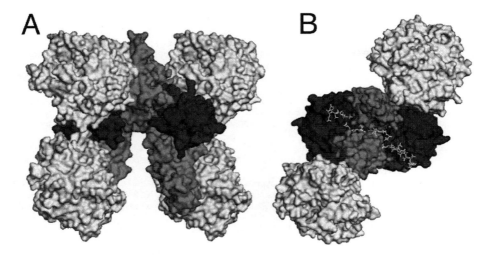

Figure 8. A proposed model of the FGF23-FGFR1c-heparin-αklotho complex. A) FGF23 is superimposed onto FGF2 in the FGF2-FGFR1c-heparin complex (PDB ID:1FQ9)[55] and is depicted in dark grey. The FGFRs from 1FQ9 are depicted in a medium tint of grey. Myrosinase (PDB ID: 1E6S),[140] a glycosidase, is used as a model for the two KL domains of αklotho and is depicted in light grey. B) The FGF23-FGFR1c-heparin-αklotho complex has been rotated 90 degrees to show the HS binding cleft in the membrane-distal portion of the complex.

assay, it was shown that FGF10 could be transformed into a functional mimic of FGF7 through a single Arg→Val mutation in the FGF10 HBS. This mutation reduced FGF10's HS affinity to FGF7-like levels and thus allowed FGF10 to diffuse through the extracellular matrix more easily and establish a gradient similar to that made by FGF7. Future studies should aim to elucidate the extent to which the differing biology of ligands within other FGF subfamilies can also be accounted for on the basis of their affinities for HS. In which case, the central role of HS affinity in differentiating the biological functions of FGFs will not be a feature only of the endocrine FGFs but of all FGFs.

REFERENCES

1. Eswarakumar VP, Lax I, Schlessinger J. Cellular signaling by fibroblast growth factor receptors. Cytokine Growth Factor Rev 2005; 16(2):139-149.
2. Johnson DE, Williams LT. Structural and functional diversity in the FGF receptor multigene family. Adv Cancer Res 1993; 60:1-41.
3. Mohammadi M, Olsen SK, Ibrahimi OA. Structural basis for fibroblast growth factor receptor activation. Cytokine Growth Factor Rev 2005; 16(2):107-137.
4. Itoh N, Ornitz DM. Evolution of the Fgf and Fgfr gene families. Trends Genet 2004;20(11):563-569.
5. Itoh N, Ornitz DM. Functional evolutionary history of the mouse Fgf gene family. Dev Dyn 2008; 237(1):18-27.
6. Popovici C, Roubin R, Coulier F et al. An evolutionary history of the FGF superfamily. Bioessays 2005; 27(8):849-857.
7. Colvin JS, Green RP, Schmahl J et al. Male-to-female sex reversal in mice lacking fibroblast growth factor 9. Cell 2001; 104(6):875-889.
8. Goriely A, Hansen RM, Taylor IB et al. Activating mutations in FGFR3 and HRAS reveal a shared genetic origin for congenital disorders and testicular tumors. Nat Genet 2009; 41(11):1247-1252.

9. Goriely A, McVean GA, Rojmyr M et al. Evidence for selective advantage of pathogenic FGFR2 mutations in the male germ line. Science 2003; 301(5633):643-646.
10. Kato S, Sekine K. FGF-FGFR signaling in vertebrate organogenesis. Cell Mol Biol (Noisy-le-grand) 1999; 45(5):631-638.
11. Niswander L, Tickle C, Vogel A et al. FGF-4 replaces the apical ectodermal ridge and directs outgrowth and patterning of the limb. Cell 1993; 75(3):579-587.
12. Baker RE, Schnell S, Maini PK. A clock and wavefront mechanism for somite formation. Dev Biol 2006; 293(1):116-126.
13. Dale JK, Malapert P, Chal J et al. Oscillations of the snail genes in the presomitic mesoderm coordinate segmental patterning and morphogenesis in vertebrate somitogenesis. Dev Cell 2006; 10(3):355-366.
14. Dubrulle J, McGrew MJ, Pourquie O. FGF signaling controls somite boundary position and regulates segmentation clock control of spatiotemporal Hox gene activation. Cell 2001; 106(2):219-232.
15. Pourquie O. The chick embryo: a leading model in somitogenesis studies. Mech Dev 2004; 121(9):1069-1079.
16. Sawada A, Shinya M, Jiang YJ et al. Fgf/MAPK signalling is a crucial positional cue in somite boundary formation. Development 2001; 128(23):4873-4880.
17. Colvin JS, White AC, Pratt SJ et al. Lung hypoplasia and neonatal death in Fgf9-null mice identify this gene as an essential regulator of lung mesenchyme. Development 2001; 128(11):2095-2106.
18. Itoh N. The Fgf families in humans, mice and zebrafish: their evolutional processes and roles in development, metabolism and disease. Biol Pharm Bull 2007; 30(10):1819-1825.
19. Lu SY, Sheikh F, Sheppard PC et al. FGF-16 is required for embryonic heart development. Biochem Biophys Res Commun 2008; 373(2):270-274.
20. Sugi Y, Ito N, Szebenyi G et al. Fibroblast growth factor (FGF)-4 can induce proliferation of cardiac cushion mesenchymal cells during early valve leaflet formation. Dev Biol 2003; 258(2):252-263.
21. O'Leary DD, Chou SJ, Sahara S. Area patterning of the mammalian cortex. Neuron 2007; 56(2):252-269.
22. Kharitonenkov A, Shiyanova TL, Koester A et al. FGF-21 as a novel metabolic regulator. J Clin Invest 2005; 115(6):1627-1635.
23. Fu L, John LM, Adams SH et al. Fibroblast growth factor 19 increases metabolic rate and reverses dietary and leptin-deficient diabetes. Endocrinology 2004; 145(6):2594-2603.
24. Holt JA, Luo G, Billin AN et al. Definition of a novel growth factor-dependent signal cascade for the suppression of bile acid biosynthesis. Genes Dev 2003; 17(13):1581-1591.
25. Tomlinson E, Fu L, John L et al. Transgenic mice expressing human fibroblast growth factor-19 display increased metabolic rate and decreased adiposity. Endocrinology 2002; 143(5):1741-1747.
26. Yu X, White KE. FGF23 and disorders of phosphate homeostasis. Cytokine Growth Factor Rev 2005; 16(2):221-232.
27. Goetz R, Dover K, Laezza F et al. Crystal structure of a fibroblast growth factor homologous factor (FHF) defines a conserved surface on FHFs for binding and modulation of voltage-gated sodium channels. J Biol Chem 2009; 284(26):17883-17896.
28. Goldfarb M. Fibroblast growth factor homologous factors: evolution, structure and function. Cytokine Growth Factor Rev 2005; 16(2):215-220.
29. Olsen SK, Garbi M, Zampieri N et al. Fibroblast growth factor (FGF) homologous factors share structural but not functional homology with FGFs. J Biol Chem 2003; 278(36):34226-34236.
30. Eriksson AE, Cousens LS, Weaver LH et al. Three-dimensional structure of human basic fibroblast growth factor. Proc Natl Acad Sci USA 1991; 88(8):3441-3445.
31. Zhang JD, Cousens LS, Barr PJ et al. Three-dimensional structure of human basic fibroblast growth factor, a structural homolog of interleukin 1 beta. Proc Natl Acad Sci USA 1991; 88(8):3446-3450.
32. Zhu X, Komiya H, Chirino A et al. Three-dimensional structures of acidic and basic fibroblast growth factors. Science 1991; 251(4989):90-93.
33. Olsen SK, Li JY, Bromleigh C et al. Structural basis by which alternative splicing modulates the organizer activity of FGF8 in the brain. Genes Dev 2006; 20(2):185-198.
34. Plotnikov AN, Hubbard SR, Schlessinger J et al. Crystal structures of two FGF-FGFR complexes reveal the determinants of ligand-receptor specificity. Cell 2000; 101(4):413-424.
35. Yeh BK, Igarashi M, Eliseenkova AV et al. Structural basis by which alternative splicing confers specificity in fibroblast growth factor receptors. Proc Natl Acad Sci USA 2003; 100(5):2266-2271.
36. Lee PL, Johnson DE, Cousens LS et al. Purification and complementary DNA cloning of a receptor for basic fibroblast growth factor. Science 1989; 245(4913):57-60.
37. Plotnikov AN, Schlessinger J, Hubbard SR et al. Structural basis for FGF receptor dimerization and activation. Cell 1999; 98(5):641-650.
38. Stauber DJ, DiGabriele AD, Hendrickson WA. Structural interactions of fibroblast growth factor receptor with its ligands. Proc Natl Acad Sci USA 2000; 97(1):49-54.
39. Olsen SK, Ibrahimi OA, Raucci A et al. Insights into the molecular basis for fibroblast growth factor receptor autoinhibition and ligand-binding promiscuity. Proc Natl Acad Sci USA 2004; 101(4):935-940.

THE STRUCTURAL BIOLOGY OF THE FGF19 SUBFAMILY

40. Wang F, Kan M, Yan G et al. Alternately spliced NH2-terminal immunoglobulin-like Loop I in the ectodomain of the fibroblast growth factor (FGF) receptor 1 lowers affinity for both heparin and FGF-1. J Biol Chem 1995; 270(17):10231-10235.
41. Johnson DE, Lu J, Chen H et al. The human fibroblast growth factor receptor genes: a common structural arrangement underlies the mechanisms for generating receptor forms that differ in their third immunoglobulin domain. Mol Cell Biol 1991; 11(9):4627-4634.
42. Miki T, Bottaro DP, Fleming TP et al. Determination of ligand-binding specificity by alternative splicing: two distinct growth factor receptors encoded by a single gene. Proc Natl Acad Sci USA 1992; 89(1):246-250.
43. Orr-Urtreger A, Bedford MT, Burakova T et al. Developmental localization of the splicing alternatives of fibroblast growth factor receptor-2 (FGFR2). Dev Biol 1993; 158(2):475-486.
44. Forsberg E, Kjellen L. Heparan sulfate: lessons from knockout mice. J Clin Invest 2001; 108(2):175-180.
45. Lin X, Buff EM, Perrimon N et al. Heparan sulfate proteoglycans are essential for FGF receptor signaling during Drosophila embryonic development. Development 1999; 126(17):3715-3723.
46. Ornitz DM, Yayon A, Flanagan JG et al. Heparin is required for cell-free binding of basic fibroblast growth factor to a soluble receptor and for mitogenesis in whole cells. Mol Cell Biol 1992; 12(1):240-247.
47. Rapraeger AC, Krufka A, Olwin BB. Requirement of heparan sulfate for bFGF-mediated fibroblast growth and myoblast differentiation. Science 1991; 252(5013):1705-1708.
48. Yayon A, Klagsbrun M, Esko JD et al. Cell surface, heparin-like molecules are required for binding of basic fibroblast growth factor to its high affinity receptor. Cell 1991; 64(4):841-848.
49. Whitelock JM, Iozzo RV. Heparan sulfate: a complex polymer charged with biological activity. Chem Rev 2005; 105(7):2745-2764.
50. Mohammadi M, Olsen SK, Goetz R. A protein canyon in the FGF-FGF receptor dimer selects from an a la carte menu of heparan sulfate motifs. Curr Opin Struct Biol 2005; 15(5):506-516.
51. Makarenkova HP, Hoffman MP, Beenken A et al. Differential interactions of FGFs with heparan sulfate control gradient formation and branching morphogenesis.Sci Signal 2009; 2(88):ra55.
52. Hacker U, Nybakken K, Perrimon N. Heparan sulphate proteoglycans: the sweet side of development. Nat Rev Mol Cell Biol 2005; 6(7):530-541.
53. Saksela O, Moscatelli D, Sommer A et al. Endothelial cell-derived heparan sulfate binds basic fibroblast growth factor and protects it from proteolytic degradation. J Cell Biol 1988; 107(2):743-751.
54. Ashikari-Hada S, Habuchi H, Kariya Y et al. Characterization of growth factor-binding structures in heparin/heparan sulfate using an octasaccharide library. J Biol Chem 2004; 279(13):12346-12354.
55. Schlessinger J, Plotnikov AN, Ibrahimi OA et al. Crystal structure of a ternary FGF-FGFR-heparin complex reveals a dual role for heparin in FGFR binding and dimerization. Mol Cell 2000; 6(3):743-750.
56. Goetz R, Beenken A, Ibrahimi OA et al. Molecular insights into the klotho-dependent, endocrine mode of action of fibroblast growth factor 19 subfamily members. Mol Cell Biol 2007; 27(9):3417-3428.
57. Chen H, Ma J, Li W et al. A molecular brake in the kinase hinge region regulates the activity of receptor tyrosine kinases. Mol Cell 2007; 27(5):717-730.
58. Chen H, Xu CF, Ma J et al. A crystallographic snapshot of tyrosine trans-phosphorylation in action. Proc Natl Acad Sci USA 2008; 105(50):19660-19665.
59. Furdui CM, Lew ED, Schlessinger J et al. Autophosphorylation of FGFR1 kinase is mediated by a sequential and precisely ordered reaction. Mol Cell 2006; 21(5):711-717.
60. Mohammadi M, Dikic I, Sorokin A et al. Identification of six novel autophosphorylation sites on fibroblast growth factor receptor 1 and elucidation of their importance in receptor activation and signal transduction. Mol Cell Biol 1996; 16(3):977-989.
61. Mohammadi M, Dionne CA, Li W et al. Point mutation in FGF receptor eliminates phosphatidylinositol hydrolysis without affecting mitogenesis. Nature 1992; 358(6388):681-684.
62. Peters KG, Marie J, Wilson E et al. Point mutation of an FGF receptor abolishes phosphatidylinositol turnover and Ca2+ flux but not mitogenesis. Nature 1992; 358(6388):678-681.
63. Seo JH, Suenaga A, Hatakeyama M et al. Structural and functional basis of a role for CRKL in a fibroblast growth factor 8-induced feed-forward loop. Mol Cell Biol 2009; 29(11):3076-3087.
64. Divecha N, Irvine RF. Phospholipid signaling. Cell 1995; 80(2):269-278.
65. Huang J, Mohammadi M, Rodrigues GA et al. Reduced activation of RAF-1 and MAP kinase by a fibroblast growth factor receptor mutant deficient in stimulation of phosphatidylinositol hydrolysis. J Biol Chem 1995;270(10):5065-5072.
66. Schlessinger J. Cell signaling by receptor tyrosine kinases. Cell 2000; 103(2):211-225.
67. Kouhara H, Hadari YR, Spivak-Kroizman T et al. A lipid-anchored Grb2-binding protein that links FGF-receptor activation to the Ras/MAPK signaling pathway. Cell 1997; 89(5):693-702.
68. Dhalluin C, Yan KS, Plotnikova O et al. Structural basis of SNT PTB domain interactions with distinct neurotrophic receptors. Mol Cell 2000; 6(4):921-929.
69. Ong SH, Guy GR, Hadari YR et al. FRS2 proteins recruit intracellular signaling pathways by binding to diverse targets on fibroblast growth factor and nerve growth factor receptors. Mol Cell Biol 2000; 20(3):979-989.

70. Ong SH, Hadari YR, Gotoh N et al. Stimulation of phosphatidylinositol 3-kinase by fibroblast growth factor receptors is mediated by coordinated recruitment of multiple docking proteins. Proc Natl Acad Sci USA 2001; 98(11):6074-6079.
71. Hadari YR, Kouhara H, Lax I et al. Binding of Shp2 tyrosine phosphatase to FRS2 is essential for fibroblast growth factor-induced PC12 cell differentiation. Mol Cell Biol 1998; 18(7):3966-3973.
72. Dailey L, Ambrosetti D, Mansukhani A et al. Mechanisms underlying differential responses to FGF signaling. Cytokine Growth Factor Rev 2005; 16(2):233-247.
73. Bellosta P, Iwahori A, Plotnikov AN et al. Identification of receptor and heparin binding sites in fibroblast growth factor 4 by structure-based mutagenesis. Mol Cell Biol 2001; 21(17):5946-5957.
74. Kalinina J, Byron SA, Makarenkova HP et al. Homodimerization Controls the FGF9 Subfamily's Receptor Binding and Heparan Sulfate Dependent Diffusion in the Extracellular Matrix. Mol Cell Biol 2009.
75. Osslund TD, Syed R, Singer E et al. Correlation between the 1.6 A crystal structure and mutational analysis of keratinocyte growth factor. Protein Sci 1998; 7(8):1681-1690.
76. Plotnikov AN, Eliseenkova AV, Ibrahimi OA et al. Crystal structure of fibroblast growth factor 9 reveals regions implicated in dimerization and autoinhibition. J Biol Chem 2001; 276(6):4322-4329.
77. Luo Y, Lu W, Mohamedali KA et al. The glycine box: a determinant of specificity for fibroblast growth factor. Biochemistry 1998; 37(47):16506-16515.
78. DiGabriele AD, Lax I, Chen DI et al. Structure of a heparin-linked biologically active dimer of fibroblast growth factor. Nature 1998; 393(6687):812-817.
79. Faham S, Hileman RE, Fromm JR et al. Heparin structure and interactions with basic fibroblast growth factor. Science 1996; 271(5252):1116-1120.
80. Zhu X, Hsu BT, Rees DC. Structural studies of the binding of the anti-ulcer drug sucrose octasulfate to acidic fibroblast growth factor. Structure 1993; 1(1):27-34.
81. Hogan BL, Yingling JM. Epithelial/mesenchymal interactions and branching morphogenesis of the lung. Curr Opin Genet Dev 1998; 8(4):481-486.
82. Jin C, Wang F, Wu X et al. Directionally specific paracrine communication mediated by epithelial FGF9 to stromal FGFR3 in two-compartment premalignant prostate tumors. Cancer Res 2004; 64(13):4555-4562.
83. Pirvola U, Zhang X, Mantela J et al. Fgf9 signaling regulates inner ear morphogenesis through epithelial-mesenchymal interactions. Dev Biol 2004; 273(2):350-360.
84. Xu X, Weinstein M, Li C et al. Fibroblast growth factor receptor 2 (FGFR2)-mediated reciprocal regulation loop between FGF8 and FGF10 is essential for limb induction. Development 1998; 125(4):753-765.
85. Zhang X, Stappenbeck TS, White AC et al. Reciprocal epithelial-mesenchymal FGF signaling is required for cecal development. Development 2006; 133(1):173-180.
86. Bellusci S, Grindley J, Emoto H et al. Fibroblast growth factor 10 (FGF10) and branching morphogenesis in the embryonic mouse lung. Development 1997; 124(23):4867-4878.
87. Hoffman MP, Kidder BL, Steinberg ZL et al. Gene expression profiles of mouse submandibular gland development: FGFR1 regulates branching morphogenesis in vitro through BMP- and FGF-dependent mechanisms. Development 2002; 129(24):5767-5778.
88. Izvolsky KI, Shoykhet D, Yang Y et al. Heparan sulfate-FGF10 interactions during lung morphogenesis. Dev Biol 2003; 258(1):185-200.
89. Makarenkova HP, Ito M, Govindarajan V et al. FGF10 is an inducer and Pax6 a competence factor for lacrimal gland development. Development 2000; 127(12):2563-2572.
90. Sekine K, Ohuchi H, Fujiwara M et al. Fgf10 is essential for limb and lung formation. Nat Genet 1999; 21(1):138-141.
91. Liu A, Joyner AL. Early anterior/posterior patterning of the midbrain and cerebellum. Annu Rev Neurosci 2001; 24:869-896.
92. Sun X, Mariani FV, Martin GR. Functions of FGF signalling from the apical ectodermal ridge in limb development. Nature 2002; 418(6897):501-508.
93. Autosomal dominant hypophosphataemic rickets is associated with mutations in FGF23. Nat Genet 2000; 26(3):345-348.
94. Goetz R, Nakada Y, Hu MC et al. Isolated C-terminal tail of FGF23 alleviates hypophosphatemia by inhibiting FGF23-FGFR-Klotho complex formation. Proc Natl Acad Sci USA 2009.
95. Shimada T, Muto T, Urakawa I et al. Mutant FGF-23 responsible for autosomal dominant hypophosphatemic rickets is resistant to proteolytic cleavage and causes hypophosphatemia in vivo. Endocrinology 2002; 143(8):3179-3182.
96. Shimada T, Mizutani S, Muto T et al. Cloning and characterization of FGF23 as a causative factor of tumor-induced osteomalacia. Proc Natl Acad Sci USA 2001; 98(11):6500-6505.
97. Shiraki-Iida T, Aizawa H, Matsumura Y et al. Structure of the mouse klotho gene and its two transcripts encoding membrane and secreted protein. FEBS Lett 1998; 424(1-2):6-10.
98. Kharitonenkov A, Dunbar JD, Bina HA et al. FGF-21/FGF-21 receptor interaction and activation is determined by betaKlotho. J Cell Physiol 2008; 215(1):1-7.

THE STRUCTURAL BIOLOGY OF THE FGF19 SUBFAMILY

99. Kurosu H, Choi M, Ogawa Y et al. Tissue-specific expression of betaKlotho and fibroblast growth factor (FGF) receptor isoforms determines metabolic activity of FGF19 and FGF21. J Biol Chem 2007; 282(37):26687-26695.
100. Lin BC, Wang M, Blackmore C et al. Liver-specific activities of FGF19 require Klotho beta. J Biol Chem 2007; 282(37):27277-27284.
101. Ogawa Y, Kurosu H, Yamamoto M et al. BetaKlotho is required for metabolic activity of fibroblast growth factor 21. Proc Natl Acad Sci USA 2007; 104(18):7432-7437.
102. Suzuki M, Uehara Y, Motomura-Matsuzaka K et al. betaKlotho is required for fibroblast growth factor (FGF) 21 signaling through FGF receptor (FGFR) 1c and FGFR3c. Mol Endocrinol 2008; 22(4):1006-1014.
103. Wu X, Ge H, Gupte J et al. Co-receptor requirements for fibroblast growth factor-19 signaling. J Biol Chem 2007; 282(40):29069-29072.
104. Kurosu H, Ogawa Y, Miyoshi M et al. Regulation of fibroblast growth factor-23 signaling by klotho. J Biol Chem 2006; 281(10):6120-6123.
105. Urakawa I, Yamazaki Y, Shimada T et al. Klotho converts canonical FGF receptor into a specific receptor for FGF23. Nature 2006; 444(7120):770-774.
106. Kuro-o M, Matsumura Y, Aizawa H et al. Mutation of the mouse klotho gene leads to a syndrome resembling ageing. Nature 1997; 390(6655):45-51.
107. Shimada T, Kakitani M, Yamazaki Y et al. Targeted ablation of Fgf23 demonstrates an essential physiological role of FGF23 in phosphate and vitamin D metabolism. J Clin Invest 2004; 113(4):561-568.
108. Inagaki T, Choi M, Moschetta A et al. Fibroblast growth factor 15 functions as an enterohepatic signal to regulate bile acid homeostasis. Cell Metab 2005; 2(4):217-225.
109. Ito S, Fujimori T, Furuya A et al. Impaired negative feedback suppression of bile acid synthesis in mice lacking betaKlotho. J Clin Invest 2005; 115(8):2202-2208.
110. Yu C, Wang F, Kan M et al. Elevated cholesterol metabolism and bile acid synthesis in mice lacking membrane tyrosine kinase receptor FGFR4. J Biol Chem 2000; 275(20):15482-15489.
111. Gattineni J, Bates C, Twombley K et al. FGF23 decreases renal NaPi-2a and NaPi-2c expression and induces hypophosphatemia in vivo predominantly via FGF receptor 1. Am J Physiol Renal Physiol 2009; 297(2):F282-291.
112. Bai XY, Miao D, Goltzman D et al. The autosomal dominant hypophosphatemic rickets R176Q mutation in fibroblast growth factor 23 resists proteolytic cleavage and enhances in vivo biological potency. J Biol Chem 2003; 278(11):9843-9849.
113. Fukumoto S. Physiological regulation and disorders of phosphate metabolism—pivotal role of fibroblast growth factor 23. Intern Med 2008; 47(5):337-343.
114. Larsson T, Marsell R, Schipani E et al. Transgenic mice expressing fibroblast growth factor 23 under the control of the alpha1(I) collagen promoter exhibit growth retardation, osteomalacia and disturbed phosphate homeostasis. Endocrinology 2004; 145(7):3087-3094.
115. Liu S, Guo R, Simpson LG et al. Regulation of fibroblastic growth factor 23 expression but not degradation by PHEX. J Biol Chem 2003; 278(39):37419-37426.
116. Riminucci M, Collins MT, Fedarko NS et al. FGF-23 in fibrous dysplasia of bone and its relationship to renal phosphate wasting. J Clin Invest 2003; 112(5):683-692.
117. Saito H, Kusano K, Kinosaki M et al. Human fibroblast growth factor-23 mutants suppress Na+-dependent phosphate cotransport activity and 1alpha,25-dihydroxyvitamin D3 production. J Biol Chem 2003; 278(4):2206-2211.
118. Segawa H, Kawakami E, Kaneko I et al. Effect of hydrolysis-resistant FGF23-R179Q on dietary phosphate regulation of the renal type-II Na/Pi transporter. Pflugers Arch 2003; 446(5):585-592.
119. Lundasen T, Galman C, Angelin B et al. Circulating intestinal fibroblast growth factor 19 has a pronounced diurnal variation and modulates hepatic bile acid synthesis in man. J Intern Med 2006; 260(6):530-536.
120. Choi M, Moschetta A, Bookout AL et al. Identification of a hormonal basis for gallbladder filling. Nat Med 2006; 12(11):1253-1255.
121. Nishimura T, Nakatake Y, Konishi M et al. Identification of a novel FGF, FGF-21, preferentially expressed in the liver. Biochim Biophys Acta 2000; 1492(1):203-206.
122. Badman MK, Pissios P, Kennedy AR et al. Hepatic fibroblast growth factor 21 is regulated by PPARalpha and is a key mediator of hepatic lipid metabolism in ketotic states. Cell Metab 2007; 5(6):426-437.
123. Galman C, Lundasen T, Kharitonenkov A et al. The circulating metabolic regulator FGF21 is induced by prolonged fasting and PPARalpha activation in man. Cell Metab 2008; 8(2):169-174.
124. Inagaki T, Dutchak P, Zhao G et al. Endocrine regulation of the fasting response by PPARalpha-mediated induction of fibroblast growth factor 21. Cell Metab 2007; 5(6):415-425.
125. Palou M, Priego T, Sanchez J et al. Sequential changes in the expression of genes involved in lipid metabolism in adipose tissue and liver in response to fasting. Pflugers Arch 2008; 456(5):825-836.
126. Coskun T, Bina HA, Schneider MA et al. Fibroblast growth factor 21 corrects obesity in mice. Endocrinology 2008; 149(12): 6018-27.

127. Kharitonenkov A, Wroblewski VJ, Koester A et al. The metabolic state of diabetic monkeys is regulated by fibroblast growth factor-21. Endocrinology 2007; 148(2):774-781.
128. Wente W, Efanov AM, Brenner M et al. Fibroblast growth factor-21 improves pancreatic beta-cell function and survival by activation of extracellular signal-regulated kinase 1/2 and Akt signaling pathways. Diabetes 2006; 55(9):2470-2478.
129. Kharitonenkov A, Shanafelt AB. Fibroblast growth factor-21 as a therapeutic agent for metabolic diseases. BioDrugs 2008; 22(1):37-44.
130. Jonsson KB, Zahradnik R, Larsson T et al. Fibroblast growth factor 23 in oncogenic osteomalacia and X-linked hypophosphatemia. N Engl J Med 2003; 348(17):1656-1663.
131. Yamazaki Y, Okazaki R, Shibata M et al. Increased circulatory level of biologically active full-length FGF-23 in patients with hypophosphatemic rickets/osteomalacia. J Clin Endocrinol Metab 2002; 87(11):4957-4960.
132. Araya K, Fukumoto S, Backenroth R et al. A novel mutation in fibroblast growth factor 23 gene as a cause of tumoral calcinosis. J Clin Endocrinol Metab 2005; 90(10):5523-5527.
133. Benet-Pages A, Orlik P, Strom TM et al. An FGF23 missense mutation causes familial tumoral calcinosis with hyperphosphatemia. Hum Mol Genet 2005; 14(3):385-390.
134. Larsson T, Yu X, Davis SI et al. A novel recessive mutation in fibroblast growth factor-23 causes familial tumoral calcinosis. J Clin Endocrinol Metab 2005; 90(4):2424-2427.
135. Lyles KW, Burkes EJ, Ellis GJ et al. Genetic transmission of tumoral calcinosis: autosomal dominant with variable clinical expressivity. J Clin Endocrinol Metab 1985; 60(6):1093-1096.
136. Gutierrez OM, Mannstadt M, Isakova T et al. Fibroblast growth factor 23 and mortality among patients undergoing hemodialysis. N Engl J Med 2008; 359(6):584-592.
137. Harmer NJ, Pellegrini L, Chirgadze D et al. The crystal structure of fibroblast growth factor (FGF) 19 reveals novel features of the FGF family and offers a structural basis for its unusual receptor affinity. Biochemistry 2004; 43(3):629-640.
138. Wu X, Ge H, Lemon B et al. FGF19 induced hepatocyte proliferation is mediated through FGFR4 activation. J Biol Chem 2009.
139. Zhu H, Ramnarayan K, Anchin J et al. Glu-96 of basic fibroblast growth factor is essential for high affinity receptor binding. Identification by structure-based site-directed mutagenesis. J Biol Chem 1995; 270(37):21869-21874.
140. Burmeister WP, Cottaz S, Rollin P et al. High resolution X-ray crystallography shows that ascorbate is a cofactor for myrosinase and substitutes for the function of the catalytic base. J Biol Chem 2000; 275(50):39385-39393.

CHAPTER 2

KLOTHO AND βKLOTHO

Makoto Kuro-o

Department of Pathology, University of Texas Southwestern Medical Center at Dallas, Dallas, Texas, USA.
Email—makoto.kuro-o@utsouthwestern.edu

Abstract: Endocrine fibroblast growth factors (FGFs) have been recognized as hormones that regulate a variety of metabolic processes. FGF19 is secreted from intestine upon feeding and acts on liver to suppress bile acid synthesis. FGF21 is secreted from liver upon fasting and acts on adipose tissue to promote lipolysis and responses to fasting. FGF23 is secreted from bone and acts on kidney to inhibit phosphate reabsorption and vitamin D synthesis. One critical feature of endocrine FGFs is that they require the Klotho gene family of transmembrane proteins as coreceptors to bind their cognate FGF receptors and exert their biological activities. This chapter overviews function of Klotho family proteins as obligate coreceptors for endocrine FGFs and discusses potential link between Klothos and age-related diseases.

INTRODUCTION

In Greek mythology, three daughters of Zeus determine the longevity allocated to each person at birth. They are Klotho, Lachesis and Atropos who spins, measures and cuts the thread of life, respectively. The *klotho* gene, named after the spinner, was identified in 1997 as a gene mutated in a mouse strain that exhibited short life span and complex phenotypes resembling human premature aging syndrome.[1] This mouse strain (the *klotho* mouse) was serendipitously generated as a by-product during an attempt to make transgenic mice by conventional pronuclear microinjection of recombinant DNA.[2] Since exogenous DNA or transgene injected into fertilized mouse eggs is integrated randomly into the mouse genome, it occasionally disrupts an endogenous mouse gene(s) at the integration site (insertional mutation), which can result in unexpected phenotypes. The *klotho* mouse was one of the transgenic mouse lines that carried an expression construct for Na-proton exchanger-I,[2] but it did not express the transgene and was supposed to be of no use.

Endocrine FGFs and Klothos, edited by Makoto Kuro-o.
©2012 Landes Bioscience and Springer Science+Business Media.

However, this line displayed multiple aging-like phenotypes when homozygous for the transgene, suggesting that insertional mutation disrupted a gene that might be involved in the suppression of aging. Analysis of the *klotho* mouse genome revealed that ~10 copies of the transgene were integrated in tandem at a single locus on chromosome 5 and disrupted a promoter region of an unknown gene, which later was identified as the *klotho* gene.[1]

The *klotho* gene is composed of 5 exons[3,4] and encodes a Type-I single-pass transmembrane protein of 1,014-amino acid-long. The intracellular domain is short (10-amino acid-long) and has no known functional domains. The extracellular domain is composed of two internal repeats, termed KL1 and KL2, with weak homology to each other. Each domain has homology to Family 1 glycosidases, including lactose-phlorizin hydrolase of mammals and β-glucosidases of bacteria and plants.[1,5] These enzymes have exoglycosidase activity that hydrolyzes β-glucosidic linkage in saccharides, glycoproteins and glycolipids. However, recombinant Klotho protein did not have β-glucosidase-like enzymatic activity, probably because critical amino acid residues in putative active centers of Klotho protein diverge from those conserved in the β-glucosidase family of enzymes.[1,5]

The *klotho* gene is expressed in limited tissues and cell types. The highest expression is observed in distal convoluted tubules in the kidney and choroid plexus in the brain.[1] Lower expression is detected in several endocrine organs such as hypothalamus, pituitary gland, parathyroid gland,[6] pancreas, ovary, testis, placenta[1] and in some non-endocrine tissues such as inner ear[7] and breast epithelial cells.[8] Klotho gene expression is strongly suppressed in the *klotho* mouse due to the insertional mutation at the 5' flanking region of the *klotho* gene.[1] Thus, the *klotho* mouse is not a null but a severe hypomorph strain. A null strain for the *klotho* gene (*Klotho*[−/−] mouse) was generated by conventional gene targeting and exhibited phenotypes identical with those observed in the original *klotho* mouse.[9] Therefore, the *Klotho*[−/−] mouse and the original *klotho* mouse are collectively described as the *Klotho*[−/−] mouse in this chapter.

PHENOTYPES OF *KLOTHO*[−/−] MICE

Klotho[−/−] mice can be obtained from mating pairs of *Klotho*[+/−] mice in an expected Mendelian ratio, indicating that Klotho deficiency does not cause embryonic lethality. *Klotho*[−/−] mice appear normal and are indistinguishable from their wild-type littermates until 3 weeks of age, when they begin to show decreased spontaneous activity and die prematurely around 9 weeks of age. The aging-like phenotypes of *Klotho*[−/−] mice include:

Growth arrest. *Klotho*[−/−] mice stop growing around 4-5 weeks of age and barely gain body weight after that until they die around 9 weeks of age. Although growth hormone (GH) producing cells in pituitary gland are small and atrophic,[1] GH-deficiency is not a likely cause of growth arrest in *Klotho*[−/−] mice, because *Klotho*[−/−] mice do not grow even after the treatment with human GH.[10]

Hypogonadotropic hypogonadism. Both males and females of *Klotho*[−/−] mice are infertile.[1] Testis of *Klotho*[−/−] mice exhibits marked atrophy and contains no mature sperm. Seminiferous tubules are small and spermatogenesis does not progress beyond the pachytene stage. Female *Klotho*[−/−] mice never experience vaginal opening. Their ovaries and uteri are extremely atrophic. Folliculogenesis does not progress beyond the secondary stage. No corpus luteum is observed, indicating that ovulation never occurs. The impaired follicle maturation is not due to a primary defect in the ovary but due to a defect in pituitary and/or hypothalamus function (hypogonadotropic hypogonadism),

KLOTHO AND βKLOTHO

because (1) expression of lutenizing hormone (LH) and follicular stimulating hormone (FSH) in pituitary gland is significantly decreased in both males and females in *Klotho*[-/-] mice, (2) administration of FSH or gonadotropin releasing hormone (GnRH) partially restore follicular maturation and growth of uteri and (3) ovaries from *Klotho*[-/-] mice can ovulate and function normally when transplanted to wild-type female mice.[11] These observations suggest that Klotho is involved in the functional regulation of pituitary gland and/or hypothalamus that is required for inducing puberty.

Premature thymic involution. Thymus of *Klotho*[-/-] mice appears normal in size up to 4 weeks of age and thereafter undergoes rapid involution, resulting in invisible thymus by 8 weeks of age.[1] Because hematopoietic stem cells from *Klotho*[-/-] mice can differentiate into normal lymphoid cells when transplanted into SCID mice,[12] thymopoietic insufficiency in *Klotho*[-/-] mice is not due to an intrinsic defect in lymphohematopoietic progenitors. Rather, it is due to a defect in thymic epithelial cells that support proliferation and survival of lymphocytes. This pathophysiology is similar to that observed in thymic involution associated with normal aging. Thymic involution can be rescued by injection of KGF (keratinocyte growth factor, a.k.a. FGF7) that induces proliferation of thymic epithelial cells in *Klotho*[-/-] mice as well as in naturally aged wild-type mice.[12]

Tissue atrophy. In addition to reproductive organs and thymus, skin and intestine exhibit marked atrophy in *Klotho*[-/-] mice. Histology of the skin shows reduced number of hair follicles and reduced thickness of dermal and epidermal layers. No subcutaneous fat is observed.[1] Histology of the intestine shows reduced thickness of the mucosal layer and reduced height of intestinal villi.[13] Skin and intestine are high-turnover tissues containing stem cells that continuously regenerate epithelial cells. These highly proliferative cells can be labeled with nucleoside analog bromodeoxyuridine (BrdU). The number of BrdU-positive cells is significantly decreased in skin and intestine in *Klotho*[-/-] mice,[14] suggesting that atrophy of these tissues may be attributed to reduced number and/or impaired proliferation of stem cells.

Lymphocytopenia. Bone marrow is another highly proliferative tissue besides skin and intestine. Flow cytometric analysis of bone marrow cells revealed selective reduction of B-lymphoid lineage cells in *Klotho*[-/-] mice.[15] B-lymphocytoepenia is one of the characteristic features in age-related changes in the immune system that potentially contribute to increased morbidity to infectious diseases in the aged population.[16] Because B-lymphoid lineage cells also give rise to osteoclasts,[15] reduction of this lineage cells in bone marrow may also contribute to bone phenotypes in *Klotho*[-/-] mice.

Low-turnover osteopenia. Overall bone mineral density (BMD) of leg bones (femur and tibia) in *Klotho*[-/-] mice is lower than that in wild-type mice.[17] This is primarily due to decreased cortical bone thickness. Histomorphometric analysis shows significant decrease in bone formation rate associated with decreased number of both osteoclasts and osteoblasts. The decrease in bone formation exceeds that in bone resorption, resulting in a net bone loss. These are characteristic features of low-turnover osteopenia observed in senile osteoporosis in the aged or metabolic bone disease associated with chronic kidney disease. In contrast to decreased thickness of cortical bones, trabecular bones are rather increased in vertebrae and the metaphysis of tibia and femur.[17-19] The mechanism by which Klotho deficiency affects cortical and trabecular bones in a different way remains to be determined.

Aging lung. Aging lung is defined as age-related morphological and functional changes of the lung, including alveolar enlargement, decreased elastic recoil and increased residual volume.[20] These features are also observed in pulmonary emphysema, but by definition, pulmonary emphysema must be accompanied by alveolar wall destruction. Histological analysis of the lung of *Klotho*[-/-] mice shows progressive emphysematous changes including destruction of alveolar walls, enlargement of air spaces and reduction of elastic recoil.[21] Thus, the lung phenotype of *Klotho*[-/-] mice is consistent with pulmonary emphysema. In addition to these histological changes, the respiratory function of *Klotho*[-/-] mice is also consistent with emphysema characterized by increased compliance and expiratory time.[22] A mathematical model suggests that the pattern of alveolar destruction in *Klotho*[-/-] mice is consistent with a random destruction caused by a systemic factor(s) rather than a correlated destruction caused by a local factor(s) such as smoking.[23] However, some investigators did not observe alveolar wall destruction[22] and concluded that the lung phenotype of *Klotho*[-/-] mice was not pulmonary emphysema but aging lung. It is possible that whether or not alveolar wall destruction takes place in *Klotho*[-/-] mice may depend on age, genetic background and/or environmental factors.

Neurodegeneration. Cognitive function of *Klotho*[-/-] mice is impaired as determined by novel-object recognition and conditioned-fear tests.[24] Hippocampus neurons of *Klotho*[-/-] mice exhibit increased apoptosis and increased oxidative damages to lipid and DNA. Furthermore, treatment of *Klotho*[-/-] mice with an antioxidant α-tocopherol improves cognitive function, attenuates oxidative damage and reduces apoptosis in hippocampus. In addition, anterior horn cells (AHCs) in the spinal cord are degenerated in *Klotho*[-/-] mice. This is similar to the finding observed in patients with amyotrophic lateral sclerosis (ALS) with motor neuron degeneration. AHCs of *Klotho*[-/-] mice show significant decrease in cytoplasmic RNA, ribosomes and rough endoplasmic reticulum and accumulation of neurofilaments as those of ALS patients.[25] Recent studies have established that increased oxidative stress is a common mechanism underlying many neurodegenerative disorders including Alzheimer's disease, Parkinson's disease and ALS.[26] These observations suggest that Klotho protein may be involved in the regulation of oxidative stress[27] and protect neurons from oxidative damage.

Hearing disorder. Klotho expression is detected in the stria vascularis and spiral ligament of the inner ear. Although no morphological abnormalities are observed in the inner ear of *Klotho*[-/-] mice, they have significantly higher threshold for auditory brainstem response than wild-type mice, indicating the existence of hearing disorder.[7] Function of Klotho in the inner ear and mechanism behind the hearing disorder remains to be determined.

Vascular calcification. *Klotho*[-/-] mice develop severe vascular calcification in aorta and arteries in the kidney. Vascular calcification occurs in the media with little or no intimal thickening or accumulation of foam cells. Thus, these vascular changes are not atherosclerosis but Mönckeberg-type arteriosclerosis often observed in the aged as well as in patients with diabetes and chronic kidney disease in humans.[1] Vascular calcification is primarily caused by dysregulated phosphate and vitamin D homeostasis due to Klotho deficiency, which is relevant to fundamental function of Klotho protein and discussed later.

Aging phenotypes not observed in Klotho[-/-] *mice.* There is no report thus far indicating that *Klotho*[-/-] mice have an increased incidence of malignant tumors or increased amyloid plaques or neurofibrillary tangles in the brain, which are common features of human aging.

It should be noted that *Klotho*[-/-] mice develop multiple phenotypes commonly observed in naturally aged wild-type mice in an accelerated manner (decreased spontaneous activity, thymic involution, B lymphocytopenia, decreased BrdU-positive cells in hair follicles, etc.), suggesting that the aging process itself may be accelerated in *Klotho*[-/-] mice. In addition, they develop phenotypes never observed in aged mice but commonly observed

KLOTHO AND βKLOTHO

in aged humans (vascular calcification, osteopenia, hypogonadism, etc). Thus, *Klotho*[-/-] mice may be viewed not only as a mouse model of accelerated aging but also as a mouse model of human premature aging syndromes. *Klotho*[-/-] mice have been used as one of the best characterized mammalian models for human aging that manifests multiple aging-like phenotypes in a single individual.

KLOTHO PROTEIN FUNCTION

The clue to understanding of Klotho protein function was the fact that mice lacking fibroblast growth factor-23 (FGF23) exhibited phenotypes identical to those observed in *Klotho*[-/-] mice.[28] FGF23 is a bone-derived hormone and was originally identified as the gene mutated in autosomal dominant hypophosphatemic rickets (ADHR), a rare hereditary disorder that exhibits renal phosphate wasting and defects in bone mineralization (discussed in detail in Chapter 3).[29] Patients with ADHR carry mutations in the *FGF23* gene that confer resistance to proteolytic degradation of FGF23 protein.[30] As a result, blood FGF23 levels are inappropriately high in ADHR patients. FGF23 acts on kidney to induce phosphate excretion into urine (phosphaturia) and suppress synthesis of 1,25-dihydroxyvitamin D_3 (calcitriol or active form of vitamin D) (discussed in detail in Chapters 3 and 4), which explain why ADHR patients develop phosphate-wasting phenotypes including hypophosphatemia and defects in bone mineralization.

To understand physiological roles of FGF23, mice lacking FGF23 were generated by targeted gene disruption.[31] As expected, *Fgf23*[-/-] mice developed phosphate retention phenotypes. Specifically, *Fgf23*[-/-] mice exhibited hyperphosphatemia due to impaired renal phosphate excretion and hypervitaminosis D (high serum calcitriol levels), which also increased serum calcium levels. As a result, *Fgf23*[-/-] mice developed extensive ectopic calcification of various soft tissues, most notably in blood vessels. In addition to these predictable phenotypes, *Fgf23*[-/-] mice unexpectedly developed multiple aging-like phenotypes as observed in *Klotho*[-/-] mice, including short life span, hypogonadotropic hypogonadism, multiple organ atrophy and emphysematous lung.[32] On the other hand, *Klotho*[-/-] mice had been known to exhibit high serum levels of phosphate, calcium and calcitriol.[9] The striking similarity in phenotypes displayed by *Fgf23*[-/-] mice and *Klotho*[-/-] mice raised the possibility that FGF23 and Klotho might function in an identical signal transduction pathway.

Because FGF23 has very low affinity to any known FGF receptors (FGFRs),[33] identity of physiological receptors for FGF23 was unclear until it was reported in 2006 that Klotho protein functions as an obligate coreceptor for FGF23.[28] Klotho protein forms a constitutive binary complex with FGF receptor-1c (FGFR1c), -3c (FGFR3c) and -4 (FGFR4) on the cell surface. When bound to Klotho, these FGFRs acquire the ability to bind to FGF23 with high affinity.[28,34] Thus, FGF23 requires Klotho to bind and activate cognate FGFRs. These observations were reconfirmed later by several laboratories.[35-37] The fact that Klotho functions as the obligate coreceptor for FGF23 explains why *Fgf23*[-/-] mice and *Klotho*[-/-] mice exhibit identical phenotypes. Also, the fact that Klotho is expressed primarily in kidney explains why FGF23 can identify kidney as its target organ among many other tissues that express FGFRs. The bone-kidney endocrine axis mediated by FGF23 and Klotho has emerged as an indispensable machinery for maintaining phosphate homeostasis, because defects in either hormone (FGF23) or receptor (Klotho) result in hyperphosphatemia in mice[1,31,32] and in humans.[38]

PHOSPHATE HOMEOSTASIS

Our knowledge on endocrine regulation of phosphate homeostasis has advanced enormously since the discovery of FGF23 and Klotho. Ten years ago, vitamin D (calcitriol) and parathyroid hormone (PTH) were two major phosphate-regulating hormones. Calcitriol is primarily secreted from kidney and acts on intestine to increase absorption of phosphate and calcium from diet. When reduction of serum calcium level is detected by calcium-sensing receptor in parathyroid gland,[39] PTH is secreted and acts on kidney to promote calcitriol synthesis. Thus, PTH increases absorption of both calcium and phosphate from intestine. However, PTH does not induce a net positive phosphate balance, because PTH induces phosphaturia at the same time. Calcitriol in turn suppresses PTH production and secretion, thereby closing a negative feedback loop. This parathyroid-kidney endocrine axis mediated by PTH and calcitriol was the sole established mechanism for phosphate homeostasis at that time.

The discovery of FGF23 and Klotho has transformed this traditional view and added new dimensions to the endocrine regulation of phosphate homeostasis (Fig. 1). When phosphate is in excess, FGF23 is secreted from bone and acts on kidney where Klotho is expressed. FGF23 promotes phosphaturia and suppress calcitriol synthesis, thereby inducing negative phosphate balance. On the other hand, calcitriol induces FGF23 expression in bone,[40] indicating the existence of a negative feedback loop between bone (FGF23) and kidney (calcitriol). Importantly, parathyroid gland is one of the few organs

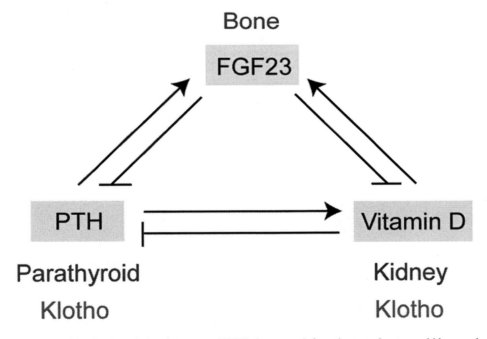

Figure 1. Phosphate-regulating hormones. FGF23 is secreted from bone and acts on kidney and pararthyroid gland where Klotho is expressed to suppress vitamin D and PTH, respectively. Vitamin D and PTH in turn increases FGF23 expression in bone.

KLOTHO AND βKLOTHO

that express Klotho endogenously, suggesting that parathyroid is also a target organ of FGF23. In fact, FGF23 acts on parathyroid gland to suppress PTH production and secretion.[6] PTH may stimulate FGF23 production and/or secretion, because parathyroidectomy significantly reduces serum FGF23 level in rats.[41] Thus, another feedback loop may exist between bone (FGF23) and parathyroid (PTH). The discovery of FGF23 and Klotho has identified novel endocrine axes between bone, intestine, parathyroid gland and kidney that coordinately regulate phosphate homeostasis.

A POTENTIAL LINK BETWEEN PHOSPHATE AND AGING

Several laboratories successfully rescued multiple aging-like phenotypes observed in *Fgf23*[-/-] mice and/or *Klotho*[-/-] mice by manipulating serum levels of phosphate, calcium and/or calcitriol using various genetic and dietary interventions (Table 1) (also see Chapter 5). First, disruption of the *Cyp27b1* gene rescued many phenotypes in *Fgf23*[-/-] mice[32] and *Klotho*[-/-] mice.[42] The *Cyp27b1* gene encodes 1α-hydroxylase, the enzyme that converts inactive vitamin D (25-hydroxyvitamin D_3) to active vitamin D (calcitriol). Disruption of the *Cyp27b1* gene depleted calcitriol, resulting in low blood phosphate and calcium levels. Vitamin D deficient diet also induced similar conditions and rescued *Fgf23*[-/-] mice and *Klotho*[-/-] mice as well.[9] Second, disruption of vitamin D receptor (VDR) also rescued *Fgf23*[-/-] phenotypes.[43] *VDR*[-/-] mice exhibited low blood phosphate and calcium levels, but serum calcitriol levels were increased due to end-organ resistance. Third, disruption of the *Slc34a1* gene (encoding Type 2a sodium phosphate cotransporter or Npt2a) rescued not only hyperphosphatemia but also multiple aging-like phenotypes in *Klotho*[-/-] mice.[44] Npt2a is expressed on the brush border membrane of renal proximal tubules in rodents and mediates phosphate reabsorption in kidney. Thus, inactivation of Npt2a abolished phosphate reabsorption in the kidney and reduced serum phosphate levels. Renal phosphate-wasting induced compensatory increase in serum calcitriol levels to maximize phosphate absorption from intestine. Because calcitriol also increases calcium absorption from intestine, disruption of Npt2a resulted in low phosphate, high calcitriol and high calcium in blood. Low phosphate diet also induced similar conditions and rescued *Fgf23*[-/-] and *Klotho*[-/-] phenotypes.[45,46] Thus, it is low serum phosphate levels, but not calcium or calcitriol, that is consistently associated with the rescue of aging-like phenotypes in *Fgf23*[-/-] and *Klotho*[-/-] mice. These observations have raised the possibility

Table 1. Interventions that rescue phenotypes of *Fgf23*[-/-] mice and/or *Klotho*[-/-] mice and resulting changes in serum phosphate, vitamin D and calcium levels

Interventions for Rescue	Serum Levels		
	Phosphate	Vitamin D	Calcium
Low phosphate diet	↓	↑	↑
Low vitamin D diet	↓	↓	↓
1α-hydroxylase knockout	↓	↓	↓
Vitamin D receptor knockout	↓	↑	↓
Na-phosphate cotransporter Type 2a knockout	↓	↑	↑

that phosphate is responsible for the aging-like phenotypes.[47] Although phosphate is an essential nutrient, it can be toxic when overloaded to the cell.

Several lines of indirect evidence also support the potential link between phosphate and aging. (1) Phosphate restriction extends life span in yeast.[48,49] (2) Transgenic mice that over-express Klotho have longer life span and lower serum phosphate levels due to higher renal phosphate excretion than wild-type mice.[50,51] (3) Serum phosphate levels inversely correlate with longevities of various mammalian species.[47,52] 4) Epidemiological studies indicated that every 1.0 mg/dl increase in serum phosphate level was independently associated with 23% increase in all-cause mortality even when serum phosphate is within normal range.[53] The mechanism by which phosphate affects aging processes remains to be determined.

SECRETED KLOTHO

The extracellular domain of Klotho protein is clipped on the cell surface by membrane-anchored proteases, including ADAM10, ADAM17 and BACE1[54,55] and released into blood, urine and cerebrospinal fluid.[50,56] Thus, Klotho protein exists in two forms; membrane Klotho and secreted Klotho. Membrane Klotho functions as the obligate coreceptor for FGF23. However, secreted Klotho is unable to function as a decoy receptor for FGF23, because it is the Klotho-FGFR complex, but not FGFR or Klotho alone, that can bind to FGF23 with high affinity. In addition, unlike membrane Klotho, secreted Klotho cannot efficiently support FGF23-induced activation of FGF signaling,[28] although secreted Klotho and FGF23 indeed forms ternary complexes with FGFR in vitro.[57] Then, what is the function of secreted Klotho? Function of secreted Klotho, if any, must be independent of FGF23.

Recent studies have demonstrated that secreted Klotho functions as a humoral factor that regulates activity of several ion channels, ion transporters and growth factor receptors on the cell surface. Specifically, secreted Klotho protein activates calcium channel TRPV5 (transient receptor potential cation channel, subfamily V, member 5)[58,59] and potassium channel ROMK1 (potassium inwardly-rectifying channel, subfamily J, member 1),[60] whereas secreted Klotho inhibits Npt2a (Type 2a Na-phosphate cotransporter).[51] These seemingly promiscuous activities of secreted Klotho may stem from its putative enzymatic activity as glycosidase. Recently, it was reported that secreted Klotho protein might have sialidase activity and removed terminal sialic acids from sugar chains of multiple cell-surface ion channels, including TRPV5 and ROMK1, to alter their cell-surface abundance[58,60] (see Chapter 7). Removal of terminal sialic acids from N-linked glycans exposes underlying galactose residues. Terminal galactose is a ligand for galectin-1, a lectin abundant in extracellular matrices. Thus, secreted Klotho induces interaction between galectin-1 and glycans on TRPV5, thereby trapping TRPV5 on the cell surface and increasing the number of cell-surface TRPV5. As a result, TRPV5-mediated calcium current is increased.[58] ROMK1-mediated potassium current is also increased by the same mechanism.[60] On the other hand, secreted Klotho decreases the number of cell-surface Npt2a.[51] Klotho-mediated modification of N-linked glycans on cell-surface Npt2a not only attenuates its transporter activity but also promotes its internalization and proteolytic degradation, thereby reducing cellular phosphate uptake in renal proximal tubular cells.

KLOTHO AND βKLOTHO   33

Secreted Klotho, when applied to cultured cells, suppresses activation of insulin/IGF-1 receptor,[8,50] although the precise mechanism remains to be determined. *Klotho*[-/-] mice are hypoglycemic and extremely sensitive to insulin.[43,61] In contrast, Klotho-overexpressing transgenic mice are moderately resistant to insulin and IGF-1, although they maintain normal fasting blood glucose levels and are not diabetic.[50] These observations are consistent with the ability of Klotho to inhibit activation of insulin/IGF-1 receptor. It is possible that secreted Klotho may modify N-linked glycans on insulin/IGF-1 receptors, which inhibits their activity and/or alters cell surface abundance. Of note, it has been known that treatment of cultured cells with bacterial sialidase induces insulin resistance.[62] In addition, transgenic mice that over-express a membrane-anchored sialidase (NEU3) develop glucose intolerance due to increased insulin resistance.[63] These observations suggest a potential link between insulin resistance and removal of cell surface sialic acids. The ability of secreted Klotho to inhibit insulin/IGF-1 signaling may partly contribute to the anti-aging properties of Klotho, since accumulating genetic evidence has indicated that moderate inhibition of insulin-like signaling pathway is one of the evolutionarily conserved mechanisms for suppressing aging.[64]

Secreted Klotho is also reported to bind to Wnt3 and inhibit activation of canonical Wnt signaling induced by Wnt3, which may contribute to reduced number of stem cells in the skin and intestine in *Klotho*[-/-] mice.[14] Thus, secreted Klotho protein has emerged as a novel endocrine regulator of glycoprotein function on the cell surface.

βKLOTHO

The *βklotho* gene was identified in 2000 based on sequence similarity with the *klotho* gene.[65] The *βklotho* gene also encoded a single-pass transmembrane protein, but tissue distribution of βKlotho was very different from that of Klotho. βKlotho was expressed predominantly in liver and white adipose tissue. At that time, there was no clue to βKlotho protein function. Five years later, mice lacking βKlotho were generated.[66] *βKlotho*[-/-] mice were grossly normal in appearance, but they exhibited increased bile acid synthesis due to increased expression of the *Cyp7a1* gene, which encodes the rate-limiting enzyme of bile acid synthesis in the liver. These phenotypes are reminiscent of those observed in mice lacking FGFR4[67] or FGF15.[68] FGF15 is expressed primarily in intestinal epithelial cells. FGF15 expression and secretion are increased in response to bile acid released into the intestinal lumen. Secreted FGF15 enters portal circulation and reaches to the liver, where it down-regulates expression of the *Cyp7a1* gene to close a negative feedback loop for bile acid synthesis[68] (see Chapter 10). FGFR4 is the most abundant FGFR isoform in the liver. These observations have led us to hypothesize that βKlotho may form a binary complex with FGFR4 to function as the obligate coreceptor for FGF15. Three groups independently demonstrated that this was indeed the case.[69-71] βKlotho forms a binary complex with FGFR4 and is required for FGF19 (the human ortholog of mouse FGF15) to bind FGFR4 with high affinity and exert its biological activity. This explains why defects in FGF15, FGFR4, or βKlotho result in identical phenotypic consequences (increased bile acid synthesis in the liver).

βKlotho also forms a binary complex with FGFR1c to function as the obligate coreceptor for FGF21.[72,73] FGF21 was originally identified as a liver-derived hormone that stimulates glucose uptake in adipocytes[74] (see Chapter 11). Unlike insulin, FGF21 increases glucose uptake through increasing expression of glucose transporter-1 and therefore requires several hours to exert this activity. Recent studies have revealed that FGF21 can mediate many other metabolic processes, including promotion of lipolysis in white adipose tissue and ketogenesis in the liver.[75] Furthermore, FGF21 can induce torpor in mice, a short-term hibernation state in which animals can save energy by reducing physical activity and body temperature.[75] All these changes can be regarded as an adaptation for fasting and in fact, FGF21 expression is induced by fasting.[75]

Although both FGF19 and FGF21 require βKlotho for high affinity binding to FGFR1c and FGFR4, they differ in the ability to activate FGFR4; FGF19 can signal through both βKlotho-FGFR1c and βKlotho-FGFR4 complexes, whereas FGF21 can signal only through the βKlotho-FGFR1c complex and not through the βKlotho-FGFR4 complex.[69] Because liver predominantly expresses FGFR4, FGF21 cannot activate FGF signaling in the liver, even though FGF21 indeed binds to the βKlotho-FGFR4 complex.[69] This may prevent FGF21 from acting as an autocrine factor in the liver. In contrast, FGF21 can act specifically on white adipose tissue where FGFR1 and βKlotho are co-expressed.[69] These studies have established a novel concept that tissue-specific expression of the Klotho gene family members determines target organs of endorine FGFs (Fig. 2). This may represent a new mechanism for regulating ligand-receptor interaction in target organs. The Klotho gene family (Klotho and βKlotho) may have evolved in the regulation of tissue-specific activity of the endocrine FGFs (FGF15/19, FGF21 and FGF23).

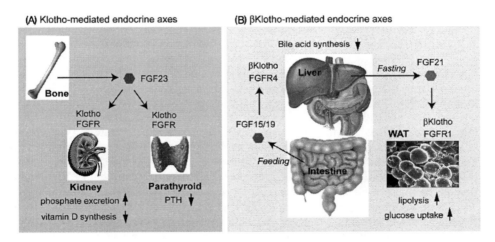

Figure 2. Endocrine axes mediated by the FGF and Klotho families. A) Klotho-mediated endocrine axes. FGF23 is secreted from bone and acts on kidney and parathyroid gland where Klotho and FGFR1c/3c/4 are co-expressed. FGF23 suppresses phosphate reabsorption and vitamin D (calcitriol) synthesis in the kidney. FGF23 also suppresses PTH expression and secretion in parathyroid gland. B) βKlotho-mediated endocrine axes. FGF15/19 is secreted from intestine upon feeding and acts on liver where βKlotho and FGFR4 are co-expressed. FGF15/19 suppresses bile acid synthesis in the liver. FGF21 is secreted from liver upon fasting and acts on white adipose tissue (WAT) where βKlotho and FGFR1c are co-expressed. FGF21 promotes lipolysis and glucose uptake in WAT. Reprinted with permission from Kuro-o M, Trends Endocrinol Metab 2008; 19:239-245. ©2008 Elsevier.

KLPH

A third member of the Klotho gene family, the *Klph* gene, was also identified based on amino acid sequence similarity with Klotho.[76] The *Klph* gene also encodes a single-pass transmembrane protein and is expressed in kidney and adipose tissues and most notably, in eyes.[77] KLPH protein binds to FGFR1b, 1c, 2c and 4. When expressed in HEK293 cells, KLPH supported activation of FGF signaling by FGF19.[77] Targeted disruption of the *Klph* gene in mice will provide insights into its physiological function.

NUCLEAR RECEPTOR REGULATION OF ENDOCRINE FGFs

Another important feature of endocrine FGFs is that their expression is primarily regulated by nuclear receptors and lipophyoic ligands (Fig. 3).

The FGF23-Klotho system. Administration of calcitriol (active form of vitamin D) increases expression of FGF23 in the bone in a vitamin D receptor (VDR) dependent manner, leading to increase in serum levels of FGF23 in mice.[40] FGF23 secreted from bone acts on kidney to suppress *Cyp27b1* and increase *Cyp24* expression.[40,78,79] The *Cyp27b1* gene encodes 1α-hydroxylase that synthesizes calcitriol, whereas the *Cyp24* gene encodes 24-hydroxylase that inactivates calcitriol.[80] Therefore, FGF23 reduces serum calcitriol levels, which in turn suppresses FGF23 expression in the bone and closes a negative feedback loop in the regulation of vitamin D homeostasis.[9] This also explains why *Fgf23*[-/-] and *Klotho*[-/-] mice have paradoxically high levels of serum calcitriol despite the fact that their blood phosphate and calcium levels are elevated.

The FGF15/19-βKlotho system. Expression of FGF15/19 (FGF15 is the mouse ortholog of human FGF19) is regulated by bile acid and its nuclear receptor FXR (farnesoid X receptor) in the small intestine.[81-83] In the initial step of enterohepatic circulation, bile acid released into intestinal lumen enters intestinal epithelial cells and binds to FXR. The ligand-bound FXR forms a heterodimer with retinoid X receptors (RXRs) and functions as a transcription factor that transactivates expression of FGF15/19. FGF15/19 secreted from intestine then acts on the liver via the βKlotho-FGFR4 complex to suppress expression of the *Cyp7a1* gene.[68,69] Because the *Cyp7a1* gene encodes the rate-limiting enzyme in the cascade of bile acid synthesis from cholesterol, suppression of *Cyp7a1* expression by FGF15/19 closes a negative feedback loop in the regulation of bile acid homeostasis. It is possible that FGF15/19 may be indirectly involved in cholesterol metabolism through regulating bile acid homeostasis.[84]

The FGF21-βKlotho system. Expression of FGF21 in the liver is regulated by peroxisome proliferator-activated receptor-α (PPARα).[75,85] PPARα, when bound to it ligands (fatty acids), forms a heterodimer with RXR and binds to PPARα-response elements in the promoter region of the *FGF21* gene to increase FGF21 expression. Because PPARα is activated upon fasting, expression of *FGF21* mRNA in the liver and serum levels of FGF21 are increased upon fasting as well.[75] PPARα is required to induce metabolic adaptation in the liver and other tissues in response to fasting, which includes promotion of gluconeogenesis, ketogenesis and fatty acid oxidation.[86-88] The ability of PPARα to induce these fasting responses is largely mediated by FGF21, because knockdown of FGF21 in the liver results in impaired ketogenesis and fatty acid oxidation in fasted mice.[85]

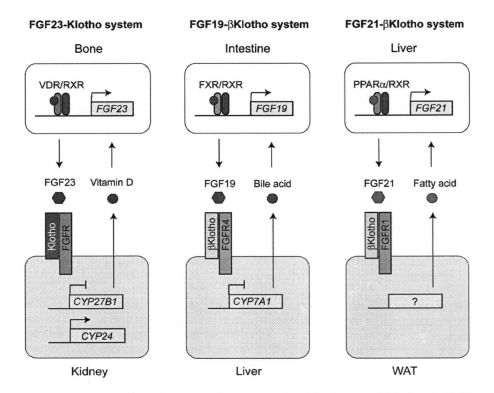

Figure 3. Feedback loops of the endocrine axes mediated by FGFs and Klothos. Left) The bone-kidney endocrine axis. Active form of vitamin D (calcitriol) binds to vitamin D receptor (VDR). The ligand-bound VDR forms a heterodimer with retinoid X receptor (RXR) and functions as a transcription factor that increases expression of FGF23 in osteocytes. FGF23 activates FGFR1c, 3c and/or 4 bound by Klotho in renal tubular cells and suppress expression of the *CYP27B1* gene and increases expression of the *CYP24* gene, thereby reducing serum calcitriol levels. Center) The intestine-liver endocrine axis. Bile acid binds to farnesoid X receptor (FXR). The ligand-bound FXR forms a heterodimer with RXR and functions as a transcription factor that increases expression of FGF19 in intestinal epithelial cells. FGF19 activates FGFR4 bound by βKlotho in hepatocytes and suppresses expression of the *CYP7A1* gene and reduces bile acid synthesis in the liver. Right) The liver-WAT (white adipose tissue) endocrine axis. Fatty acids bind to peroxisome proliferator-activated receptor α (PPARα). The ligand-bound PPARα forms a heterodimer with RXR and functions as a transcription factor that increases expression of FGF21 in hepatocytes. FGF21 activates FGFR1c bound by βKlotho in adipocytes and promotes lipolysis. A CYP gene(s) regulated by the FGF21-βKlotho system in adipocytes remains to be identified. Reprinted with permission from Kuro-o M, Trends Endocrinol Metab 2008; 19:239-245. ©2008 Elsevier.

CONCLUSION

Discovery of Klotho has unveiled the molecular mechanism by which endocrine FGFs exert their specific biological activity in their target organs. It has also identified novel endocrine axes with conserved feedback loop structure that is composed of lipophylic ligands (vitamin D, bile acid, fatty acids), nuclear receptors (VDR, FXR, PPARα) and cytochrome P450 family enzymes (Cyp27b1, Cyp24, Cyp7a1). These

KLOTHO AND βKLOTHO

endocrine axes mediated by endocrine FGFs and Klothos have opened a new research field in endocrinology, which potentially provides new insights into physiology and pathophysiology of multiple metabolic processes, including mineral metabolism, bile acid metabolism and fatty acid metabolism.

REFERENCES

1. Kuro-o M, Matsumura Y, Aizawa H et al. Mutation of the mouse klotho gene leads to a syndrome resembling ageing. Nature 1997; 390(6655):45-51.
2. Kuro-o M, Hanaoka K, Hiroi Y et al. Salt-sensitive hypertension in transgenic mice overexpressing Na(+)-proton exchanger. Circ Res 1995; 76(1):148-153.
3. Matsumura Y, Aizawa H, Shiraki-Iida T et al. Identification of the human klotho gene and its two transcripts encoding membrane and secreted klotho protein. Biochem Biophys Res Commun 1998; 242(3):626-630.
4. Shiraki-Iida T, Aizawa H, Matsumura Y et al. Structure of the mouse klotho gene and its two transcripts encoding membrane and secreted protein. FEBS Lett 1998; 424(1-2):6-10.
5. Mian IS. Sequence, structural, functional and phylogenetic analyses of three glycosidase families. Blood Cells Mol Dis 1998; 24(2):83-100.
6. Ben-Dov IZ, Galitzer H, Lavi-Moshayoff V et al. The parathyroid is a target organ for FGF23 in rats. J Clin Invest 2007; 117(12):4003-4008.
7. Kamemori M, Ohyama Y, Kurabayashi M et al. Expression of Klotho protein in the inner ear. Hear Res 2002; 171(1-2):103-110.
8. Wolf I, LevanonCohen S, Bose S et al. Klotho: a tumor suppressor and a modulator of the IGF-1 and FGF pathways in human breast cancer. Oncogene 2008; 27(56):7094-7105.
9. Tsujikawa H, Kurotaki Y, Fujimori T et al. Klotho, a gene related to a syndrome resembling human premature aging, functions in a negative regulatory circuit of vitamin D endocrine system. Mol Endocrinol 2003; 17(12):2393-2403.
10. Kashimada K, Yamashita T, Tsuji K et al. Defects in growth and bone metabolism in klotho mutant mice are resistant to GH treatment. J Endocrinol 2002; 174(3):403-410.
11. Toyama R, Fujimori T, Nabeshima Y et al. Impaired regulation of gonadotropins leads to the atrophy of the female reproductive system in klotho-deficient mice. Endocrinology 2006; 147(1):120-129.
12. Min D, Panoskaltsis-Mortari A, Kuro-o M et al. Sustained thymopoiesis and improvement in functional immunity induced by exogenous KGF administration in murine models of aging. Blood 2007; 109(6):2529-2537.
13. Nakatani T, Sarraj B, Ohnishi M et al. In vivo genetic evidence for klotho-dependent, fibroblast growth factor 23 (Fgf23)-mediated regulation of systemic phosphate homeostasis. FASEB J 2009; 23(2):433-441.
14. Liu H, Fergusson MM, Castilho RM et al. Augmented Wnt signaling in a mammalian model of accelerated aging. Science 2007; 317(5839):803-806.
15. Manabe N, Kawaguchi H, Chikuda H et al. Connection between B lymphocyte and osteoclast differentiation pathways. J Immunol 2001; 167(5):2625-2631.
16. Linton PJ, Dorshkind K. Age-related changes in lymphocyte development and function. Nat Immunol 2004; 5(2):133-139.
17. Kawaguchi H, Manabe N, Miyaura C et al. Independent impairment of osteoblast and osteoclast differentiation in klotho mouse exhibiting low-turnover osteopenia. J Clin Invest 1999; 104(3):229-237.
18. Yamashita T, Nifuji A, Furuya K et al. Elongation of the epiphyseal trabecular bone in transgenic mice carrying a klotho gene locus mutation that leads to a syndrome resembling aging. J Endocrinol 1998; 159(1):1-8.
19. Yamashita T, Yamashita M, Noda M et al. High-resolution micro-computed tomography analyses of the abnormal trabecular bone structures in klotho gene mutant mice. J Endocrinol 2000; 164(2):239-245.
20. Fukuchi Y. The aging lung and chronic obstructive pulmonary disease: similarity and difference. Proc Am Thorac Soc 2009; 6(7):570-572.
21. Suga T, Kurabayashi M, Sando Y et al. Disruption of the klotho gene causes pulmonary emphysema in mice. Defect in maintenance of pulmonary integrity during postnatal life. Am J Respir Cell Mol Biol 2000; 22(1):26-33.
22. Ishii M, Yamaguchi Y, Yamamoto H et al. Airspace enlargement with airway cell apoptosis in klotho mice: a model of aging lung. J Gerontol A Biol Sci Med Sci 2008; 63(12):1289-1298.
23. Sato A, Hirai T, Imura A et al. Morphological mechanism of the development of pulmonary emphysema in klotho mice. Proc Natl Acad Sci USA 2007; 104(7):2361-2365.
24. Nagai T, Yamada K, Kim HC et al. Cognition impairment in the genetic model of aging klotho gene mutant mice: a role of oxidative stress. FASEB J 2003; 17(1):50-52.

25. Anamizu Y, Kawaguchi H, Seichi A et al. Klotho insufficiency causes decrease of ribosomal RNA gene transcription activity, cytoplasmic RNA and rough ER in the spinal anterior horn cells. Acta Neuropathol (Berl) 2005; 109(5):457-466.
26. Lin MT, Beal MF. Mitochondrial dysfunction and oxidative stress in neurodegenerative diseases. Nature 2006; 443(7113):787-795.
27. Yamamoto M, Clark JD, Pastor JV et al. Regulation of oxidative stress by the anti-aging hormone Klotho. J Biol Chem 2005; 280(45):38029-38034.
28. Kurosu H, Ogawa Y, Miyoshi M et al. Regulation of fibroblast growth factor-23 signaling by klotho. J Biol Chem 2006; 281(10):6120-6123.
29. White KE, Evans WE, O'Rlordan JLH et al. Autosomal dominant hypophosphataemic rickets is associated with mutations in FGF23. Nat Genet 2000; 26(3):345-348.
30. White KE, Carn G, Lorenz-Depiereux B et al. Autosomal-dominant hypophosphatemic rickets (ADHR) mutations stabilize FGF-23. Kidney Int 2001; 60(6):2079-2086.
31. Shimada T, Kakitani M, Yamazaki Y et al. Targeted ablation of Fgf23 demonstrates an essential physiological role of FGF23 in phosphate and vitamin D metabolism. J Clin Invest 2004; 113(4):561-568.
32. Razzaque MS, Sitara D, Taguchi T et al. Premature aging-like phenotype in fibroblast growth factor 23 null mice is a vitamin D-mediated process. FASEB J 2006; 20(6):720-722.
33. Yu X, Ibrahimi OA, Goetz R et al. Analysis of the biochemical mechanisms for the endocrine actions of fibroblast growth factor-23. Endocrinology 2005; 146(11):4647-4656.
34. Goetz R, Beenken A, Ibrahimi OA et al. Molecular Insights into the Klotho-Dependent, Endocrine Mode of Action of FGF19 Subfamily Members. Mol Cell Biol 2007; 27:3417-3428.
35. Urakawa I, Yamazaki Y, Shimada T et al. Klotho converts canonical FGF receptor into a specific receptor for FGF23. Nature 2006; 444(7120):770-774.
36. Tomiyama K, Maeda R, Urakawa I et al. Relevant use of Klotho in FGF19 subfamily signaling system in vivo. Proc Natl Acad Sci USA 2010; 107(4):1666-1671.
37. Farrow EG, Davis SI, Summers LJ et al. Initial FGF23-mediated signaling occurs in the distal convoluted tubule. J Am Soc Nephrol 2009; 20(5):955-960.
38. Ichikawa S, Imel EA, Kreiter ML et al. A homozygous missense mutation in human KLOTHO causes severe tumoral calcinosis. J Clin Invest 2007; 117:2692-2701.
39. Brown EM, Gamba G, Riccardi D et al. Cloning and characterization of an extracellular Ca(2+)-sensing receptor from bovine parathyroid. Nature 1993; 366(6455):575-580.
40. Shimada T, Hasegawa H, Yamazaki Y et al. FGF-23 is a potent regulator of vitamin D metabolism and phosphate homeostasis. J Bone Miner Res 2004; 19(3):429-435.
41. Saji F, Shiizaki K, Shimada S et al. Regulation of fibroblast growth factor 23 production in bone in uremic rats. Nephron Physiol 2009; 111(4):59-66.
42. Ohnishi M, Nakatani T, Lanske B, Razzaque MS. Reversal of mineral ion homeostasis and soft-tissue calcification of klotho knockout mice by deletion of vitamin D 1alpha-hydroxylase. Kidney Int 2009; 75:1166-1172.
43. Hesse M, Frohlich LF, Zeitz U et al. Ablation of vitamin D signaling rescues bone, mineral and glucose homeostasis in Fgf-23 deficient mice. Matrix Biol 2007; 26(2):75-84.
44. Ohnishi M, Nakatani T, Lanske B et al. In vivo genetic evidence for suppressing vascular and soft-tissue calcification through the reduction of serum phosphate levels, even in the presence of high serum calcium and 1,25-dihydroxyvitamin d levels. Circ Cardiovasc Genet 2009; 2(6):583-590.
45. Stubbs JR, Liu S, Tang W et al. Role of Hyperphosphatemia and 1,25-Dihydroxyvitamin D in Vascular Calcification and Mortality in Fibroblastic Growth Factor 23 Null Mice. J Am Soc Nephrol 2007; 18(7):2116-2124.
46. Morishita K, Shirai A, Kubota M et al. The progression of aging in klotho mutant mice can be modified by dietary phosphorus and zinc. J Nutr 2001; 131(12):3182-3188.
47. Kuro-o M. A potential link between phosphate and aging—lessons from Klotho-deficient mice. Mech Ageing Dev 2010; 131(4):270-275.
48. Brauer MJ, Huttenhower C, Airoldi EM et al. Coordination of growth rate, cell cycle, stress response and metabolic activity in yeast. Mol Biol Cell 2008; 19(1):352-367.
49. Boer VM, de Winde JH, Pronk JT et al. The genome-wide transcriptional responses of Saccharomyces cerevisiae grown on glucose in aerobic chemostat cultures limited for carbon, nitrogen, phosphorus, or sulfur. J Biol Chem 2003; 278(5):3265-3274.
50. Kurosu H, Yamamoto M, Clark JD et al. Suppression of Aging in Mice by the Hormone Klotho. Science 2005; 309(5742):1829-1833.
51. Hu MC, Shi M, Zhang J et al. Klotho: a novel phosphaturic substance acting as an autocrine enzyme in the renal proximal tubule. FASEB J 2010.
52. Kuro-o M. Klotho and aging. Biochim Biophys Acta 2009; 1790(10):1049-1058.

KLOTHO AND βKLOTHO

53. Kestenbaum B, Sampson JN, Rudser KD et al. Serum phosphate levels and mortality risk among people with chronic kidney disease. J Am Soc Nephrol 2005; 16(2):520-528.

54. Chen CD, Podvin S, Gillespie E et al. Insulin stimulates the cleavage and release of the extracellular domain of Klotho by ADAM10 and ADAM17. Proc Natl Acad Sci USA 2007; 104(50):19796-19801.

55. Bloch L, Sineshchekova O, Reichenbach D et al. Klotho is a substrate for alpha-, beta- and gamma-secretase. FEBS Lett 2009; 583(19):3221-3224.

56. Imura A, Iwano A, Tohyama O et al. Secreted Klotho protein in sera and CSF: implication for posttranslational cleavage in release of Klotho protein from cell membrane. FEBS Lett 2004; 565(1-3):143-147.

57. Goetz R, Nakada Y, Hu MC et al. Isolated C-terminal tail of FGF23 alleviates hypophosphatemia by inhibiting FGF23-FGFR-Klotho complex formation. Proc Natl Acad Sci USA 2010; 107(1):407-412.

58. Cha SK, Ortega B, Kurosu H et al. Removal of sialic acid involving Klotho causes cell-surface retention of TRPV5 channel via binding to galectin-1. Proc Natl Acad Sci USA 2008; 105(28):9805-9810.

59. Chang Q, Hoefs S, van der Kemp AW et al. The beta-glucuronidase klotho hydrolyzes and activates the TRPV5 channel. Science 2005; 310(5747):490-493.

60. Cha SK, Hu MC, Kurosu H et al. Regulation of ROMK1 channel and renal K+ excretion by Klotho. Mol Pharmacol 2009; 76(1):38-46.

61. Utsugi T, Ohno T, Ohyama Y et al. Decreased insulin production and increased insulin sensitivity in the klotho mutant mouse, a novel animal model for human aging. Metabolism 2000; 49(9):1118-1123.

62. Salhanick AI, Amatruda JM. Role of sialic acid in insulin action and the insulin resistance of diabetes mellitus. Am J Physiol 1988; 255(2 Pt 1):E173-E179.

63. Sasaki A, Hata K, Suzuki S et al. Overexpression of plasma membrane-associated sialidase attenuates insulin signaling in transgenic mice. J Biol Chem 2003; 278(30):27896-27902.

64. Tatar M, Bartke A, Antebi A. The Endocrine Regulation of Aging by Insulin-like Signals. Science 2003; 299(5611):1346-1351.

65. Ito S, Kinoshita S, Shiraishi N et al. Molecular cloning and expression analyses of mouse betaklotho, which encodes a novel Klotho family protein. Mech Dev 2000; 98(1-2):115-119.

66. Ito S, Fujimori T, Furuya A et al. Impaired negative feedback suppression of bile acid synthesis in mice lacking betaKlotho. J Clin Invest 2005; 115(8):2202-2208.

67. Yu C, Wang F, Kan M et al. Elevated cholesterol metabolism and bile acid synthesis in mice lacking membrane tyrosine kinase receptor FGFR4. J Biol Chem 2000; 275(20):15482-15489.

68. Inagaki T, Choi M, Moschetta A et al. Fibroblast growth factor 15 functions as an enterohepatic signal to regulate bile acid homeostasis. Cell Metab 2005; 2(4):217-225.

69. Kurosu H, Choi M, Ogawa Y et al. Tissue-specific expression of betaKlotho and fibroblast growth factor (FGF) receptor isoforms determines metabolic activity of FGF19 and FGF21. J Biol Chem 2007; 282(37):26687-26695.

70. Lin BC, Wang M, Blackmore C et al. Liver-specific Activities of FGF19 Require Klotho beta. J Biol Chem 2007; 282(37):27277-27284.

71. Wu X, Ge H, Gupte J et al. Co-receptor requirements for fibroblast growth factor-19 signaling. J Biol Chem 2007; 282(40):29069-29072.

72. Ogawa Y, Kurosu H, Yamamoto M et al. betaKlotho is required for metabolic activity of fibroblast growth factor 21. Proc Natl Acad Sci USA 2007; 104(18):7432-7437.

73. Kharitonenkov A, Dunbar JD, Bina HA et al. FGF-21/FGF-21 receptor interaction and activation is determined by betaKlotho. J Cell Physiol 2007; 215(1):1-7.

74. Kharitonenkov A, Shiyanova TL, Koester A et al. FGF-21 as a novel metabolic regulator. J Clin Invest 2005; 115(6):1627-1635.

75. Inagaki T, Dutchak P, Zhao G et al. Endocrine Regulation of the Fasting Response by PPARalpha-Mediated Induction of Fibroblast Growth Factor 21. Cell Metab 2007; 5(6):415-425.

76. Ito S, Fujimori T, Hayashizaki Y et al. Identification of a novel mouse membrane-bound family 1 glycosidase-like protein, which carries an atypical active site structure. Biochim Biophys Acta 2002; 1576(3):341-345.

77. Tacer KF, Bookout AL, Ding X et al. Comprehensive Expression Atlas of the Fibroblast Growth Factor System in Adult Mouse. Mol Endocrinol 2010; In press.

78. Shimada T, Mizutani S, Muto T et al. Cloning and characterization of FGF23 as a causative factor of tumor-induced osteomalacia. Proc Natl Acad Sci USA 2001; 98(11):6500-6505.

79. Bai XY, Miao D, Goltzman D et al. The autosomal dominant hypophosphatemic rickets R176Q mutation in fibroblast growth factor 23 resists proteolytic cleavage and enhances in vivo biological potency. J Biol Chem 2003; 278(11):9843-9849.

80. Dusso AS, Brown AJ, Slatopolsky E. Vitamin D. Am J Physiol Renal Physiol 2005; 289(1):F8-28.

81. Goodwin B, Jones SA, Price RR et al. A regulatory cascade of the nuclear receptors FXR, SHP-1 and LRH-1 represses bile acid biosynthesis. Mol Cell 2000; 6(3):517-526.

82. Lu TT, Makishima M, Repa JJ et al. Molecular basis for feedback regulation of bile acid synthesis by nuclear receptors. Mol Cell 2000; 6(3):507-515.
83. Kalaany NY, Mangelsdorf DJ. LXRS and FXR: the yin and yang of cholesterol and fat metabolism. Annu Rev Physiol 2006; 68:159-191.
84. Kuipers F, Stroeve JH, Caron S et al. Bile acids, farnesoid X receptor, atherosclerosis and metabolic control. Curr Opin Lipidol 2007; 18(3):289-297.
85. Badman MK, Pissios P, Kennedy AR et al. Hepatic Fibroblast Growth Factor 21 Is Regulated by PPARalpha and Is a Key Mediator of Hepatic Lipid Metabolism in Ketotic States. Cell Metab 2007; 5(6):426-437.
86. Lefebvre P, Chinetti G, Fruchart JC et al. Sorting out the roles of PPAR alpha in energy metabolism and vascular homeostasis. J Clin Invest 2006; 116(3):571-580.
87. Kersten S, Seydoux J, Peters JM et al. Peroxisome proliferator-activated receptor alpha mediates the adaptive response to fasting. journal of clinical investigation 1999; 103(11):1489-1498.
88. Kersten S, Desvergne B, Wahli W. Roles of PPARs in health and disease. Nature 2000; 405(6785):421-424.

CHAPTER 3

FGF23 AND SYNDROMES OF ABNORMAL RENAL PHOSPHATE HANDLING

Clemens Bergwitz[1] and Harald Jüppner*[,1,2]

[1]Endocrine Unit and [2]Pediatric Nephrology Unit, Massachusetts General Hospital and Harvard Medical School, Boston, Massachusetts, USA.
*Corresponding Author: Harald Jüppner—Email: hjueppner@partners.org

Abstract: Fibroblast growth factor 23 (FGF23) is part of a previously unrecognized hormonal bone-parathyroid-kidney axis, which is modulated by 1,25(OH)$_2$-vitamin D (1,25(OH)$_2$D), dietary and circulating phosphate and possibly PTH. FGF23 was discovered as the humoral factor in tumors that causes hypophosphatemia and osteomalacia and through the identification of a mutant form of FGF23 that leads to autosomal dominant hypophosphatemic rickets (ADHR), a rare genetic disorder. FGF23 appears to be mainly secreted by osteocytes where its expression is up-regulated by 1,25(OH)$_2$D and probably by increased serum phosphate levels. Its synthesis and secretion is reduced through yet unknown mechanisms that involve the phosphate-regulating gene with homologies to endopeptidases on the X chromosome (PHEX), dentin matrix protein 1 (DMP1) and ecto-nucleotide pyrophosphatase/phosphodiesterase 1 (ENPP1). Consequently, loss-of-function mutations in these genes underlie hypophosphatemic disorders that are either X-linked or autosomal recessive. Impaired O-glycosylation of FGF23 due to the lack of UDP-N-acetyl-alpha-D-galactosamine:polypeptide N-acetylgalactosaminyl-transferase 3 (GALNT3) or due to certain homozygous FGF23 mutations results in reduced secretion of intact FGF23 and leads to familial hyperphosphatemic tumoral calcinosis. FGF23 acts through FGF-receptors and the coreceptor Klotho to reduce 1,25(OH)$_2$D synthesis in the kidney and probably the synthesis of parathyroid hormone (PTH) by the parathyroid glands. It furthermore synergizes with PTH to increase renal phosphate excretion by reducing expression of the sodium-phosphate cotransporters NaPi-IIa and NaPi-IIc in the proximal tubules. Loss-of-function mutations in these two transporters lead to autosomal recessive Fanconi syndrome or to hereditary hypophosphatemic rickets with hypercalciuria, respectively.

Endocrine FGFs and Klothos, edited by Makoto Kuro-o.
©2012 Landes Bioscience and Springer Science+Business Media.

INTRODUCTION

Identification of the genetic causes of rare familial disorders of renal phosphate handling has provided, over the past decade, important novel insights into the regulation of mammalian phosphate homeostasis. This chapter will provide an update on the current knowledge of the pathophysiology, the clinical presentation, diagnostic evaluation and therapy of FGF23-dependent and -independent disorders of phosphate homeostasis and tissue mineralization.

AUTOSOMAL DOMINANT HYPOPHOSPHATEMIC RICKETS (ADHR, OMIM 193100)

Genetics of ADHR

In 1971 Bianchine et al described a small family, in which male-to-male transmission suggested an autosomal dominant form of hypophosphatemic rickets (ADHR, for a full list of abbreviations see Table 1).[1] Subsequently, Econs and McEnery identified a large family, in which numerous members are affected by ADHR.[2] The authors were able to map the disorder to a locus on chromosome 12p13,[3] which ultimately allowed identification of the genetic mutation leading to this rare inherited form of hypophosphatemia.[4] The incidence of ADHR is unknown; thus far, only a few families have been described in which hypophosphatemia follows an autosomal dominant trait.[4-6] ADHR thus appears to be much less frequent than X-linked hypophosphatemia (XLH), which affects approximately 1:20,000 births.[7] The ADHR patients described to date carry a heterozygous mutation in fibroblast growth factor 23 (FGF23, chromosome 12p13.3) at amino acid positions 176 or 179, a site for cleavage by subtilisin-like proptein convertases (SPC)[4-6] (Fig. 1). Although the exact mechanism is unknown, it appears that ADHR mutations enhance resistance towards SPC leading to deminished intracellular cleavage of intact FGF23. The resulting secretion of inappropriate amounts of biologically active FGF23 suggests that the identified amino acid changes are "gain-of-function" mutations, which lead to renal phosphate wasting, possibly combined with an abnormal feed-back regulation of FGF23 synthesis, as discussed below.

The phosphaturic action of FGF23 requires the coreceptor alpha Klotho (KL, also see below). A genetic defect leading to a disorder resembling ADHR has recently been reported in a single sporadic case of hypophosphatemia.[8] The patient, a 13 month old girl, presented with rickets due to excessive renal phosphate excretion and hyperparathyroidism and cytogenetic studies revealed a de novo chromosomal translocation with the breakpoint being located adjacent to the gene encoding KL. As a result, plasma levels of soluble KL and KL-associated beta-glucuronidase activity were increased, along with increased levels of immunoreactive FGF23. It remains uncertain, however, whether the elevated levels of FGF23 and/or PTH are solely responsible for the increased urinary excretion of phosphate, or whether the elevated levels of soluble Klotho contribute to the abnormal renal handling of this mineral.

FGF23 Synthesis and Secretion

The main sources of FGF23 are osteocytes and osteoblasts in the skeleton, but low levels of uncertain biological significance can be detected in the ventrolateral thalamic nucleus,

FGF23 AND SYNDROMES OF ABNORMAL RENAL PHOSPHATE HANDLING
43

Table 1. Abbreviations

Key-Term/ Acronym	Definition
GALNT3	UDP-N-Acetyl-α-D-Galactosamine:Polypeptide N-Acetylgalactosaminyl-transferease, isoform 3 appears to be important for O-glycosylation of FGF23
FGF23	Fibroblast growth Factor 23
KL	Alpha Klotho
PHEX	Phosphate-regulating gene with Homologies to Endopeptidases on the X chromosome
FGF23	Fibroblast Growth Factor 23
DMP1	Dentin Matrix Acidic Phosphoprotein 1
MEPE	Matrix extracellular phosphoprotein
FGFR	Fibroblast Growth Factor Receptor
ARHP	Autosomal Recessive Hypophosphatemia
HHRH	Hereditary Hypophosphatemic Rickets with Hypercalciuria
PTHR1	PTH/PTHrP Receptor
VDR	Vitamin D receptor, forms heterodimer with RXR
HFTC	Hyperphosphatemic Familial Tumoral Calcinosis
XLH	X-linked Dominant hypophosphatemia
ADHR	Autosomal Dominant Hypophosphatemic Rickets
SLC34	Solute Carrier Family 34 (sodium-phosphate cotransporter) (also referred to as NaPi-II or NPT2), members 1 (NaPi-IIa) and 3 (NaPi-IIc) are expressed in the proximal renal tubule, member 2 (NaPi-IIb) is expressed in the intestine
CYP27B1	1-alpha-hydroyxylase, vitamin D-activating enzyme, which is expressed in the renal proximal tubule
TRPV	Transient receptor potential cation channel, subfamily V, members 5 and 6 are calcium-selective
PMCA	plasma membrane Ca^{2+} ATPase
Pit-2	Solute Carrier Family 20 (sodium-phosphate cotransporter), member 2

the thymus, the small intestine and the heart.[9] FGF23 synthesis is stimulated by dietary phosphate[10-12] and the application of 1,25(OH)2D.[13,14] Conversely, the phosphate-regulating gene with homologies to endopeptidases on the X chromosome (PHEX), dentin matrix protein 1 (DMP1) and ecto-nucleotide pyrophosphatase/phosphodiesterase 1(ENPP1) appear to have an important role in suppressing FGF23 synthesis. PHEX and DMP1 were shown to be genetically up-stream of FGF23, as the lack of either of these two proteins leads to an up-regulation of FGF23 expression in bone, most likely through indirect mechanisms.[15] After cleavage of the signal sequence comprising 24 amino acids and O-glycosylation by UDP-N-acetyl-alpha-D-galactosamine:polypeptide N-acetylgalactosaminyl-transferase 3 (GALNT3), mature FGF23(25-251) is secreted into the circulation. O-glycosylation of FGF23 occurs in the 162-228 region[16] and this posttranslational modification, which may involve residue 178, appears to protect FGF23 from cleavage by subtilisin-like proprotein convertases as shown with recombinant peptides in vitro.[17] O-glycosylation is also essential for secretion and processing of intact FGF23 in CHO cells.[17,18] Furthermore, in HEK293 cells, expression of GALNT3 is stimulated by extracellular phosphate and suppressed by extracellular calcium and 1,25(OH)2D.[19] This suggests that GALNT3 may be an important component of an as-of-yet incompletely understood circuit regulating FGF23 secretion in

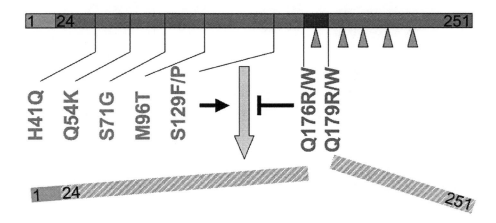

Figure 1. Known mutations in FGF23. Human mutations leading to hyperphosphatemic tumoral calcinosis Type 2 (HFTC2, red) are located in the first half of the FGF23 molecule. These impair, directly or indirectly, O-glycosylation (solid arrows) in the second half of FGF23 and thereby prevent secretion of intact FGF23. Human FGF23 mutations leading to autosomal dominant hypophosphatemic rickets (ADHR, green) change the subtilisin-furin cleavage motif RXXR in amino acid positions 176-179 (blue box) and lead to increased secretion and stability of FGF23 that is independent of O-glycosylation. A color version of this figure is available online at http://www.landesbioscience.com/curie.

vivo, which may be disrupted by ADHR mutations. The two FGF23 amino acid residues that are mutated in patients affected by ADHR, R176 or R179, constitute, as mentioned above, the site for cleavage by subtilisin/furin-like endopeptidases. Just like O-linked glycosylation of residue T178 in wild-type FGF23,[17] mutation of either residue 176 or 179 appears to protect FGF23 against intracellular proteolytic cleavage and degradation,[20] leading to the secretion of intact FGF23, although it remains unclear, how production of the ADHR mutant escapes feed-back regulation by phosphate and 1,25(OH)$_2$D.

FGF23 Mode of Action

The phosphaturic actions of FGF23 require KL as a coreceptor. Consequently, a homozygous inactivating KL mutation was shown to lead to FGF23 resistance in a patient with familial hyperphosphatemic tumoral calcinosis Type 3 (HFTC3, see below)[21] and mice that are null for Kl, the murine ortholog of KL, show a corresponding phenotype.[22,23] Immunoprecipitation experiments, surface plasmon resonance (SPR) spectroscopy and functional assays measuring the mitogenic response of BaF3 cells or activation of the MAPK-pathway in HEK293 cells have shown that KL, a protein with a single membrane-spanning domain, forms a ternary complex with FGF23 in conjunction with FGFR1c and other FGF receptors.[24,25] Recent work using neutralizing anti-FGF23 antibodies indicates that the N-terminal portion of FGF23 interacts with FGFR1c, while the C-terminus binds to KL and both interactions appear to be important for bioactivity in vitro and in vivo.[26] Low affinity binding of FGF23 to FGFR1c and other FGF receptors in the absence of KL has been shown to occur in vitro.[27-29] However, this coreceptor appears to be essential for high-affinity binding of FGF23 to the receptor and its phosphate-regulating effects in vivo. The ablation of either FGFR3 or FGFR4, individually or in combination, was unable to rescue the hyp phenotype in mice suggesting that these two FGF receptors are not involved in mediating the renal effects of FGF23.[30]

The site of FGF23 action in the kidney is still controversial. While FGF23 decreases NPT2a and NPT2c expression[31-33] and CYP27B1 activity in the proximal tubules,[34-36] the coreceptor KL is expressed mainly in the distal tubules. In addition, mice injected with recombinant FGF23 show phosphorylation of MAPK and an up-regulation of the early growth-response gene 1 (Egr-1) in the distal tubules.[37] These findings suggest either that FGF23 uses a noncanonical signal-transduction pathway in the proximal tubules or that it induces the secretion of an "intermediary phosphatonin" from the distal tubules, which acts in a paracrine fashion on the proximal tubules. KL may furthermore have ligand-independent actions by regulating expression of the calcium channels TRPV5[38] and the potassium channel ROMK in the distal tubules.[39]

Although the transgenic overexpression of soluble KL can prolong the life span of mice,[24,40] no data on FGF23 levels or mineral ion homeostasis have been reported to date in these animals. Increased FGF23 levels in the patient described by Brownstein et al (see above) suggests that KL may interfere with the normal feed-back inhibition of FGF23 synthesis and/or secretion; however, the exact mechanism of hypophosphatemia in this patient remains unclear.

Clinical Findings in ADHR

Chronic hypophosphatemia in ADHR can lead to abnormal bone growth and mineralization, i.e., rickets in growing children and osteomalacia in adults. As in nutritional rickets, osteoid is undermineralized, leading to a blurring of the microtrabecular architecture and pseudo-fractures (Looser zones) on radiographs. The clinical consequences of rickets or osteomalacia in children or adults, respectively, are bone pain and impaired mechanical properties of the affected bones leading to deformities of the lower extremities, often leading to characteristic wind-swept deformities. Lack of chondrocyte apoptosis in the growing skeleton furthermore leads to an expansion of the epiphyses in children, giving rise to swollen wrists and rachitic rosary.[41] Serum biochemical findings indicative of rickets and osteomalacia include elevated bone-specific alkaline phosphatase, osteocalcin, procollagen, pyridinoline cross-links as well as N- and C-telopeptides.[42,43] Chronic hypophosphatemia also leads to muscle weakness, which when compared to the effects on the skeleton, is less well understood and may be related to the role of phosphate in intracellular signal transduction and synthesis of ATP or creatine phosphate.[42,43]

While rickets or osteomalacia can be observed to variable degrees in all hypophosphatemic disorders, including ADHR, subtle but important differences can guide the differential diagnostic and therapeutic decisions. When compared to other types of hypophosphatemic rickets, ADHR appears to have a quite variable clinical phenotype. The kindred described by Econs et al[2] can be divided into two subgroups. Group 1 consisted of nine female patients who presented with renal phosphate-wasting during adolescence and adulthood. These individuals presented with bone pain, weakness and insufficiency fractures, but without deformities of the lower extremities. Group 2 consisted of nine male and female patients, with onset of symptoms, including bone deformities, during childhood. Two of these nine individuals lost the renal phosphate-wasting defect later in life, which is consistent with observations in the affected members of two other unrelated kindreds, in which five of eight carriers of a heterozygous, "activating" FGF23 mutation had lost the renal phosphate-wasting defect later in life.[5,6] Enthesopathies, which refer to painful or indolent mineral deposits near the insertion sites of tendons usually at the spine and lower extremities, can occur in patients with XLH, ADHR, or autosomal recessive hypophosphatemia (ARHP) (see below) and are readily identified on radiographs.[44] The development of these lesions may involve FGFR3 and Klotho, which are expressed in fibrocartilage cells of the entheses.[45] Dental cysts or dental dysplasia, respectively, were described in patients with XLH and ADHR, but these abnormalities do not appear to be as common in ADHR as in XLH.[2,5] Likewise, midfacial hypoplasia and frontal bossing may also be observed in ADHR.[5] Although these findings may be related to the severity of hypophosphatemia, therapy with $1,25(OH)_2D$ and phosphate supplements often cannot reverse or prevent these changes. Thus, local effects of FGF23 excess and activation of canonical FGF receptor signaling (Klotho-independent) or other factors up-stream of FGF23 may play a role.[44,46]

Diagnostic Evaluation of ADHR and Other Disorders of Phosphate Homeostasis

The clinical presentation of ADHR and other hypophosphatemic disorders includes renal phosphate wasting, leading to rickets and/or osteomalacia and abnormal vitamin D metabolism. The diagnosis therefore depends on careful evaluation of phosphate homeostasis, which can be challenging since serum phosphate concentrations are influenced by the time of day, relationship to meals and age of the subject and none of the methods for determination of tubular phosphate reabsorption are entirely satisfying. To determine the cause of abnormal serum phosphate levels in a patient, who has normal parathyroid and renal function, we generally first assess his or her tubular reabsorption for phosphate (%TRP). For this purpose the patient is asked to collect a 3-hour timed urine for measurement of phosphate and creatinine along with the corresponding serum parameters after 8 hours of fasting. %TRP is calculated according to Formula 1 and the tubular maximum of reabsorption for phosphate (TmP/GFR) is derived from a nomogram, which was devised by Walton and Bijvoet[47] to correct for the nonlinear relationships of %TRP and TmP/GFR when %TRP is higher 80%.

Formula 1: %TRP = 100x(1 – (U-P x S-creat)/(S-P x U-creat)) using timed 3-hour urine and blood creatinine and phosphate concentrations

TmP/GFR reflects the threshold of the serum phosphate concentration above which phosphate is no longer fully reclaimed from the glomerular filtrate in the proximal tubules. While the TmP/GFR derived from the Walton and Bijvoet nomogram is generally sufficient in adults, the nomogram does not accommodate the higher normal range of serum phosphate values in newborns and toddlers.[48] Thus, calculating TP/GFR provides a more accurate assessment of renal phosphate handling in the pediatric population[49] (Formula 2).

Formula 2: TP/GFR = S-P − (U-P x S-creat/U-creat), using simultaneous urine and blood creatinine and phosphate concentrations

Inappropriately low %TRP in the setting of hypophosphatemia is suggestive of a proximal renal tubular defect as the underlying cause, which can be further classified by determining the patient's vitamin D status: In patients with ADHR, %TRP and $1,25(OH)_2D$ levels are concordantly (inappropriately) reduced, suggesting excess FGF23 activity. In contrast, appropriately elevated $1,25(OH)_2D$ levels suggest an FGF23-independent, renal tubular defect leading to abnormal serum phosphate levels, which can for example be seen in hereditary hypophosphatemic rickets with hypercalciuria (HHRH).[50] If hypophosphatemia occurs without an obvious increase in urinary phosphate excretion, nutritional deficiencies, malabsorption, or liver disease should be considered; in rare cases, a primary intestinal defect involving reduced NPT2b expression has been described in pulmonary alveolar microlithiasis (PAM)[51] (see Table 2).

Excess production of $1,25(OH)_2D$ due to FGF23-independent hypophosphatemia may lead to increased absorption of calcium in the gut, resulting in hypercalciuria and some suppression of PTH production. Measuring the levels of $1,25(OH)_2D$ and PTH, as well as serum and urinary calcium, may therefore help to distinguish FGF23-dependent from FGF23-independent forms of hypophosphatemia. It is important to keep in mind that vitamin D deficiency and secondary hyperparathyroidism may mask the findings as described for HHRH and serum and urinary studies need to be repeated after repletion of vitamin D stores.[52]

Measurement of FGF23 Levels

Circulating FGF23 levels can be determined from EDTA plasma or serum using several commercially available enzyme-linked immunometric assays.[53,54] The intact FGF23 assay (referred to as iFGF23 assay) uses antibodies directed against N-terminal and C-terminal portions of the peptide for capture and detection, respectively (Kainos, Tokyo, Japan). In contrast, the C-terminal FGF23 assay (referred to as cFGF23 assay) (Immutopics, Inc., San Clemente, CA) uses two antibodies directed against distinct epitopes within the C-terminal region of FGF23 and thus detects intact FGF23 as well as C-terminal fragments. Both assays can help establish the diagnosis of FGF23-dependent disorders of phosphate homeostasis, when FGF23 levels are elevated above the normal range or are inappropriately normal.[55,56] They can also help establishing the diagnosis of FGF23-dependent hypophosphatemic disorders such as HHRH or Fanconi syndrome, albeit with significant differences in sensitivity, which is currently best for the iFGF23 assay.[54,57] The measurement of FGF23 with both assays can help establish the diagnosis of tumoral calcinosis (HFTC), since homozygous inactivating mutations in GALNT3 (HFTC1) or FGF23 (HFTC2) result in low levels of intact FGF23, yet often significant elevations of C-terminal FGF23 fragments while both intact and C-terminal FGF23 are elevated in HFTC3 (see below for details).[21,58-60]

Treatment of ADHR

The clinical course of ADHR is usually comparable to mild forms of XLH, which will be described in more detail below. As a result phosphate and $1,25(OH)_2D$ supplementation are often required only during skeletal growth in childhood and adolescence. These therapeutic interventions provide symptomatic relieve and improves the bone abnormalities, but are

Table 2. Serum biochemical findings in disorders of phosphate homeostasis

Parameter	Hypophosphatemia		Hyperphosphatemia	
	FGF23-dependent	FGF23-independent	FGF23-deficient	FGF23-resistant
Acquired	TIO, postrenal transplant	Posthepatectomy	NA	NA
Inherited	XLH, ADHR, ARHP, OGD, OSD, FD/MAS, NF1+2	HHRH, Fanconi with hypercalciuria	HFTC1, 2	HFTC3
S-Ca	NL	NL	NL to HIGH	NL to HIGH
S-PTH	NL to HIGH	NL to LOW	NL to LOW	NL to HIGH
S-1,25(OH)$_2$D	NL to LOW	HIGH	HIGH	HIGH
U-Ca	NL to LOW	HIGH	NL to HIGH	NL to HIGH
S-P	LOW	LOW	HIGH	HIGH
S-FGF23	NL to HIGH	LOW to NL	LOW	HIGH
U-P	HIGH	HIGH	LOW	LOW
Current Rx:	Phosphate and 1,25(OH)$_2$D replacement	Phosphate replacement only	Phosphate binders, acetazolamide	Phosphate binders, acetazolamide

usually unable to normalize serum phosphate levels. Treatment can be complicated by the development of secondary hyperparathyroidism, hypercalciuria and nephrocalcinosis.[44,46] Thus, treatment needs to be monitored carefully for these complications and the use of calcium-sensing receptor agonists such as cinacalcet, which has been successfully used to normalize parathyroid hormone secretion and to reduce the magnitude of phosphaturia in XLH patients,[61,62] may have a role in ADHR as well. We prefer potassium-containing phosphate supplements over sodium-containing phosphate supplements since the former seem to induce less sodium-related phosphaturia, although formal studies to support this practice are still missing. Thiazide diuretics may be helpful in slowing the progression of nephrocalcinosis.[63] Anti-FGF23 antibodies hold promise to become a therapeutic option for humans with ADHR since treatment has been successful in hyp mice, which like patients with ADHR have high circulating FGF23 levels.[26,64]

OTHER DISORDERS OF RENAL PHOSPHATE EXCRETION

Acquired Hypophosphatemic Disorders: Tumor Induced Osteomalacia (TIO)

Tumor-induced osteomalacia (TIO), also referred to as oncogenic osteomalacia (OOM),[65] is an acquired disorder of FGF23 excess,[66] or possibly FGF7 excess.[67] Tumors secreting the phosphaturic factor are usually benign mixed connective tissue tumors. Other factors such as matrix extracellular protein (MEPE)[68] or secreted frizzled related protein 4 (sFRP4)[69] were also isolated from TIO tumors and may contribute to the abnormal regulation of renal phosphate handling.[70] Tumor-induced osteomalacia is a relatively rare condition, with only slightly more than one hundred cases described in

FGF23 AND SYNDROMES OF ABNORMAL RENAL PHOSPHATE HANDLING

the literature to date.[65] Drezner reviewed 120 cases of tumor-induced osteomalacia and identified four distinct morphologic patterns:[1] primitive-appearing, mixed connective tissue tumors;[2] osteoblastoma-like tumors;[3] non-ossifying fibroma-like tumors; and[4] ossifying fibroma-like tumors. Hypophosphatemia was also described in patients with widespread fibrous dysplasia of bone, neurofibromatosis and linear nevus sebaceous syndrome (see further below) and concurrent with breast carcinoma, prostate carcinoma, oat cell carcinoma, small cell carcinoma, multiple myeloma and chronic lymphocytic leukemia. Proof of a causal relationship has been that removal of the tumor resulted in appropriate biochemical and radiographic improvements; however, since most cases were reported before the discovery of FGF23 (and of MEPE and sFRP-4), the phosphaturic factor secreted by these previously reported tumors has not been determined. The tumors are frequently located in the visceral scull or in the tendons of hands and feet and may only be a few millimeters large and indolent. They commonly escape detection by physical exam and computed tomography scans and may require more sensitive techniques for localization including whole-body octreotide-scans,[71] PET-CT scans using $[^{18}F]$-FDG,[72] or $[^{68}Ga]$-DOTANOC[73] as tracers. Selective vein sampling[74] can permit localization, even in individuals with only mildly elevated circulating FGF23 levels.[75-77] Therapy consists of surgical tumor excision, once its location has been revealed, which usually results in normalization of serum phosphate levels within 24 hrs. In those patients, where localization of the tumor is impossible or if tumor resection is incomplete, symptomatic therapy, as described in more detail above, should be continued.

Other Acquired Syndromes of Renal Phosphate Wasting

Another increasingly recognized acquired syndrome of renal phosphate wasting is postrenal transplant hypophosphatemia, which often cannot be attributed to tertiary hyperparathyroidism alone.[78,79] Bhan et al[80] and Pande et al[81] recently showed that posttransplant hypophosphatemia correlates inversely with serum FGF23 levels and coined the term "tertiary hyperphosphatoninism" due to persistent production of FGF23, which persists longer than would be expected from the half-life of the hormone.[80,82] Hypophosphatemia in the setting of inappropriate renal phosphate excretion has also been recognized with severe burn-injuries[83,84] and after partial hepatectomy,[85] although these forms of hypophosphatemia appear to be independent of FGF23.[86,87]

INHERITED HYPOPHOSPHATEMIC DISORDERS OTHER THAN ADHR (SEE FIG. 2)

X-Linked Autosomal Hypophosphatemia (XLH, OMIM 307800)

XLH, the most common form of hypophosphatemia, was first recognized by Albright and coworkers in 1937.[88] Lack of male-to-male transmission was observed by Winters in 1958,[89] which suggested an X-linked mode of inheritance. Using a positional cloning approach, the genetic defect was ultimately identified in 1995[90] and a large number of different loss-of-function mutations in PHEX, phosphate-regulating gene with homologies to endopeptidases on the X chromosome, have since been reported.[91] Deletion of the Phex gene in hyp-mice leads to increased FGF23 gene transcription in osteocytes resulting

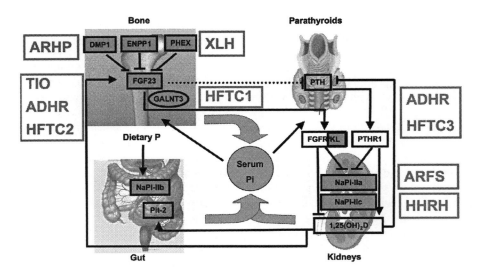

Figure 2. Disorders of phosphate homeostasis. FGF23 expression in bone is up-regulated by an increase in dietary phosphate intake and by 1,25(OH)$_2$D and down-regulated, through yet unknown mechanisms, by PHEX, DMP1, ENPP1 and probably several additional proteins. FGF23 acts through one or more FGF receptors, with Klotho as coreceptor, to inhibit renal phosphate re-absorption and to decrease circulating 1,25(OH)$_2$D levels and possibly to inhibit PTH secretion by the parathyroid glands (dashed line). The net effect of these PTH-dependent actions is a decrease in serum phosphate levels, yet an increase in serum 1,25(OH)$_2$D levels. 1-alpha hydroxylase activity is also up-regulated by low serum phosphate levels and down-regulated by increased serum calcium and phosphate levels and by increased FGF23 levels. 1,25(OH)$_2$D acts through VDR/RXR heterodimers to enhance the intestinal absorption of phosphate and to stimulate FGF23 synthesis and secretion by osteocytes; it furthermore inhibits PTH synthesis and secretion by the parathyroid glands. The net 1,25(OH)$_2$D effect is an increase in serum phosphate levels. The disorders affecting phosphate homeostasis are indicated in green. Please see the text for a detailed description. A color version of this figure is available online at http://www.landesbioscience.com/curie.

in elevated circulating levels of FGF23 and thus renal phosphate wasting,[92,93] which is similar to findings in human XLH patients.[53,54] It was therefore concluded that PHEX may be involved in the feed-back regulation of FGF23 secretion,[94] which may explain why males and females are equally affected, although the exact mechanism remains unclear.

Although some patients have normal growth, most XLH patients show stunted growth despite treatment with phosphate and active vitamin D analogs.[95] Additional clinical features include craniosynostosis, frontal bossing and mid-facial hypoplasia as described above.[44,46] Owen et al recently examined female monozygotic twins with a documented PHEX mutation; one twin had overt XLH with hypophosphatemia leading to abnormal growth and rickets, while the other displayed normal renal phosphate handling, normal growth and no evidence for rickets. The authors suggested that nonrandom expression of the normal PHEX allele in critical tissues may be responsible for the discordant XLH phenotype.[96]

As outlined above for ADHR, treatment of XLH consists of oral phosphate supplementation and active vitamin D analogs, which provides symptomatic relieve and improves the bone abnormalities, but is usually unable to normalize serum phosphate levels. Treatment may be complicated by the development of secondary hyperparathyroidism, hypercalciuria and nephrocalcinosis[44,46] and needs to be monitored carefully for these complications. The calcium-sensing receptor agonist cinacalcet has been successfully used to normalize parathyroid hormone secretion and to reduce the magnitude of phosphaturia in XLH patients, who had developed severe secondary or tertiary hyperparathyroidism.[61,62] However, parathyroidectomy may ultimately be required, which results generally in improved phosphatemic control, although the postoperative course may be complicated by severe hungry bone syndrome.[97] Thiazide diuretics may be helpful in slowing the progression of nephrocalcinosis.[63] Replacement of the membrane-anchored PHEX with a soluble form of PHEX did not prove effective to reverse hypophosphatemia in hyp mice, while treatment with anti-FGF23 antibodies has been successful in these animals and holds promise to become a therapeutic option for humans with XLH.[26,64] Growth hormone therapy was reported to improve linear growth in some patients, although it remains unclear, whether the observed improvements were in part attributable to an increase in tubular reabsorption of phosphate during growth hormone treatment.[98]

Autosomal Recessive Hypophosphatemia (ARHP, OMIM 241520)

ARHP is caused by homozygous, presumably loss-of-function mutations in DMP1[99,100] or in ecto-nucleotide pyrophosphatase/phosphodiesterase 1 (ENPP1).[101,102] Intact DMP1 is cleaved into a 35 and a 57 kDa fragment, possibly by bone morphogenic protein 1 (BMP1),[103] which, in turn, is activated by a complex consisting of the subtilisin propeptide convertase SPC2 and the co-activator 7B2.[104] Transgenic over-expression of the C-terminal 57 kDa DMP1 fragment is both necessary and sufficient to rescue the bone phenotype (and probably the hypophosphatemia resulting from increased FGF23 secretion) of DMP1-null mice.[105] This DMP1 fragment appears to have nuclear effects, which are required to prevent excess FGF23 secretion, renal phosphate wasting and hypophosphatemia.[105] Thus, the C-terminal fragment may be required for suppression and/or feed back regulation of FGF23 gene transcription and/or FGF23 secretion.[99] ENPP1 is a membrane-bound ecto-enzyme responsible for the generation of the mineralization inhibitor pyrophosphate (PPi).[106] Loss-of function mutations in this enzyme are the cause of generalized arterial calcifications of infancy (GACI),[107,108] which was associated in some patients with mild hypophosphatemia.[109] Recently, two groups identified homozygous loss-of-function mutations in ENPP1 in kindreds with hypophosphatemic rickets due to FGF23-dependent renal phosphate-wasting.[101,102] Further evaluation of the phenotypic expression of ENPP1 mutations indicated that members of the same kindred, carrying the same homozygous mutation can present either with arterial calcifications or with rickets. Furthermore, subtle changes, i.e., mild hypophosphatemia was observed in individuals with GACI and thickening of the cardiac valves in individuals with rickets were observed, suggesting the presence of modifiers (genetic or environmental) that underlie the variable expressivity of the disorders. The cause of FGF23 excess may be directly related to lack of PPi production, or due to accumulation of precursors, such as ATP in the extracellular space. However, presence of mild hyperphosphatemia in individuals suffering from hypophosphatasia, which is caused by loss-of-function mutations in the PPi-degrading enzyme, namely tissue nonspecific alkaline phosphatase (TNALP),[110] suggests that PPi may directly or indirectly suppress FGF23 production.

Treatment of ARHP is currently symptomatic and relies, like the treatment of XLH and ADHR, on oral phosphate supplementation and repletion of $1,25(OH)_2D$ to prevent the development of hyperparathyroidism. In the future, it may be possible to treat these groups of patients with monoclonal, inactivating antibodies against FGF23.

Hereditary Hypophosphatemic Rickets with Hypercalciuria (HHRH, OMIM 241530)

HHRH is a rare disorder with autosomal recessive inheritance that was first described in 1985 in a large consanguineous Bedouin kindred.[50] Unlike patients with XLH, individuals affected by HHRH do not develop dental abscesses or craniofacial abnormalities (i.e., craniosynostosis, frontal bossing, scaphocephaly, Chiari I malformation)[111,112] and different from patients affected by XLH, ADHR and ARHR, HHRH patients show suppressed or low-normal FGF23 levels.[113] This reduction in FGF23, combined with the hypophosphatemia, contributes to the compensatory increase in the plasma level of $1,25(OH)_2D$ typically observed in HHRH. This appropriate rise in the concentration of the biologically active form of vitamin D results in absorptive hypercalciuria, the cardinal feature that distinguishes HHRH from most other Mendelian hypophosphatemic disorders. Measurement of $1,25(OH)_2D$ and urinary calcium excretion are thus essential for establishing the diagnosis of HHRH, although both may be normal, if vitamin D deficiency and/or secondary hyperparathyroidism are present.[52]

HHRH is caused by homozygous or compound heterozygous loss-of-function mutations in NaPi-IIc/SLC34A3.[113-115] Expressivity can be quite variable and may be affected like in other hypophosphatemic forms of rickets by vitamin D status. Some heterozygous individuals may have an increased urinary calcium excretion and occasionally some of the above biochemical features of HHRH, while bone changes are generally missing. Likewise, individuals with two mutated SLC34A3 alleles can initially present with renal stones alone even in the absence of clinical symptoms associated with rickets or osteomalacia.[52,116]

In contrast to patients with XLH, ADHR or ARHP, who are usually treated with multiple daily doses of oral phosphate and high doses of calcitriol (i.e.,$1,25(OH)_2D$), the effective therapy of individuals affected by HHRH consists of oral phosphate supplementation alone. The prescription of biologically active vitamin D analogs is contraindicated and may lead to hypercalcemia, hypercalciuria, nephrocalcinosis and possibly renal insufficiency.[52,116]

Hypophosphatemia with Osteoporosis and Nephrolithiasis Due to NaPi-IIa (OMIM 612286) or NHERF-1 Mutations (OMIM 604990)

Prie et al investigated a heterogeneous group of patients with idiopathic hypercalciuria, osteoporosis and renal stones. Using a candidate gene approach they found in 2 of 20 individuals heterozygous for nonsynonymous SNPs in NaPi-IIa/SLC34A1.[117] Consistent with their findings, Lapointe et al identified additional heterozygous mutations in other patients affected by calcium nephrolithiasis and a renal phosphate leak.[118] These latter NaPi-IIa mutations did not lead to functional abnormalities of sodium-dependent phosphate cotransport, when tested in vitro. In contrast, Prie et al presented experimental in vitro evidence for dominant negative effects of the NaPi-IIa/SLC34A1 alterations on proximal renal tubular phosphate reabsorption; these findings, however, were challenged by others.[119] The same group more recently identified heterozygous, nonsynonymous amino acid changes in NHERF-1; three different changes were identified in 7 of 94 individuals with idiopathic hypercalciuria, osteoporosis and renal stones.[120] However, two of those amino acid changes are listed in the NCBI dbSNP database as low frequency polymorphisms,[121] raising the possibility that these are either not contributing to the clinical and laboratory

FGF23 AND SYNDROMES OF ABNORMAL RENAL PHOSPHATE HANDLING   **53**

findings in the investigated cohort or that these changes may lead more frequently to hypophosphatemia, albeit without readily appreciated clinical abnormalities. Further studies are thus required to prove that mutations in NaPi-IIa or NHERF-1 are indeed responsible for idiopathic hypercalciuria, osteoporosis and renal stones.

Autosomal Recessive Fanconi Syndrome (ARFS)

In 1988, Tieder et al described a consanguineous Arab kindred with childhood rickets and defective proximal tubular handing of phosphate, amino acids and glucose consistent with renal Fanconi syndrome.[122] Distinct from other forms of Fanconi, their patients also had elevated 1,25(OH)$_2$D levels and absorptive hypercalciuria. Recently, homozygosity mapping of this kindred showed linkage of the disease to chromosome 5q35.1-3 and subsequent sequence analysis of the SLC34A1/NaPi-IIa gene in the linked interval revealed a homozygous duplication g.2061_2081dup (p.I154_V160dup).[123] Expression of the mutant NaPi-IIa protein in *Xenopus* oocytes and *Opossum* kidney cells showed complete loss-of-function and lack of membrane insertion, respectively. The two patients described in the 1988 report, who are now 39 and 43 years old, continue to have low TmP/GFR and their FGF23 and PTH levels were recently shown to be low-normal (despite impaired renal function), suggesting that their hypophosphatemia is FGF23- and PTH-independent. However, their previously documented absorptive hypercalciuria due to increased 1,25(OH)$_2$D levels, had normalized in the setting of vitamin D deficiency. Although symptoms of rickets were present in childhood, both patients have been relatively asymptomatic during adulthood and discontinued phosphate supplementation. Both homozygous carriers developed CKD Stage 2-3 in their 30s, which is in contrast to the other familial hypophosphatemia disorders described above. Heterozygous carriers of p.I154_V160dup had normal renal function and no evidence for renal phosphate leak or proximal tubulopathy, arguing strongly against dominant negative effects of the mutant NaPi-IIa. However, it is still possible that intracellular accumulation of the mutant transporters in homozygous carriers may be involved in the pathogenesis of their proximal tubular defect, since no comparable abnormalities were observed in mice that are null for the murine ortholog of NaPi-IIa.[124, 125]

Other Inherited Forms of Renal Phosphate Wasting

Osteoglophonic dysplasia (OGD, OMIM166250) is an autosomal dominant disorder caused by activating missense mutations in the gene encoding fibroblast growth factor receptor-1 (FGFR1).[126,127] Affected individuals may develop hypophosphatemia and renal phosphate wasting due to an increased production of FGF23.[128,129] Other clinical features include those seen in other syndromes that are caused by fibroblast growth factor receptor-1 mutations, such as craniosynostosis, midfacial hypoplasia, prognatism and rizomelic chondrodysplasia. In addition, individuals affected by OGD have symmetrical radiolucent metaphyseal defects, which may be the source of their excess FGF23 production, although direct histological evidence to support this hypothesis is to date lacking.

Opsismodysplasia (OSD, OMIM 258480)[130] is an autosomal recessive skeletal dysplasia that is characterized by a delay in epiphyseal ossification, platyspondyly, metaphyseal cupping, resulting in brachydactyly with short metacarpals and phalanges. The genetic defect is unknown. Like OGD, opsismodysplasia can go along with FGF23 excess leading to renal phosphate wasting.[131]

Linear nevus sebaceous syndrome (LNSS)/epidermal nevus syndrome (ENS) (OMIM163200) is characterized by sebaceous nevi, often in the face, abnormalities of the central nervous system, ocular anomalies, including coloboma and skeletal defects.[132-135] Most patients with LNSS or ENS carry mosaic FGFR3 mutations.[136] Some patients develop hypophosphatemic rickets[137,138] and recently some nevi were shown to secrete FGF23 thus providing an explanation for the underlying renal phosphate wasting.[139-141] However, it is unknown whether the FGFR3 mutations alone or additional unknown somatic mutations lead to the increase in FGF23 production and thus renal phosphate-wasting.

Fibrous dysplasia (FD)/McCune Albright syndrome (MAS) (OMIM 174800) is caused by somatic activating missense mutations in the alpha subunit of the stimulatory G-protein (encoded by GNAS).[142-144] The classical triad of MAS includes polyostotic FD, café-au-lait spots and precocious puberty. However, a number of other endocrine disorders such as thyrotoxicosis, pituitary gigantism and Cushing syndrome are often present as well.[144] The nonmineralizing bone lesions of FD/MAS may secrete FGF23, which can lead to hypophosphatemic rickets or osteomalacia.[145-148] FGF23-mediated hypophosphatemia can also be observed in Jansen's metaphyseal chondrodysplasia (OMIM 156400), which is caused by heterozygous activating PTH/PTHrP receptor mutations and may be, like in FD/MAS, a consequence of agonist-independent activation of the cAMP/PKA signaling pathway.[149] Finally, hypophosphatemia leading to osteomalacia has been described in some individuals with neurofibromatosis 1 and 2,[150,151] although the mechanism remains to be clarified.

HYPERPHOSPHATEMIC DISORDERS

Hyperphosphatemic Familial Tumoral calcinosis, HFTC (OMIM 211900)

Tumoral calcinosis is a clinically and genetically heterogeneous group of disorders first described by Giard (1898)[152] and then by Duret (1899).[153] Tumoral calcinosis is characterized by calcium-phosphate deposits in different tissues, but osteogenic cells and matrix formation are absent, which distinguishes this disorders from heterotopic ossification. For the purpose of this chapter, the hyperphosphatemic forms of familial tumoral calcinosis shall be classified as Type 1-3 (HFTC1-3), which all follow an autosomal recessive mode of inheritance and all furthermore show inappropriately enhanced renal tubular absorption of phosphate leading to hyperphosphatemia as a common laboratory feature. The activity of renal 1-alpha-hydroxylase is increased resulting in elevated serum $1,25(OH)_2D$ levels and thus increased intestinal absorption of calcium (and phosphate), suppression of parathyroid hormone production and hypercalciuria. The increased serum calcium-phosphate product results in the characteristic abnormal tissue mineralization observed in tumoral calcinosis.

Ectopic tissue mineralization in HFTC is often seen in juxtaarticular muscular and subcutaneous tissues. Patients with tumoral calcinosis often also show dental pulp stones, which may lead to a complete obliteration of the dental pulp cavities. Other clinical features, which may constitute the only clinical evidence for tumoral calcinosis, can include eye-lid calcifications, vascular calcifications and/or nephrocalcinosis. There can also be mineralization of the juxtaarticular bone marrow cavities, however, the remaining skeleton often shows low bone mineral density due to a mineralization defect, which at the moment is only poorly understood.[154,155]

FGF23 AND SYNDROMES OF ABNORMAL RENAL PHOSPHATE HANDLING   55

Familial tumoral calcinosis Type 1 (HFTC1) is caused by homozygous loss-of-function mutations in the gene encoding UDP-N-acetyl-alpha-D-galactosamine: polypeptide N-acetylgalactosaminyl-transferase 3 (GALNT3).[58,156] GALNT3 is responsible for O-glycosylation and proper secretion of intact FGF23.[17] Patients with HFTC1 characteristically have low or undetectable intact, but increased C-terminal FGF23 levels. HFTC1 is allelic with the hyperostosis-hyperphosphatemia syndrome (HSS; OMIM 610233). Patients affected by HSS show, besides the characteristic biochemical abnormalities in serum and urine, recurrent albeit transient painful swellings of the long bones and associated radiographic findings that are consistent with periosteal reaction and cortical hyperostosis.[157,158] Similar or identical GALNT3 mutations thus can lead to HSS (bone abnormalities without skin involvement) or to HFTC1, but the genetic modifiers responsible for these differences in disease presentation are currently unknown.

HFTC2 is caused by mutations affecting both alleles of FGF23, which reduce the circulating levels of bioactive intact FGF23, just like in HFTC1 (Fig. 1).[156,159] C-terminal FGF23 fragments are secreted, however and high levels of this fragment are often detected in the circulation.[60] The mechanism by which HPTC2 mutations may lead to loss-of-function of FGF23 is unclear. Consistent with the patients' low circulating intact FGF23 levels, we recently showed that HEK293 or COS-7 cells expressing FGF23 with proline at position 129 ([P129]FGF23) (or with other mutations identified in HFTC2 patients), secrete the intact hormone only poorly into the medium. Interestingly, total lysates of HEK293 or COS-7 cells expressing [P129]FGF23 (and other FGF23 mutants) revealed a 25 kDa form of FGF23 that was also detected in cells expressing wild-type FGF23. However, in contrast to cells expressing wild-type FGF23, a 32 kDa protein species was missing in cells expressing FGF23 mutants.

The 32 kDa protein species of FGF23 is detected with antibodies raised against N-terminal and C-terminal portions of FGF23. Thus full-length FGF23 is larger than predicted from the primary amino acid sequence of mature FGF23(25-251). This suggested that wild-type FGF23 is modified and that this modification is absent in patients carrying one of the FGF23 mutations leading to tumoral calcinosis. Enzymatic deglycosylation of wild-type FGF23 suggested that the modification consists of O-linked glycans (Fig. 3A) and that these modifications are absent in [P129]FGF23 (Fig. 3B) as well as in [G71] FGF23 and [F129]FGF23, other mutant forms of FGF23 that cause tumoral calcinosis. These findings suggest that a similar mechanism, namely lack of O-glycosylation, underlies the poor secretion observed of these FGF23 mutants. Both, HFTC1 (GALNT3 mutations) and HFTC2 (FGF23 mutations), thus appear to be disorders of abnormal O-glycosylation of FGF23.

The mechanism by which the FGF23 mutations identified in HFTC2 lead to impaired O-glycosylation of FGF23 remains unclear. However, previous reports have indicated that O-glycosylation occurs within the C-terminal portion of FGF23, i.e., the 168-228 region,[16] while all HFTC2 mutations identified to date reside in the N-terminal portion of FGF23; some of these mutations furthermore do not affect potential O-glycosylation sites (H41Q,[160] Q54K).[161] Thus, HFTC2 mutations likely cause misfolding, which may delay or impair O-glycosylation, as previously suggested.[60,159]

The lack of O-glycosylation can be the consequence or the cause of poor secretion of the mutant FGF23 by HEK293 cells; the former scenario would likely result in the secretion of large quantities of an unmodified 25 kDa FGF23 into the medium of cells transfected with plasmids encoding the mutant forms of FGF23. However, even after concentrating 50-fold the conditioned medium from HEK293 expressing the mutant FGF23,

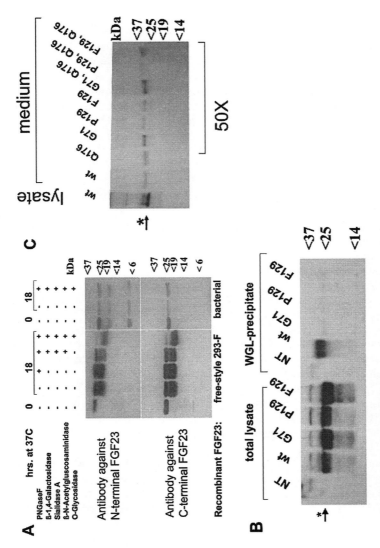

Figure 3. Tumoral calcinosis mutations impair O-glycosylation and secretion of FGF23. A) Enzymatic deglycosylation of recombinant FGF23: N-linked and O-linked carbohydrates were removed under denaturing conditions using approximately 1 μg of recombinant FGF23 and the E-DEGLY kit (Sigma). The molecular shift observed with O-glycosidase indicates that recombinant FGF23 is O-glycosylated. B) Total lysates from HEK293 cells expressing wild-type FGF23(1-251) (wt) or FGF23 with mutations at residues 71 or 129 (G71, P129, or F129) show equal expression of the 25 kDa form of FGF23; the 32 kDa form of FGF23 is detected only in the glycoprotein fraction (star); NT, nontransfected. C) The 32 kDa, presumably glycosylated form of FGF23, but not the more abundant 25 kDa protein species is detected after concentrating 50-fold conditioned media from HEK293 cells expressing wild-type and mutant FGF23; this suggests that FGF23 carrying one of the mutations that cause tumoral calcinosis can undergo limited O-glycosylation thus allowing secretion (star); see text for details. Reproduced with permission from Bergwitz C et al. J Clin Endocrinol Metab 2009; 94:4267-4274.[172] ©2009 The Endocrine Society.

FGF23 AND SYNDROMES OF ABNORMAL RENAL PHOSPHATE HANDLING

the 25 kDa protein species could not be detected. Instead, small amounts of the 32 kDa protein species were observed in concentrated medium, indicating that O-glycosylation can occur, albeit inefficiently for FGF23 mutants and that these mutants can be secreted once glycosylated (Fig. 3C).

Recently, a 13 year-old girl with a disorder resembling HFTC1 and HFTC2, was reported, who had extremely high circulating levels of intact and C-terminal FGF23.[21] Radiographs of the patient showed osteopenia, patchy sclerosis in the hands, feet, long bones and calvaria, intracranial calcifications and calcifications of the dura and carotid arteries. Interestingly and distinct from the first two forms of HFTC, she had elevated PTH levels due to four-gland parathyroid hyperplasia. Very high circulating levels of intact FGF23 had also been observed in mice that are null for klotho (KL), the coreceptor for FGF23 (see above). The authors therefore decided to analyze the gene encoding KL, which led to the discovery of a homozygous missense mutation in the second putative beta-glycosidase domain that is presumably inactivating, leading to end-organ resistance to FGF23.[21]

Like humans with HFTC, mice that are null for FGF23,[162,163] GALNT3,[164] KL[22] suffer from hyperphosphatemia and increased $1,25(OH)_2D$ levels due to the loss of biologically active FGF23 or due to end-organ resistance to FGF23, respectively. Consistent with an increased intestinal absorption of calcium from the gut, mice with these genetic modifications also display mild hypercalcemia, suppressed PTH levels and hypercalciuria. The resulting increase in the calcium-phosphorus product in each of these disorders is thought to cause the characteristic tissue calcifications. The homozygous ablation of Npt2a,[154] the VDR,[163] or CYP27B1[165] rescues the serum-biochemical abnormalities of FGF23-null mice and the homozygous ablation of CYP27B1 was also shown to rescue the klotho-null mice.[166] Likewise, diets low in phosphate or low in vitamin D[167] can normalize the changes in mineral ion homeostasis of FGF23- and klotho-null animals, although mineralization defects in the skeleton may persist.[154,162,168]

Heterozygous carriers of the genetic mutations that cause HFTC1-3 may have mildly abnormal blood chemistries,[156] but do not develop significant calcified lesions and they do not require specific treatment. Treatment of homozygous individuals currently relies on minimizing the intestinal absorption of phosphate through appropriate binders such as aluminum hydroxide or sevelamer[60] and on inhibiting renal phosphate reabsorption with acetazolamide;[169] in one study treatment with PTH was temporarily able to induce phosphaturia.[170] In the future, patients affected by HFTC may be treated with calcilytic agents that reduce the activity of the calcium-sensing receptor and thus stimulate endogenous PTH secretion[171] or with recombinant FGF23, if this becomes available for use in humans.

CONCLUSION

FGF23 was identified initially as the hypophosphatemic factor that is abundantly expressed in TIO tumors and that is mutated in individuals affected by ADHR. This phosphaturic hormone was subsequently found to be readily detectable in the circulation of healthy individuals and research over the past years strongly suggests that it is an important regulator of normal phosphate homeostasis. Dysregulation of FGF23 occurs in more than 13 acquired and inherited disorders of phosphate homeostasis and it is emerging as a predictor of disease progression in chronic kidney disease and of mortality of incident and long-term dialysis patients. Further investigations are required to understand the regulation of FGF23 expression through changes in dietary intake and/or serum phosphate levels and through PTH and 1,25(OH)$_2$D. It will also be important to understand which FGF receptors mediate the actions of FGF23 in parathyroids and kidney and how PHEX, DMP1, ENPP1 and most likely several other proteins contribute to the regulation of FGF23 synthesis and secretion.

REFERENCES

1. Bianchine JW, Stambler AA, Harrison HE. Familial hypophosphatemic rickets showing autosomal dominant inheritance. Birth Defects Orig Artic Ser 1971; 7:287-295.
2. Econs MJ, McEnery PT. Autosomal dominant hypophosphatemic rickets/osteomalacia: clinical characterization of a novel renal phosphate-wasting disorder. J Clin Endocrinol Metab 1997; 82:674-681.
3. Econs M, McEnery P, Lennon F et al. Autosomal dominant hypophosphatemic rickets is linked to chromosome 12p13. J Clin Invest 1997; 100:2653-2657.
4. ADHR Consortium T. White KE, Evans WE et al. Autosomal dominant hypophosphataemic rickets is associated with mutations in FGF23. Nat Genet 2000; 26:345-348.
5. Gribaa M, Younes M, Bouyacoub Y et al. An autosomal dominant hypophosphatemic rickets phenotype in a Tunisian family caused by a new FGF23 missense mutation. J Bone Miner Metab 2009; 28:111-115.
6. Kruse K, Woelfel D, Strom TM. Loss of renal phosphate wasting in a child with autosomal dominant hypophosphatemic rickets caused by a FGF23 mutation. Horm Res 2001; 55:305-308.
7. Burnett CH, Dent CE, Harper C et al. Vitamin D-resistant rickets. Analysis of twenty-four pedigrees with hereditary and sporadic cases. Am J Med 1964; 36:222-232.
8. Brownstein CA, Adler F, Nelson-Williams C et al. A translocation causing increased alpha-klotho level results in hypophosphatemic rickets and hyperparathyroidism. Proc Natl Acad Sci USA 2008; 105:3455-3460.
9. Yamashita T, Yoshioka M, Itoh N. Identification of a novel fibroblast growth factor, FGF-23, preferentially expressed in the ventrolateral thalamic nucleus of the brain. Biochem Biophys Res Commun 2000; 277:494-498.
10. Burnett SM, Gunawardene SC, Bringhurst FR et al. Regulation of C-terminal and intact FGF-23 by dietary phosphate in men and women. J Bone Miner Res 2006; 21:1187-1196.
11. Perwad F, Zhang MY, Tenenhouse HS et al. Fibroblast growth factor 23 impairs phosphorus and vitamin D metabolism in vivo and suppresses 25-hydroxyvitamin D-1alpha-hydroxylase expression in vitro. Am J Physiol Renal Physiol 2007; 293:F1577-F1583.
12. Nagano N, Miyata S, Abe M et al. Effect of manipulating serum phosphorus with phosphate binder on circulating PTH and FGF23 in renal failure rats. Kidney Int 2006; 69:531-537.
13. Nishi H, Nii-Kono T, Nakanishi S et al. Intravenous calcitriol therapy increases serum concentrations of fibroblast growth factor-23 in dialysis patients with secondary hyperparathyroidism. Nephron Clin Pract 2005; 101:c94-c99.
14. Kolek OI, Hines ER, Jones MD et al. 1alpha,25-Dihydroxyvitamin D3 upregulates FGF23 gene expression in bone: the final link in a renal-gastrointestinal-skeletal axis that controls phosphate transport. Am J Physiol Gastrointest Liver Physiol 2005; 289:G1036-G1042.
15. Quarles LD. Endocrine functions of bone in mineral metabolism regulation. J Clin Invest 2008; 118:3820-3828.
16. Shimada T, Muto T, Urakawa I et al. Mutant FGF-23 responsible for autosomal dominant hypophosphatemic rickets is resistant to proteolytic cleavage and causes hypophosphatemia in vivo. Endocrinology 2002; 143:3179-3182.
17. Kato K, Jeanneau C, Tarp MA et al. Polypeptide GalNAc-transferase T3 and familial tumoral calcinosis. Secretion of fibroblast growth factor 23 requires O-glycosylation. J Biol Chem 2006; 281:18370-18377.
18. Frishberg Y, Ito N, Rinat C et al. Hyperostosis-hyperphosphatemia syndrome: a congenital disorder of O-glycosylation associated with augmented processing of fibroblast growth factor 23. J Bone Miner Res 2007; 22:235-242.
19. Chefetz I, Kohno K, Izumi H et al. GALNT3, a gene associated with hyperphosphatemic familial tumoral calcinosis, is transcriptionally regulated by extracellular phosphate and modulates matrix metalloproteinase activity. Biochim Biophys Acta 2009; 1792:61-67.
20. White KE, Carn G, Lorenz-Depiereux B et al. Autosomal-dominant hypophosphatemic rickets (ADHR) mutations stabilize FGF-23. Kidney Int 2001; 60:2079-2086.
21. Ichikawa S, Imel EA, Kreiter ML et al. A homozygous missense mutation in human KLOTHO causes severe tumoral calcinosis. J Clin Invest 2007; 117:2684-2691.

FGF23 AND SYNDROMES OF ABNORMAL RENAL PHOSPHATE HANDLING

22. Kuro-o M, Matsumura Y, Aizawa H et al. Mutation of the mouse klotho gene leads to a syndrome resembling ageing. Nature 1997; 390:45-51.
23. Bai X, Dinghong Q, Miao D et al. Klotho ablation converts the biochemical and skeletal alterations in FGF23 (R176Q) transgenic mice to a Klotho-deficient phenotype. Am J Physiol Endocrinol Metab 2009; 296:E79-E88.
24. Kurosu H, Yamamoto M, Clark JD et al. Suppression of aging in mice by the hormone Klotho. Science 2005; 309:1829-1833.
25. Urakawa I, Yamazaki Y, Shimada T et al. Klotho converts canonical FGF receptor into a specific receptor for FGF23. Nature 2006; 444:770-774.
26. Yamazaki Y, Tamada T, Kasai N et al. Anti-FGF23 neutralizing antibodies show the physiological role and structural features of FGF23. J Bone Miner Res 2008; 23:1509-1518.
27. Zhang X, Ibrahimi OA, Olsen SK et al. Receptor specificity of the fibroblast growth factor family. The complete mammalian FGF family. J Biol Chem 2006; 281:15694-15700.
28. Yu X, White KE. Fibroblast growth factor 23 and its receptors. Ther Apher Dial 2005; 9:308-312.
29. Yamashita T, Konishi M, Miyake A et al. Fibroblast growth factor (FGF)-23 inhibits renal phosphate reabsorption by activation of the mitogen-activated protein kinase pathway. J Biol Chem 2002; 277:28265-28270.
30. Liu S, Zhou J, Tang W et al. Pathogenic role of Fgf23 in Dmp1-null mice. Am J Physiol Endocrinol Metab 2008; 295:E254-E261.
31. Segawa H, Yamanaka S, Ohno Y et al. Correlation between hyperphosphatemia and type II Na-Pi cotransporter activity in klotho mice. Am J Physiol Renal Physiol 2007; 292:F769-F779.
32. Baum M, Schiavi S, Dwarakanath V et al. Effect of fibroblast growth factor-23 on phosphate transport in proximal tubules. Kidney Int 2005; 68:1148-1153.
33. Segawa H, Onitsuka A, Aranami F et al. Npt2a and Npt2c in Mice Play Distinct and Synergistic Roles in Inorganic Phosphate Metabolism and Skeletal Development. RENAL WEEK 2006, San Francisco, 2007; pp abstract SA-FC101.
34. Strom TM, Jüppner H. PHEX, FGF23, DMP1 and beyond. Curr Opin Nephrol Hypertens 2008; 17:357-362.
35. Liu S, Gupta A, Quarles LD. Emerging role of fibroblast growth factor 23 in a bone-kidney axis regulating systemic phosphate homeostasis and extracellular matrix mineralization. Curr Opin Nephrol Hypertens 2007; 16:329-335.
36. Fukumoto S, Yamashita T. FGF23 is a hormone-regulating phosphate metabolism—unique biological characteristics of FGF23. Bone 2007; 40:1190-1195.
37. Farrow EG, Davis SI, Summers LJ et al. Initial FGF23-mediated signaling occurs in the distal convoluted tubule. J Am Soc Nephrol 2009; 20:955-960.
38. Chang Q, Hoefs S, van der Kemp AW et al. The beta-glucuronidase klotho hydrolyzes and activates the TRPV5 channel. Science 2005; 310:490-493.
39. Cha SK, Hu MC, Kurosu H et al. Regulation of renal outer medullary potassium channel and renal K($^+$) excretion by Klotho. Mol Pharmacol 2009; 76:38-46.
40. Masuda H, Chikuda H, Suga T et al. Regulation of multiple ageing-like phenotypes by inducible klotho gene expression in klotho mutant mice. Mech Ageing Dev 2005; 126:1274-12783.
41. Donohue MM, Demay MB. Rickets in VDR null mice is secondary to decreased apoptosis of hypertrophic chondrocytes. Endocrinology 2002; 143:3691-3694.
42. Narchi H, El Jamil M, Kulaylat N. Symptomatic rickets in adolescence. Arch Dis Child 2001; 84:501-503.
43. Francis RM, Selby PL. Osteomalacia. Baillieres Clin Endocrinol Metab 1997; 11:145-163.
44. Econs MJ, Samsa GP, Monger M et al. X-Linked hypophosphatemic rickets: a disease often unknown to affected patients. Bone Miner 1994; 24:17-24.
45. Liang G, Katz LD, Insogna KL et al. Survey of the enthesopathy of X-linked hypophosphatemia and its characterization in Hyp mice. Calcif Tissue Int 2009; 85:235-246.
46. DiMeglio LA, Econs MJ. Hypophosphatemic rickets. Rev Endocr Metab Disord 2001; 2:165-173.
47. Walton RJ, Bijvoet OL. Nomogram for derivation of renal threshold phosphate concentration. Lancet 1975; 2:309-310.
48. Brodehl J, Gellissen K, Weber HP. Postnatal development of tubular phosphate reabsorption. Clin Nephrol 1982; 17:163-171.
49. Alon U, Hellerstein S. Assessment and interpretation of the tubular threshold for phosphate in infants and children. Pediatr Nephrol 1994; 8:250-251.
50. Tieder M, Modai D, Samuel R et al. Hereditary hypophosphatemic rickets with hypercalciuria. N Engl J Med 1985; 312:611-617.
51. Corut A, Senyigit A, Ugur SA et al. Mutations in SLC34A2 cause pulmonary alveolar microlithiasis and are possibly associated with testicular microlithiasis. Am J Hum Genet 2006; 79:650-656.

52. Kremke B, Bergwitz C, Ahrens W et al. 2009 Hypophosphatemic rickets with hypercalciuria due to mutation in SLC34A3/NaPi-IIc can be masked by vitamin D deficiency and can be associated with renal calcifications. Exp Clin Endocrinol Diabetes 117:49-56.
53. Yamazaki Y, Okazaki R, Shibata M et al. Increased circulatory level of biologically active full-length FGF-23 in patients with hypophosphatemic rickets/osteomalacia. J Clin Endocrinol Metab 2002; 87:4957-4960.
54. Jonsson KB, Zahradnik R, Larsson T et al. Fibroblast growth factor 23 in oncogenic osteomalacia and X-linked hypophosphatemia. N Engl J Med 2003; 348:1656-1663.
55. Endo I, Fukumoto S, Ozono K et al. Clinical usefulness of measurement of fibroblast growth factor 23 (FGF23) in hypophosphatemic patients: proposal of diagnostic criteria using FGF23 measurement. Bone 2008; 42:1235-1239.
56. Imel EA, Peacock M, Pitukcheewanont P et al. Sensitivity of fibroblast growth factor 23 measurements in tumor-induced osteomalacia. J Clin Endocrinol Metab 2006; 91:2055-2061.
57. Imel EA, Hui SL, Econs MJ. FGF23 concentrations vary with disease status in autosomal dominant hypophosphatemic rickets. J Bone Miner Res 2007; 22:520-526.
58. Topaz O, Shurman DL, Bergman R et al. Mutations in GALNT3, encoding a protein involved in O-linked glycosylation, cause familial tumoral calcinosis. Nat Genet 2004; 36:579-581.
59. Ichikawa S, Guigonis V, Imel EA et al. Novel GALNT3 mutations causing hyperostosis-hyperphosphatemia syndrome result in low intact fibroblast growth factor 23 concentrations. J Clin Endocrinol Metab 2007; 92:1943-1947.
60. Larsson T, Yu X, Davis SI et al. A novel recessive mutation in fibroblast growth factor-23 causes familial tumoral calcinosis. J Clin Endocrinol Metab 2005; 90:2424-2427.
61. Geller JL, Khosravi A, Kelly MH et al. Cinacalcet in the management of tumor-induced osteomalacia. J Bone Miner Res 2007; 22:931-937.
62. Alon US, Levy-Olomucki R, Moore WV et al. Calcimimetics as an adjuvant treatment for familial hypophosphatemic rickets. Clin J Am Soc Nephrol 2008; 3:658-664.
63. Seikaly MG, Baum M. Thiazide diuretics arrest the progression of nephrocalcinosis in children with X-linked hypophosphatemia. Pediatrics 2001; 108:E6.
64. Aono Y, Yamazaki Y, Yasutake J et al. Therapeutic effects of anti-FGF23 antibodies in hypophosphatemic rickets/osteomalacia. J Bone Miner Res 2009; 24:1879-1888.
65. Drezner MK. Tumor-induced osteomalacia. Rev Endocr Metab Disord 2001; 2:175-186.
66. Shimada T, Mizutani S, Muto T et al. Cloning and characterization of FGF23 as a causative factor of tumor-induced osteomalacia. Proc Natl Acad Sci USA 2001; 98:6500-6505.
67. Carpenter TO, Ellis BK, Insogna KL et al. Fibroblast growth factor 7: an inhibitor of phosphate transport derived from oncogenic osteomalacia-causing tumors. J Clin Endocrinol Metab 2005; 90:1012-1020.
68. Rowe PS, de Zoysa PA, Dong R et al. MEPE, a new gene expressed in bone marrow and tumors causing osteomalacia. Genomics 2000; 67:54-68.
69. Jan De Beur S, Finnegan R, Vassiliadis J et al. Tumors associated with oncogenic osteomalacia express genes important in bone and mineral metabolism. J Bone Miner Res 2002; 17:1102-1110.
70. Berndt T, Craig T, Bowe A et al. Secreted frizzled-related protein 4 is a potent tumor-derived phosphaturic agent. J Clin Invest 2003; 112:785-794.
71. Nguyen BD, Wang EA. Indium-111 pentetreotide scintigraphy of mesenchymal tumor with oncogenic osteomalacia. Clin Nucl Med 1999; 24:130-131.
72. Dupond JL, Magy N, Mahammedi M et al. [Oncogenic osteomalacia: the role of the phosphatonins. Diagnostic usefulness of the Fibroblast Growth Factor 23 measurement in one patient]. Rev Med Interne 2005; 26:238-241.
73. Hesse E, Moessinger E, Rosenthal H et al. Oncogenic osteomalacia: exact tumor localization by coregistration of positron emission and computed tomography. J Bone Miner Res 2007; 22:158-162.
74. Takeuchi Y, Suzuki H, Ogura S et al. Venous sampling for fibroblast growth factor-23 confirms preoperative diagnosis of tumor-induced osteomalacia. J Clin Endocrinol Metab 2004; 89:3979-3982.
75. Westerberg PA, Olauson H, Toss G et al. Preoperative tumor localization by means of venous sampling for fibroblast growth factor-23 in a patient with tumor-induced osteomalacia. Endocr Pract 2008; 14:362-367.
76. Nasu T, Kurisu S, Matsuno S et al. Tumor-induced hypophosphatemic osteomalacia diagnosed by the combinatory procedures of magnetic resonance imaging and venous sampling for FGF23. Intern Med 2008; 47:957-961.
77. van Boekel G, Ruinemans-Koerts J, Joosten F et al. Tumor producing fibroblast growth factor 23 localized by two-staged venous sampling. Eur J Endocrinol 2008; 158:431-437.
78. Parfitt AM, Kleerekoper M, Cruz C. Reduced phosphate reabsorption unrelated to parathyroid hormone after renal transplantation: implications for the pathogenesis of hyperparathyroidism in chronic renal failure. Miner Electrolyte Metab 1986; 12:356-362.
79. Levi M. Post-transplant hypophosphatemia. Kidney Int 59:2377-2387.

FGF23 AND SYNDROMES OF ABNORMAL RENAL PHOSPHATE HANDLING 61

80. Bhan I, Shah A, Holmes J et al. Post-transplant hypophosphatemia: Tertiary 'Hyper-Phosphatoninism'? Kidney Int 2006; 70:1486-1494.
81. Pande S, Ritter CS, Rothstein M et al. FGF-23 and sFRP-4 in chronic kidney disease and postrenal transplantation. Nephron Physiol 2006; 104:p23-p32.
82. Khosravi A, Cutler CM, Kelly MH et al. Determination of the elimination half-life of fibroblast growth factor-23. J Clin Endocrinol Metab 2007; 92:2374-2377.
83. Nordstrom H, Lennquist S, Lindell B et al. Hypophosphataemia in severe burns. Acta Chir Scand 1977; 143:395-399.
84. Dickerson RN, Gervasio JM, Sherman JJ et al. A comparison of renal phosphorus regulation in thermally injured and multiple trauma patients receiving specialized nutrition support. JPEN J Parenter Enteral Nutr 2001; 25:152-159.
85. Nafidi O, Lepage R, Lapointe RW et al. Hepatic resection-related hypophosphatemia is of renal origin as manifested by isolated hyperphosphaturia. Ann Surg 2007; 245:1000-1002.
86. Nafidi O, Lapointe RW, Lepage R et al. Mechanisms of renal phosphate loss in liver resection-associated hypophosphatemia. Ann Surg 2009; 249:824-827.
87. Salem RR, Tray K. Hepatic resection-related hypophosphatemia is of renal origin as manifested by isolated hyperphosphaturia. Ann Surg 2005; 241:343-348.
88. Albright F, Butler AM, Bloomberg E. Rickets resistant to vitamin D therapy Am J Dis Child 1937; 54 529-547.
89. Winters RW, Graham JB, Williams TF et al. A genetic study of familial hypophosphatemia and vitamin D resistant rickets with a review of the literature. Medicine (Baltimore) 1958; 37:97-142.
90. HYP-Consortium. A gene (PEX) with homologies to endopeptidases is mutated in patients with X-linked hypophosphatemic rickets. Nat Genet 1995; 11:130-136.
91. Sabbagh Y, Jones AO, Tenenhouse HS. PHEXdb, a locus-specific database for mutations causing X-linked hypophosphatemia. Hum Mutat 2000; 16:1-6.
92. Liu S, Guo R, Simpson LG et al. Regulation of fibroblastic growth factor 23 expression but not degradation by PHEX. J Biol Chem 2003; 278:37419-37426.
93. Sitara D, Razzaque MS, Hesse M et al. Homozygous ablation of fibroblast growth factor-23 results in hyperphosphatemia and impaired skeletogenesis and reverses hypophosphatemia in Phex-deficient mice. Matrix Biol 2004; 23:421-432.
94. Liu S, Quarles LD. How fibroblast growth factor 23 works. J Am Soc Nephrol 2007; 18:1637-1647.
95. Makras P, Hamdy NA, Kant SG et al. Normal growth and muscle dysfunction in X-linked hypophosphatemic rickets associated with a novel mutation in the PHEX gene. J Clin Endocrinol Metab 2008; 93:1386-1389.
96. Owen CJ, Habeb A, Pearce SH et al. Discordance for X-linked hypophosphataemic rickets in identical twin girls. Horm Res 2009; 71:237-244.
97. Rivkees SA, el-Hajj-Fuleihan G, Brown EM et al. Tertiary hyperparathyroidism during high phosphate therapy of familial hypophosphatemic rickets. J Clin Endocrinol Metab 1992; 75:1514-1518.
98. Seikaly MG, Brown R, Baum M. The effect of recombinant human growth hormone in children with X-linked hypophosphatemia. Pediatrics 1997; 100:879-884.
99. Lorenz-Depiereux B, Bastepe M, Benet-Pages A et al. DMP1 mutations in autosomal recessive hypophosphatemia implicate a bone matrix protein in the regulation of phosphate homeostasis. Nat Genet 2006; 38:1248-1250.
100. Feng JQ, Ward LM, Liu S et al. Loss of DMP1 causes rickets and osteomalacia and identifies a role for osteocytes in mineral metabolism. Nat Genet 2006; 38:1310-1315.
101. Levy-Litan V, Hershkovitz E, Avizov L et al. Autosomal-recessive hypophosphatemic rickets is associated with an inactivation mutation in the ENPP1 gene. Am J Hum Genet 86:273-278.
102. Lorenz-Depiereux B, Schnabel D, Tiosano D et al. Loss-of-function ENPP1 mutations cause both generalized arterial calcification of infancy and autosomal-recessive hypophosphatemic rickets. Am J Hum Genet 86:267-272.
103. von Marschall Z, Fisher LW. Dentin matrix protein-1 isoforms promote differential cell attachment and migration. J Biol Chem 2008; 283:32730-32740.
104. Yuan B, Meudt J, Feng JO et al. 7B2 protein mediated inhibition of DMP1 cleavage in osteoblasts enhances FGF-23 production in hyp-mice. JBMR 2008; 23:s16 (abstract 1053).
105. Lu Y, Qin C, Xie Y et al. Studies of the DMP1 57-kDa functional domain both in vivo and in vitro. Cells Tissues Organs 2009; 189:175-185.
106. Terkeltaub R. Physiologic and pathologic functions of the NPP nucleotide pyrophosphatase/phosphodiesterase family focusing on NPP1 in calcification. Purinergic Signal 2006; 2:371-377.
107. Rutsch F, Vaingankar S, Johnson K et al. PC-1 nucleoside triphosphate pyrophosphohydrolase deficiency in idiopathic infantile arterial calcification. Am J Pathol 2001; 158:543-554.
108. Rutsch F, Ruf N, Vaingankar S et al. Mutations in ENPP1 are associated with 'idiopathic' infantile arterial calcification. Nat Genet 2003; 34:379-381.

109. Ramjan KA, Roscioli T, Rutsch F et al. Generalized arterial calcification of infancy: treatment with bisphosphonates. Nat Clin Pract Endocrinol Metab 2009; 5:167-172.
110. Chodirker BN, Evans JA, Seargeant LE et al. Hyperphosphatemia in infantile hypophosphatasia: implications for carrier diagnosis and screening. Am J Hum Genet 1990; 46:280-285.
111. Jones A, Tzenova J, Frappier D et al. Hereditary hypophosphatemic rickets with hypercalciuria is not caused by mutations in the Na/Pi cotransporter NPT2 gene. J Am Soc Nephrol 2001; 12:507-514.
112. Tenenhouse HS, Econs MJ. Mendelian hypophosphatemias. In: Scriver CR, Beaudet AL, Valle D, Sly WS, Vogelstein B, Childs B, Kinzler KW, eds. The Metabolic and Molecular Bases of Inherited Diseases, 8th ed. McGraw-Hill, New York; 2001; pp 5039-5067.
113. Lorenz-Depiereux B, Benet-Pages A, Eckstein G et al. Hereditary Hypophosphatemic Rickets with Hypercalciuria Is Caused by Mutations in the Sodium-Phosphate Cotransporter Gene SLC34A3. Am J Hum Genet 2006; 78:193-201.
114. Bergwitz C, Roslin NM, Tieder M et al. SLC34A3 Mutations in Patients with Hereditary Hypophosphatemic Rickets with Hypercalciuria Predict a Key Role for the Sodium-Phosphate Cotransporter NaPi-IIc in Maintaining Phosphate Homeostasis. Am J Hum Genet 2006; 78:179-192.
115. Ichikawa S, Sorenson AH, Imel EA et al. Intronic Deletions in the SLC34A3 Gene Cause Hereditary Hypophosphatemic Rickets with Hypercalciuria. J Clin Endocrinol Metab 2006; 91:4022-4027.
116. Jaureguiberry G, Carpenter TO, Forman S et al. A novel missense mutation in SLC34A3 that causes hereditary hypophosphatemic rickets with hypercalciuria in humans identifies threonine 137 as an important determinant of sodium-phosphate cotransport in NaPi-IIc. Am J Physiol Renal Physiol 2008; 295:F371-F379.
117. Prie D, Huart V, Bakouh N et al. Nephrolithiasis and osteoporosis associated with hypophosphatemia caused by mutations in the type 2a sodium-phosphate cotransporter. N Engl J Med 2002; 347:983-991.
118. Lapointe JY, Tessier J, Paquette Y et al. NPT2a gene variation in calcium nephrolithiasis with renal phosphate leak. Kidney Int 2006; 69:2261-2267.
119. Virkki LV, Forster IC, Hernando N et al. Functional characterization of two naturally occurring mutations in the human sodium-phosphate cotransporter type IIa. J Bone Miner Res 2003; 18:2135-2141.
120. Karim Z, Gerard B, Bakouh N et al. NHERF1 mutations and responsiveness of renal parathyroid hormone. N Engl J Med 2008; 359:1128-1135.
121. Bergwitz C, Bastepe M. NHERF1 Mutations and Responsiveness of Renal Parathyroid Hormone. NEJM 2008; 359:2615-2617.
122. Tieder M, Arie R, Modai D et al. Elevated serum 1,25-dihydroxyvitamin D concentrations in siblings with primary Fanconi's syndrome. N Engl J Med 1988; 319:845-849.
123. Magen D, Berger L, Coady MJ et al. A loss-of-function mutation in NaPi-IIa and renal Fanconi's syndrome. N Engl J Med 2010; 362:1102-1109.
124. Beck L, Karaplis AC, Amizuka N et al. Targeted inactivation of Npt2 in mice leads to severe renal phosphate wasting, hypercalciuria and skeletal abnormalities. Proc Natl Acad Sci USA 1998; 95:5372-5377.
125. Iwaki T, Sandoval-Cooper MJ, Tenenhouse HS et al. A missense mutation in the sodium phosphate cotransporter Slc34a1 impairs phosphate homeostasis. J Am Soc Nephrol 2008; 19:1753-1762.
126. Beighton P, Cremin BJ, Kozlowski K. Osteoglophonic dwarfism. Pediatr Radiol 1980; 10:46-50.
127. Sklower Brooks S, Kassner G, Qazi Q et al. Osteoglophonic dysplasia: review and further delineation of the syndrome. Am J Med Genet 1996; 66:154-162.
128. White KE, Cabral JM, Davis SI et al. Mutations that cause osteoglophonic dysplasia define novel roles for FGFR1 in bone elongation. Am J Hum Genet 2005; 76:361-367.
129. Farrow EG, Davis SI, Mooney SD et al. Extended mutational analyses of FGFR1 in osteoglophonic dysplasia. Am J Med Genet A 2006; 140:537-539.
130. Maroteaux P, Stanescu V, Stanescu R et al. Opsismodysplasia: a new type of chondrodysplasia with predominant involvement of the bones of the hand and the vertebrae. Am J Med Genet 1984; 19:171-182.
131. Zeger MD, Adkins D, Fordham LA et al. Hypophosphatemic rickets in opsismodysplasia. J Pediatr Endocrinol Metab 2007; 20:79-86.
132. Solomon LM, Fretzin DF, Dewald RL. The epidermal nevus syndrome. Arch Dermatol 1968; 97:273-285.
133. Gellis SS, Feingold M. Linear nevus sebaceous syndrome. Am J Dis Child 1970; 120:139-140.
134. Rogers M. Epidermal nevi and the epidermal nevus syndromes: a review of 233 cases. Pediatr Dermatol 1992; 9:342-344.
135. Menascu S, Donner EJ. Linear nevus sebaceous syndrome: case reports and review of the literature. Pediatr Neurol 2008; 38:207-210.
136. Hafner C, van Oers JM, Vogt T et al. Mosaicism of activating FGFR3 mutations in human skin causes epidermal nevi. J Clin Invest 2006; 116:2201-2207.
137. Skovby F, Svejgaard E, Moller J. Hypophosphatemic rickets in linear sebaceous nevus sequence. J Pediatr 1987; 111:855-857.
138. Zutt M, Strutz F, Happle R et al. Schimmelpenning-Feuerstein-Mims syndrome with hypophosphatemic rickets. Dermatology 2003; 207:72-76.

FGF23 AND SYNDROMES OF ABNORMAL RENAL PHOSPHATE HANDLING

139. Hoffman WH, Jüppner HW, Deyoung BR et al. Elevated fibroblast growth factor-23 in hypophosphatemic linear nevus sebaceous syndrome. Am J Med Genet A 2005; 134:233-236.
140. John M, Shah NS. Hypophosphatemic rickets with epidermal nevus syndrome. Indian Pediatr 2005; 42:611-612.
141. Hoffman WH, Jain A, Chen H et al. Matrix extracellular phosphoglycoprotein (MEPE) correlates with serum phosphorus prior to and during octreotide treatment and following excisional surgery in hypophosphatemic linear sebaceous nevus syndrome. Am J Med Genet A 2008; 146A:2164-2168.
142. Schwindinger WF, Francomano CA, Levine MA. Identification of a mutation in the gene encoding the alpha subunit of the stimulatory G protein of adenylyl cyclase in McCune-Albright syndrome. Proc Natl Acad Sci USA 1992; 89:5152-5156.
143. Weinstein LS, Shenker A, Gejman PV et al. Activating mutations of the stimulatory G protein in the McCune-Albright syndrome. New Engl J Med 1991; 325:1688-1695.
144. Weinstein L, Yu S, Warner D et al. Endocrine Manifestations of Stimulatory G Protein alpha-Subunit Mutations and the Role of Genomic Imprinting. Endocr Rev 2001; 22:675-705
145. Yamamoto T, Miyamoto KI, Ozono K et al. Hypophosphatemic rickets accompanying McCune-Albright syndrome: evidence that a humoral factor causes hypophosphatemia. J Bone Miner Metab 2001; 19:287-295.
146. Riminucci M, Collins MT, Fedarko NS et al. FGF-23 in fibrous dysplasia of bone and its relationship to renal phosphate wasting. J Clin Invest 2003; 112:683-692.
147. Collins MT, Chebli C, Jones J et al. Renal phosphate wasting in fibrous dysplasia of bone is part of a generalized renal tubular dysfunction similar to that seen in tumor-induced osteomalacia. J Bone Miner Res 2001; 16:806-813.
148. Kobayashi K, Imanishi Y, Koshiyama H et al. Expression of FGF23 is correlated with serum phosphate level in isolated fibrous dysplasia. Life Sci 2006; 78:2295-2301.
149. Brown WW, Jüppner H, Langman CB et al. Hypophosphatemia with elevations in serum fibroblast growth factor 23 in a child with Jansen's metaphyseal chondrodysplasia. J Clin Endocrinol Metab 2009; 94:17-20.
150. Haviv YS, Silver J. Late onset oncogenic osteomalacia-associated with neurofibromatosis type II. Clin Nephrol 2000; 54:429-430.
151. Konishi K, Nakamura M, Yamakawa H et al. Hypophosphatemic osteomalacia in von Recklinghausen neurofibromatosis. Am J Med Sci 1991; 301:322-328.
152. Giard JM. Sur la calcifcation hibernale. Compes Rend Seanes Soc Biol So 1998; 34:1013-1015.
153. Duret MH. Tumeurs multiples et singulieres des bourses sereuses. Bull Mem Soc Ant Paris 1899; 74:725-731.
154. Sitara D, Kim S, Razzaque MS et al. Genetic evidence of serum phosphate-independent functions of FGF-23 on bone. PLoS Genet 2008; 4:e1000154.
155. Martin A, David V, Laurence JS et al. Degradation of MEPE, DMP1 and release of SIBLING ASARM-peptides (minhibins): ASARM-peptide(s) are directly responsible for defective mineralization in HYP. Endocrinology 2008; 149:1757-1772.
156. Ichikawa S, Lyles KW, Econs MJ. A novel GALNT3 mutation in a pseudoautosomal dominant form of tumoral calcinosis: evidence that the disorder is autosomal recessive. J Clin Endocrinol Metab 2005; 90:2420-2423.
157. Frishberg Y, Topaz O, Bergman R et al. Identification of a recurrent mutation in GALNT3 demonstrates that hyperostosis-hyperphosphatemia syndrome and familial tumoral calcinosis are allelic disorders. J Mol Med 2005; 83:33-38.
158. Frishberg Y, Araya K, Rinat C et al. Sprecher E Hyperostosis-hyperphosphatemia syndrome caused by mutations in GALNT3 and associated with augmented processing of FGF-23. American Society of Nephrology, Philadelphia, 2004; pp F-P0937.
159. Benet-Pages A, Orlik P, Strom TM et al. An FGF23 missense mutation causes familial tumoral calcinosis with hyperphosphatemia. Hum Mol Genet 2005; 14:385-390.
160. Masi L, Gozzini A, Carbonell S et al. A novel recessive mutation in fibroblast growth factor-23 (FGF23) causes a tumoral calcinosis. J Bone Miner Res 2005; p S128.
161. Garringer HJ, Malekpour M, Esteghamat F et al. Molecular genetic and biochemical analyses of FGF23 mutations in familial tumoral calcinosis. Am J Physiol Endocrinol Metab 2008; 295:E929-E937.
162. Shimada T, Kakitani M, Yamazaki Y et al. Targeted ablation of Fgf23 demonstrates an essential physiological role of FGF23 in phosphate and vitamin D metabolism. J Clin Invest 2004; 113:561-568.
163. Hesse M, Fröhlich LF, Zeitz U et al. Ablation of vitamin D signaling rescues bone, mineral and glucose homeostasis in Fgf-23 deficient mice. Matrix Biol 2007; 26:75-84.
164. Ichikawa S, Sorenson AH, Austin AM et al. Ablation of the Galnt3 gene leads to low-circulating intact fibroblast growth factor 23 (Fgf23) concentrations and hyperphosphatemia despite increased Fgf23 expression. Endocrinology 2009; 150:2543-2550.
165. Razzaque MS, Sitara D, Taguchi T et al. Premature aging-like phenotype in fibroblast growth factor 23 null mice is a vitamin D-mediated process. FASEB J 2006; 20:720-722.
166. Ohnishi M, Nakatani T, Lanske B et al. Reversal of mineral ion homeostasis and soft-tissue calcification of klotho knockout mice by deletion of vitamin D 1alpha-hydroxylase. Kidney Int 2009; 75:1166-1172.

167. Stubbs JR, Liu S, Tang W et al. Role of hyperphosphatemia and 1,25-dihydroxyvitamin d in vascular calcification and mortality in fibroblastic growth factor 23 null mice. J Am Soc Nephrol 2007; 18:2116-2124.
168. Liu S, Zhou J, Tang W et al. Pathogenic role of Fgf23 in Hyp mice. Am J Physiol Endocrinol Metab 2006; 291:E38-E49.
169. Yamaguchi T, Sugimoto T, Imai Y et al. Successful treatment of hyperphosphatemic tumoral calcinosis with long-term acetazolamide. Bone 1995; 16(4 Suppl):247S-250S.
170. Lufkin EG, Wilson DM, Smith LH et al. Phosphorus excretion in tumoral calcinosis: response to parathyroid hormone and acetazolamide. J Clin Endocrinol Metab 1980; 50:648-653.
171. Gowen M, Stroup GB, Dodds RA et al. Antagonizing the parathyroid calcium receptor stimulates parathyroid hormone secretion and bone formation in osteopenic rats. J Clin Invest 2000; 105:1595-1604.
172. Bergwitz C, Banerjee S, Abu-Zahra H et al. Defective O-glycosylation due to a novel homozygous S129P mutation is associated with lack of fibroblast growth factor 23 secretion and tumoral calcinosis. J Clin Endocrinol Metab 2009; 94:4267-4274.

CHAPTER 4

EVIDENCE FOR FGF23 INVOLVEMENT IN A BONE-KIDNEY AXIS REGULATING BONE MINERALIZATION AND SYSTEMIC PHOSPHATE AND VITAMIN D HOMEOSTASIS

Aline Martin and L. Darryl Quarles*

University of Tennessee Health Science Center, Department of Medicine, Memphis, Tennessee, USA.
*Corresponding Author: L. Darryl Quarles—Email: dquarles@uthsc.edu

Abstract: Bone is involved in the maintenance of phosphate and vitamin D homeostasis via its production and secretion of FGF23 and serves as a reservoir for the storage and release of calcium and phosphate into the circulation. Alterations in mineralization of extracellular matrix and the remodeling activities of the skeleton are coupled to the kidney conservation of phosphate and production of 1,25(OH)2D via the regulation of FGF23 production by osteocytes through yet-to-be defined locally derived factors. In addition, FGF23 production is regulated by 1,25(OH)2D in a feedback loop where FGF23 stimulate Cyp24 mediated degradation of 1,25(OH)2D that serves to protect the organism from the toxic effects of vitamin D excess. In this chapter, we will review the regulation and function of FGF23.

INTRODUCTION

Coordination of calcium and phosphate homeostasis involves absorption of these minerals by the intestine, influx and efflux of calcium and phosphate from bone and their kidney excretion or reabsorption. Traditionally, the regulation of calcium and phosphate homeostasis has been viewed from the perspective of the PTH/vitamin D axis, where $1,25(OH)_2D$ predominate effect is to increase gastrointestinal calcium and phosphate aborption, and PTH principal function is to maintain serum calcium levels by increasing $1,25(OH)_2D$ production and calcium reabsorption by the kidney and calcium and phosphate efflux from bone by stimulating bone remodeling. The discovery of FGF23, which

---

Endocrine FGFs and Klothos, edited by Makoto Kuro-o.
©2012 Landes Bioscience and Springer Science+Business Media.

suppresses 1,25(OH)₂D levels via effects on the kidney to stimulate Cyp24 mediated degradation and suppress Cyp27b1-mediated production, and inhibits renal phosphate reabsorption, has lead to an expansion of the PTH/Vitamin D axis to include another hormonal axis that functions as a counter regulatory pathway to suppress 1,25(OH)₂D production and enhance renal phosphate secretion. FGF23 is predominately produced by osteocytes in bone, which places it in an optimal place to coordinate renal phosphate handling with bone mineralization and turnover (Fig. 1). Our understanding of FGF23 biology is derived from comparative analyses of hereditary diseases involving mutations of essential bone factors such as Phex, a transmembrane endopeptidase and Dmp1, an extracellular matrix SIBLING protein. Most of these diseases are characterized by bone mineralization disorders, rickets, increased bone production of FGF23 and subsequent

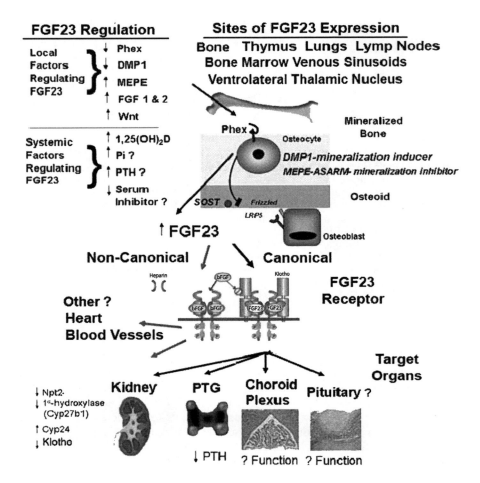

Figure 1. Regulation and function of FGF23. Local and systemic factors regulate FGF23 expression principally in osteocytes in bone. Circulating FGF23 targets koltho/FGFR complexes in a limited number of tissues, including kidney, PTG, choroid plexus and pituitary. The main physiological functions of FGF23 are to inhibit renal phosphate reabsorption and 1,25(OH)₂D production by the kidney.

FGF23 AND BONE DISORDERS

hypophosphatemia and abnormal vitamin D metabolism (XLH, ARHR). However, other disorders (McCune Albright syndrome, CKD) affecting the degree of bone remodeling and leading to different bone phenotypes are also associated with an elevation of FGF23 levels, suggesting that the status of bone mineralization and remodeling actually regulates FGF23 production. The precise mechanism in bone that regulate FGF23 are currently being elucidated. To date, it appears that Phex and Dmp1 are both essential and interconnected players in the local regulation of FGF23 production. Recent studies also propose the FGFs/FGFreceptor pathway as an important step in the signalization cascade. This chapter reviews the well-known and emerging functions of FGF23 and its regulation in response to bone modification signals.

EVOLVING INSIGHTS INTO SKELETAL FUNCTIONS

Bone Remodeling

Bone is a dynamic load bearing organ that creates the framework of the human body. Bone structural integrity is maintained due to a constant remodeling of the bone matrix, which consists of an iterative cycle of removal of mineralized tissue by the osteoclast-mediated resorption followed by its replacement with new collagenous matrix (osteoid) by osteoblast-mediated bone formation. After a delay of several days, osteoid undergoes mineralization. During the mineralization process, a subset of the osteoblasts become embedded in the matrix forming osteocytes that have dendrite-like cytoplasmic extensions forming a canalicular (neural-like) network inside the mineralized matrix. Osteocytes function as sensors and effectors maintaining skeletal homeostasis. Mineralization of osteoid is a regulated process representing a balance between inorganic factors, such as the local phosphate and pyrophosphates, which respectively promote and inhibit hydroxyapatite formation as well as extracellular matrix proteins that either facilitate or impede the mineralization process. Autocrine, paracrine and hormonal factors as well as physical and chemical processes regulate the overall remodeling process.

Bone, a Target for Systemic and Local Factors Regulating Mineral Metabolism

In addition to its scaffolding function, bone is a metabolically active organ that participates in mineral metabolism. The traditional view is that bone is a mineral reservoir, where calcium and phosphate move in and out bone from a poorly described exhalable pool, from the area surrounding osteocytes (called osteolysis) and from osteoclast-mediated resorptive sites. In this conceptualization, calcium and phosphate are in equilibrium with the systemic milieu under steady state and the influx and efflux of calcium and phosphate from bone is under control of both passive physiochemical forces and active cellular processes, such as systemic hormones and mechanical/local factors. An example of physiochemical effects is metabolic acidosis, which leads to a net loss of bone mineral. The best example an endocrine network involving bone is the PTH-Vitamin D axis. In response to hypocalcemia, the Parathyroid Gland (PTG) increases the production and secretion of PTH, which targets the renal distal tubule to decrease renal calcium excretion and the proximal tubule to inhibit phosphate reabsorption and to stimulate $1,25(OH)_2D$ production. Action of $1,25(OH)_2D$ on the small intestines increases active calcium and phosphate transport.[2] PTH also has direct effects on bone via PTH receptors in osteoblasts, resulting in increased calcium and phosphate efflux from the exchangeable bone fluid compartment[3] and through RANKL-dependent, osteoclast-mediated bone resorption of mineralized bone.[4] The direct kidney and bone effects of PTH, along with the concomitant actions of 1,25(OH)$_2$D, restore serum calcium levels to normal (Fig. 2A). The phosphaturic actions of PTH offset vitamin D–mediated gastrointestinal phosphate absorption[2] and PTH-dependent phosphate efflux from bone,[5] thereby preventing the development of hyperphosphatemia. A dramatic clinical example of the contribution of bone remodelling to systemic calcium and phosphate homeostasis is "hungry bone syndrome", where an acute lowering of PTH by parathyroidectomy in patients with hyperparathyroidism leads to hypocalcemia and hypophoshatemia due to rapid uptake of the calcium and phosphate into bone matrix as remodelling shifts from a bone resorption to a formation phase. A more recent example is that fat through leptin controls bone mass by a hypothalamic relay and sympathetic nervous system.[6] Other hormones such as estrogens and androgens, thyroid hormone and glucocorticoids, as well as the sympathetic nervous system target receptors in bone to coordinate calcium and phosphate flux in response to systemic alterations in mineral homeostasis. Finally, malignant processes such as multiple myeloma or metastatic diseases that cause increased bone resorption can lead to hypercalcemia through release of calcium from bone.

Bone as an Endocrine Organ

A new concept that has emerged is that bone is an endocrine organ that releases hormones that communicate with other organs to control systemic homeostasis. At present, FGF23 and osteocalcin are two known hormones secreted by cells in the osteoblast lineage that act systemically to respectively coordinate mineral homeostasis[1] and energy metabolism.[6]

FGF23, the Prototypic Bone Derived Hormone

FGF23 is a ~32 kDa protein with an N-terminal FGF-homology domain and a novel 71 amino acid C-terminus.[7] FGF23 is grouped with FGF19 (mouse FGF15) and FGF21[7,8] subfamily that, in contrast to the paracrine and heparin-dependent functions of canonical FGFs,[9] act as hormones.[10] FGF23 is produced in cells in the osteoblast lineage, principally the osteocyte in some diseases of FGF23 excess, such as hereditary hypophosphatemic disorders and mainly in peritrabecular osteoblasts in other disorders, such as McCune Albrights Hereditary Osteodystrophy and osteitis fibrosa in chronic kidney disease. FGF23 targets organs that co-express FGFR/Klotho complexes,[11,12] including the kidney, parathyroid gland, pituitary gland and choroid plexus.[12] α-Klotho is a Type I membrane, ß-glycosidase-like protein, which binds to FGFRs and the C-terminus of FGF23[12-15] to permit heparin independent activation of FGFRs by FGF23. The principal target for FGF23 is the kidney, where this hormone has phosphaturic effects that are mediated by inhibition of renal phosphate reabsorption and serves as a counter-regulatory factor that inhibits 1,25(OH)$_2$D production.[11,12,16] Excess FGF23 causes hypophosphatemia via inhibition of solute carrier family 34, member 1–dependent (SLC34A1-dependent) and SLC34A2-dependent phosphate transport (also known as sodium phosphate cotransporter 2 [NPT2a] or NaPi-IIa and NPT2c or NaPi-IIc, respectively). Excess FGF23 also suppresses 1,25(OH)$_2$D via inhibition of 25-hydroxyvitamin D-1α–hydroxylase (CYP27B1) that is converted to 1,25(OH)$_2$D and stimulation of 24-hydroxylase (CYP24) that inactivates

FGF23 AND BONE DISORDERS

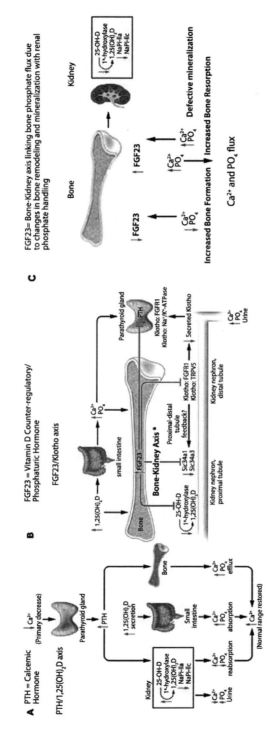

Figure 2. Comparison of PTH and FGF23 functions. A) PTH-Vitamin D axis. B) FGF23 as a Vitamin D counter regulatory hormone. C) FGF23 in a Bone-Kidney axis coordinating bone remodeling and mineralization with renal phosphate handling. Adapted with permission from Quarles LD. J Clin Invest 2008; 118(12):3820-8.[1]

1,25(OH)$_2$D in the proximal tubule of the kidney.[17-20] Although the biological actions of FGF23 are in the proximal tubule, the precise segments of the kidney and receptors that mediate the renal response to FGF23 are not entirely clear. Ex vivo studies of proximal tubular segments or cell lines derived from proximal tubules have produced variable results of FGF23-mediated inhibition of sodium-dependent phosphate transport.[19,21] The highest expression of Klotho-FGF receptor complexes are in the distal tubule,[14] suggesting a distal to proximal feedback mechanism. Although initial studies that excluded individual effects of FGFR3 and FGFR4 and by default implicated FGFR1 and the proximal tubule as the principal target in kidney for FGF23 {Liu, 2008 #4165}, we have found that combined loss of FGFR 3 and 4 partially correct the Hyp phenotype. Thus, FGFR 1, 3 and 4 may act in concert to mediate the effects of FGF23 on the kidney.

FGF23 also targets FGFR-Klotho complexes in the PTG,[22] but the physiological effects of FGF23 remains controversial.[14,23] On the one hand, there is clinical evidence that excess FGF23 stimulates PTH secretion,[12] as evidence by the strong association between elevated FGF23 levels and the severity of hyperparathyroidism in CKD. On the other hand, FGF23 has been shown to suppress PTH mRNA expression in vitro and decrease serum PTH in vivo.[24]

In addition, FGF23 is produced in the ventrolateral thalamic nucleus and may target the choroid plexus, which expresses FGFR-Klotho complexes, as well as sodium-dependent phosphate transporters. Since there is a gradient for phosphate between the Cerebral Spinal Fluid (CSF) and serum (with the phosphate concentration being significantly lower in the CSF), it is tempting to speculate that FGF23 may also regulate CSF phosphate concentrations. Another potential target of FGF23 is the pituitary gland, where upregulation of early response gene expression has been observed in response to acute FGF23 administration to mice. FGF23 null mice have severe growth retardation, suggesting that FGF23 may regulate pituitary function. There is controversy as to whether bone itself is a target for FGF23. Klotho, which is required for FGF23 actions, is not expressed in bone and the observed changes in bone in states of high and low FGF23 appear to be secondary to changes in serum phosphate and 1,25(OH)$_2$D levels. However, recent studies may have identified that FGF23 suppresses differentiation and matrix production in osteoblasts cultures.[25] FGF23 may also impact upon glucose homeostasis, thymus function, growth and aging.[26-28] At present, however, it appears that many of these abnormalities associated with FGF23 deficiency in tissues lacking Klotho represent secondary effects of elevated 1,25(OH)$_2$D and hyperphosphatemia, rather than direct effects of FGF23 on these organs.[29-31] In addition to these canonical effects mediated by FGFR-Klotho complexes, it has been suggested that high concentrations of FGF23, such as that observed in CKD, may have broader, nonspecific actions to activate FGFRs independent of membrane bound Klotho. Currently, there is no direct evidence for such off-target effects of FGF23.

GENETIC DISORDERS DEFINE FGF23 FUNCTION

Excess of FGF23

Several hereditary and acquired disorders are mediated by FGF23 leading to a common phenotype characterized by hypophosphatemia due to renal phosphate wasting and inappropriately normal 1,25(OH)$_2$D levels for the degree of hypophosphatemia, rickets/osteomalacia and the absence of hypercalciuria.

FGF23 AND BONE DISORDERS

Hereditary hypophosphatemic disorders caused by a primary increase in circulating levels of FGF23 include:

- autosomal dominant hypophosphatemic rickets (ADHR; MIM 193100),
- autosomal recessive hypophosphatemic rickets (ARHR; MIM 241520),
- X-linked hypophosphatemic rickets (XLH; MIM 307800),
- osteoglophonic dysplasia (OGD; MIM 166250).

There is a single report of a patient with hypophosphatemic rickets and hyperparathyroidism (HRH; MIM pending).[32] HRH differs from the others disorders described here in that hyperparathyroidism is a predominant feature.

Whereas the clinical features of ADHR, ARHR, XLH, OGD and HRH are caused by excess FGF23, the mechanisms whereby FGF23 levels are increased differ between these disorders. ADHR is caused by mutations (R176Q and R179W) in the RXXR furinlike cleavage domain of FGF23 that impairs proteolytic inactivation of FGF23.[33,34] Interestingly, some small integrin-binding ligand N-linked glycoproteins (SIBLINGs) can participate or even be the cause of these pathological disorders. ARHR is caused by inactivating mutations in DMP1,[35,36] a member of the SIBLING family of extracellular matrix protein that augments mineralization.[37] Loss of *DMP1* results in increased transcription of FGF23 by osteocytes.[35] Tumor-induced osteomalacia, or oncogenic osteomalacia, is a paraneoplastic syndrome of renal phosphate wasting, aberrant vitamin D metabolism and osteomalacia that is associated with elevated FGF23 levels.[38,39] The proximate cause of increased FGF23 production is not known, but this disorder is associated with increased levels of MEPE, another rmember of the SIBLING family and sFRP4, which may regulate PHEX and DMP1 metabolism, respectively.[40] XLH is caused by inactivating mutations in *PHEX*[41-43] on the X chromosome. Loss of *PHEX* also leads to increased expression of FGF23 by osteocytes.[44] Mutant Phex in Hyp mice (murine homologous for X-Linked Hypophosphatemic rickets) induces an intrinsic bone mineralization defect associated with the increased production of FGF23 by the bone and hypophosphatemia. OGD is an autosomal dominant bone dysplastic disorder caused by activating mutations in the *FGFR1* gene, suggesting FGFR1 may regulate FGF23 expression in bone and/or the renal handling of phosphate.[45] HRH also has elevated FGF23 levels, but the primary genetic abnormality is a translocation causing elevated circulating levels of Klotho.[32]

Deficiency in FGF23

In contrast, deficiency of FGF23 results in the opposite renal phenotype, consisting of hyperphosphatemia and elevated production of $1,25(OH)_2D$.[26-28,30,46,47] Hyperphosphatemic familial tumoral calcinosis (HFTC; MIM 211900) is a rare autosomal recessive disorder characterized by hyperphosphatemia, normal or elevated $1,25(OH)_2D$ levels, soft-tissue calcifications and typically massive lobulated periarticular calcifications levels.[48] In addition, *Fgf23*-null mice have soft tissue calcifications, severe growth retardation, abnormalities of bone mineralization and a shortened lifespan.[27,28,30] To date there have been three different mutations identified as leading to either decreased bioactive circulating FGF23 levels or end-organ resistance to FGF23. These include mutations of the genes encoding FGF23,[46] Klotho[49] and GalNAc transferase 3 (GALNT3),[50] a Golgi-associated enzyme that *O*-glycosylates a furin-like convertase recognition sequence in FGF23.[51] Missense mutations of FGF23 impair its secretion, leading to inadequate circulating levels of this phosphaturic factor.[46] Mutations of GALNT3 destabilize FGF23, resulting in a low-intact serum FGF23 levels but high levels of biologically inactive C-terminal FGF23 fragments. A mutation in the gene coding for Klotho (H193R) results in decreased Klotho expression and reduced FGF23-Klotho-FGFR complex formation and to end-organ resistance to FGF23.[49] In mice, inactivating mutations or deletion of *Kl* results in end-organ insensitivity to FGF23, which is characterized by hyperphosphatemia, elevated 1,25(OH)$_2$D levels, early mortality and soft-tissue calcifications,[23,49,52,53] a phenotype that resembles that of *Fgf23*-null mice.[54]

PHYSIOLOGICAL FUNCTIONS OF FGF23

Since both extremes of hypophosphatemia and hyperphosphatemia have negative biological effects,[55,56] it is not surprising that adaptive mechanisms have evolved to protect organisms from hypophosphatemia and hyperphosphatemia. The PTH-Vitamin D axis has long held the position of a phosphate regulating pathway, but the regulation and function of PTH are more consistent with its principal biological role in maintaining calcium homeostasis (Fig. 2A). In the kidney there is threshold for phosphate reabsorption in the tubules, such that phosphate above a specific concentration is excreted in the urine. While PTH regulates the tubular threshold for phosphate, other circulating factors, called phosphatonins, have been postulated to regulate phosphate homeostasis under conditions not explained by alteration in PTH or other regulators of renal tubular phosphate handling. FGF23 is the first bone fide phosphatonin to be characterized. Three major physiological functions of FGF23 have been recognized thus far.

A Counter Regulatory Hormone for 1,25(OH)$_2$D

Surprisingly, of the potential systemic factors that might be expected to regulate FGF23 expression, such as calcium, phosphate, PTH and 1,25(OH)$_2$D, only 1,25(OH)$_2$D has been shown to regulate FGF23 production (Fig. 2B). In this regard, 1,25(OH)$_2$D directly stimulates FGF23 expression in osteocytes via a vitamin D response element (VDRE) in the *Fgf23* promoter.[16] Since FGF23 targets the kidney to suppress 1,25(OH)$_2$D production, 1,25(OH)$_2$D stimulation of FGF23 closes a feedback loop.[16] In the setting of excess 1,25(OH)$_2$D, which also reduces PTH levels, FGF23-mediated phosphaturia helps prevent potential hyperphosphatemia from enhanced 1,25(OH)$_2$D-dependent gastrointestinal phosphate absorption and diminished PTH-mediated phosphaturia. 1,25(OH)$_2$D may also indirectly regulate FGF23 expression. In this regard, selective deletion of the vitamin D receptor (VDR) in cartilage produces an unidentified chondrocyte-derived inhibitor of *FGF23* transcription.[57] Thus, FGF23's ability to enhance phosphate excretion and suppress 1,25(OH)$_2$D production may have evolved to protect an organism from vitamin D intoxication.[16]

Regulation of Parathyroid Gland Function?

FGF23 targets FGFR-Klotho complexes in the PTG,[22] but the physiological effects of FGF23 remains controversial.[14,23] On the one hand, there is clinical evidence that excess FGF23 stimulates PTH secretion,[12] as evidence by the strong association between elevated FGF23 levels and the severity of hyperparathyroidism in chronic kidney disease (CKD). On the other hand, FGF23 has been shown to suppresses PTH mRNA expression in

FGF23 AND BONE DISORDERS 73

vitro and decreases serum PTH in vivo.[24] There is also controversy about how or if PTH regulates the circulating level of FGF23. PTH does not appear to directly stimulate FGF23 production, since PTH does not stimulates FGF23 in osteoblasts in vitro[16] or in mice[58] in the setting of normal renal function. Also, FGF23 is not elevated in patients with primary hyperparathyroidism.[59] However, the PTH-cyclinD1 transgenic mouse model of primary hyperparathyroidism has elevated FGF23 levels that decreased after parathyroidectomy.[60] There is also potential cross-talk between PTH and FGF23 through the indirect mechanism of PTH stimulation of $1,25(OH)_2D$.[61]

Given that FGF23 is a "phosphaturic hormone", it is expected to be regulated by serum phosphate. Extracellular phosphate, however, does not appear to directly stimulate *FGF23* mRNA levels or *FGF23* gene promoter activity in osteoblastic cultures.[16] Although phosphate loading in mice increases FGF23 levels,[62] evidence of the importance of dietary phosphate in regulating FGF23 levels in humans is also conflicting.[63,64] Phosphate may regulate FGF23 certain pathological circumstances such as CKD, where elevations in FGF23 compensate for the reduced renal clearance of phosphate and the degree of FGF23 elevation correlates with the severity of hyperphosphatemia.[39] The stimulus for increased FGF23 in CKD is not known, but the pattern of FGF23 expression in bone is distinct from hereditary disorders of FGF23 excess, suggesting other regulatory pathways, possibly from the kidney to close the bone-kidney axis loop. In addition, metabolic acidosis is another setting where phosphate appears to regulate FGF23. In this regard, low levels of FGF23 can be restored to normal after oral phosphate therapy in chronic metabolic acidosis.[65]

Pathways controlling serum calcium may also regulate FGF23. In this regard, elevated FGF23 level is associated with both low calcium intake in the absence of vitamin D deficiency;[66] whereas suppression of FGF23 is observed in response to the hypocalcemic hormone calcitonin in patients with tumor-induced osteomalacia.[60,67] However, neither extracellular calcium nor PTH directly stimulate FGF23 promoter activity in osteoblasts,[16] suggesting that their effects may also be indirect or that the response to PTH might be different according to the short vs long term exposure to PTH. We also know that the bone remodelling status varies according to the intermittent vs continuous exposure to PTH. This reinforces the hypothesis that FGF23 levels are regulated by the bone remodelling status and that PTH might indirectly regulate FGF23 through local bone remodelling factors.

FGF23 Coordinates Bone Mineralization/Turnover and Renal Phosphate Handling

The other function of FGF23 appears to be to serve as the "primary" phosphaturic hormone in a bone-kidney axis to coordinate renal phosphate handling with bone mineralization and possibly bone remodeling activity (Fig. 2C). The possibility that bone is an endocrine organ that secretes FGF23 to coordinate renal phosphate handling to match bone mineralization and turnover has arisen from the study of XLH and ARHR and metabolic bone diseases associated with elevated FGF23.[26,35,68,69]

PHEX Regulation of FGF23

XLH is caused by mutations of PHEX, or Phosphate regulating gene with homologies to endopeptidases on the X chromosome, a member of the endothelin-converting enzyme family that leads to increased FGF23 gene transcription. Conditional deletion of Phex in the osteoblast lineage in vivo is sufficient to reproduce the Hyp phenotype, suggesting bone is the main source of FGF23.[70] How Phex mutations lead to increased Fgf23 gene transcription in bone is not clear and the relative importance of Fgf23 production by osteoblasts and osteocytes is not known.

Although an initial study suggested that Phex processes FGF23,[71] subsequent studies failed to establish Phex-dependent cleavage of FGF23.[34,44,72] Screening of substrate phage libraries by us and others have identified that Phex cleaves small peptides,[73] but failed to identify a physiologically relevant substrate for Phex.[72] The ASARM peptide, a motif in MEPE and Dmp1, is substrate for Phex in vitro[74] and additional data suggest that accumulation of ASARM as a consequence of inactivation of Phex can impair mineralization.[75] Although we have shown that ASARM binds to and inhibits Phex activity against a synthetic substrate in vitro,[69] it is unlikely that ASARM from MEPE is responsible for stimulating FGF23 in Hyp, since ablation of MEPE fails to alter FGF23 expression in Hyp mice.[76]

In the osteoblastic lineage Phex and FGF23 expression are not fully concordant. FGF23 is markedly increased in terminally differentiation osteoblasts embedded in mineralization nodules, equivalent of an osteocyte embedded in bone,[41,44] whereas Phex is expressed in osteoblasts and osteocytes. Thus, Phex deficiency is necessary, but not sufficient to stimulate FGF23 (a functional Phex is missing in osteoblasts of Hyp mice, but osteoblasts do not upregulate FGF23 expression until they differentiate into osteocytes).[26] The importance of additional Fgf23 stimulating factors is illustrated by the paradox that restoration of Phex in transgenic mice using heterologous promoters fails to rescue the elevated FGF23 expression.[77-79]

DMP1 Regulation of FGF23

The breakthrough in understanding FGF23 transcriptional regulation in bone came through the comparative analysis of the Hyp mouse homologue of XLH and dentin matrix acidic phosphoprotein 1 (Dmp1) null mice. Ablation of Dmp1, an extracellular matrix SIBLING protein that regulates mineralization, markedly stimulates the transcription of FGF23 by osteocytes in mice, leading to the discovery that ARHR is caused by inactivating mutations of DMP1.[80] DMP1 is a multifunctional protein localized to the dendrites of osteocytes and its main function is to act as a nucleator of mineralization. Dmp1 is also reported translocate to the nucleus to regulate gene transcription,[81] however, we have excluded this mechanism for increased Fgf23 transcription in Dmp1 null mice, since overexpression of Dmp1 fails to regulate the Fgf23 promoter. DMP1 contains an RGD domain for integrin binding, an ASARM peptide (which might allow DMP1 to bind to PHEX), a large number of acidic domains, an N-terminal site for binding to MMP9, sites for casein kinase-2–mediated phosphorylation and conserved cleavage sites for bone morphogenetic protein 1/Tolloid-like MMPs (BMP1/Tolloid-like MMPs) and cathepsin B. DMP1 exists as a latent protein that is cleaved into 37-kDa and 57-kDa fragments by BMP1 or cathepsin B. The highly phosphorylated C-terminal 57-kDa fragment functions as a nucleator for mineralization. The NH_2-terminal fragment from DMP1 is a proteoglycan with a chondroitin sulfate chain attached through Ser74 that binds to proMMP-9 and may sequester growth factors.[82] Recent studies indicate that FGF23 is increased in callus during fracture healing, consistent with local matrix derived FGF23 stimulating factors.[83]

We have excluded several potential mechanisms whereby loss of Dmp1 or Phex leads to increased FGF23 expression in osteocytes. For example, PHEX does not cleave

FGF23 AND BONE DISORDERS

DMP1, indicating that other enzymes are required for DMP1 processing.[84] Similarly, ASARM in Dmp1 cannot be responsible for elevations of FGF23 since ablation of Dmp1 leads to increased FGF23 in Dmp1 null mice. Impaired differentiation of osteocytes, an explanation for Dmp1-dependent regulation of Fgf23 proposed by others, cannot be the reason for increased Fgf23 expression, since Fgf23 is not upregulated in earlier stages of the osteoblast lineage. Rather, assessment of bone marrow stromal cultures and explantation of either Hyp[85] or Dmp1 null bone[85] into wild-type mice demonstrates that the high expression of FGF23 in osteocytes is an intrinsic/local effect caused by the loss of either Phex or Dmp1. Moreover, there is strong evidence that the proximate cause of increased FGF23 expression in osteocytes in Hyp and Dmp1 null mice is intrinsic to the bone milieu and is mediated by the presence of unknown matrix-derived FGF23 stimulatory factors that are increased as a consequence of either Phex or Dmp1 mutations. In addition, FGFR1-dependent signaling pathways have emerged as important regulator of FGF23 expression in osteocytes. In this regard, OsteoGlophonic Dysplasia (OGD), an autosomal dominant bone dysplastic disorder caused by activating mutations in FGFR1, also has hypophosphatemia and elevated FGF23 levels.[45]

PHEX and DMP1-Dependent Defect in Mineralization: A Comparative Analysis

PHEX and DMP1 are both produced by osteoblasts and osteocytes,[26,41,44] where they control the mineralization of extracellular matrix, thereby providing signaling pathways for coordinating bone phosphate accretion with renal phosphate conservation through the regulation of Fgf23. Inactivation of Phex leads to two fundamental abnormalities in osteoblasts/osteocytes, 1) elevations of FGF23, which leads to hypophosphatemia and decreased 1,25(OH)$_2$D production and increments in a putative mineralization inhibitor called Minhibin, that leads to an intrinsic defect in the mineralization of extracellular matrix that is independent of phosphate.[86] Inactivation of either Phex or Dmp1 in mice results in nearly identical increments in FGF23 production by osteocytes and overlapping phenotypes characterized by hypophosphatemia, aberrant vitamin D metabolism and rickets/osteomalacia.[26,80,87,88] Ablation of FGF23 corrects these abnormalities in both Phex- and Dmp1-deficient mice.[26,87] However, restoring phosphate and 1.25(OH)$_2$D serum levels in Hyp does not completely correct the mineralization defect.[89]

FGF23-dependent hypophosphatemia contributes to impaired mineralization in Hyp and Dmp1 null mice, but we know that the increase in FGF23 levels is not the only cause for bone mineralization defects even if recent studies suggest that FGF23 may also have direct effects on bone to inhibit mineralization.[90] Indeed, we failed to detect Klotho expression in bone on microarray analysis, consistent with the failure to detect Klotho transcripts in bone by RT-PCR.[91] FGF23 also failed to stimulate FGF23 promoter activity in cultured osteoblasts. Consequently, it is unlikely that putative direct actions of FGF23 are mediated by FGFR-Klotho-dependent mechanisms. Rather, bone explantation studies indicate that an intrinsic defect in mineralization of extracellular matrix persists in Phex mutant mice, even when phosphate is normalized.[85] In Dmp1-null mice, the absence of the nucleation and mineral propagation functions of DMP1 likely contribute to the defective mineralization.[37,92] Indeed, Col1a-3.6 promoter-mediated overexpression of Dmp1 in osteoblast lineage is sufficient to rescue the increased Fgf23 expression and bone phenotype in Dmp1 null mice.[84] The mechanism whereby Phex mutations lead to impaired mineralization is less clear. Phex gene restoration in Hyp mice only partially corrects the bone phenotype and does not have any impact on the hypophosphatemia.[77,78]

PHEX does not cleave DMP1, indicating that other enzymes are required for DMP1 processing.[84] Another possibility is that the ASARM peptides derived from Mepe regulate mineralization in Hyp, since the ASARM peptides accumulate in Hyp bone[75] and deletion of MEPE in Hyp leads to increased mineralization of osteoblasts ex vivo.[76] One of the reason for ASARM accumulation in Hyp bone is that Phex is able to bind ASARM[75] and to cleave it[74] in normal condition. Moreover, Hyp bone is characterized by elevated proteolytic activity that is associated with the increased production of an ASARM peptide from MEPE and DMP1 and the administration of cathepsin inhibitors improves bone mineralization in Hyp mice, without correcting FGF23 expression or hypophosphatemia, indicating the separate regulation of hypophosphatemia and mineralization by PHEX.[93] However, deletion of MEPE from Hyp mice fail to rescues the rickets or osteomalacia, suggest that other factors may be important in mediating the intrinsic mineralization defect resulting from PHEX mutations.[76]

Gene expression profiling in Hyp bone also identified several previously unrecognized abnormalities in factors regulating mineralization in Hyp bone. For example, we found that the expression of matrix Gla-protein (Mgp), a inhibitor of mineralization,[94-96] was increased in Hyp bone. Thrombospondin 4 (Thbs4) is also markedly increased in Hyp bone as well as Hyp-derived cultured osteoblasts. Thbs4 is a member of the group B thrombospondin family of extracellular matrix glycoproteins containing heparin- and calcium-binding sites, as well as binding sites for collagens I, II, III and V, laminin-1, fibronectin and matrilin-2. Thbs4 is known to be expressed in osteogenic tissues,[97,98] but its physiological function in bone has not been established. However, the observations that ablation of the related Thbs3 results in accelerated endochondral ossification[99] and over expression of Thbs1 inhibits mineralization in osteoblast cultures,[100] suggest that increased Thbs4 in Hyp osteoblasts may play a role in the intrinsic mineralization defect. We found that carbonic anhydrase 12 (Car12) and 3 (Car3) as well as the sodium-dependent citrate transporter (Slc13a5), are decreased in Hyp bone. Car12 and Car 3 are members of the family of zinc metalloenzymes that catalyze the reversible hydration of carbon dioxide to bicarbonate and thereby have the potential to influence mineralization and osteoclast function by regulating the pH in the local milieu. While little is known about the membrane bound Car12 in bone, Car2, a cytosolic member of this family causes autosomal recessive osteopetrosis via disruption of osteoclast function.[101] The reduction of Slc13a5 might lead to local increases in citrate levels, which could impair mineralization, as well as deprive the cell of citrate necessary for energy metabolism in Hyp bone. With regards to known regulators of mineralization, we found no significant changes in Dmp1, Enpp1 and Tnap in Hyp bone. Ank (1.7) and Enpp6 (1.7) were respectively increased and decreased in Hyp bone.

Integrative Model Explaining PHEX and DMP1 Regulation of FGF23

Based on our current understanding of PHEX and DMP1 structure and function, we proposed a model that outlines a possible mechanism for local regulation of FGF23 expression in osteocytes by PHEX and DMP1. The model predicts that in wild-type bone DMP1 and PHEX suppress FGF23 transcription, possibly through sequestration of putative FGF23 stimulating factors. The model also predicts that Dmp1 binds to Phex via its ASARM motif and integrins via its RGD sequence located in the 57 kDa C-terminal fragment to form a PHEX-DMP1-integrin "ternary complex", required for suppression of FGF23 transcription. Recent studies found that $\alpha v\beta 3$ integrins bind to FGF1 leading

FGF23 AND BONE DISORDERS

to FGFR activation.[102] The ASARM motif in SIBLING protein MEPE binds to PHEX and inhibit its activity[103,104] and DMP1, which also contains the ASARM motif, inhibits PHEX activity, providing indirect evidence for DMP1-PHEX interactions. The C terminus of DMP1, which contains RGD integrin-binding and ASARM domains, is sufficient to rescue the phenotype of *Dmp1*-null mice,[105] suggesting that this region holds the key to understanding FGF23 regulation.

In contrast, inactivating mutations of PHEX and DMP1 increase FGF23 expression in this model via indirect effects mediated by the production and accumulation in the bone matrix of an unidentified PHEX substrate or interacting protein that acts as an FGF23-stimulating factor. We know that DMP1 is not a PHEX substrate. Impaired processing of DMP1 by other enzymes, such as BMP1, would also be predicted stimulate FGF23, by creating a functional DMP1-deficient state equivalent to *Dmp1*-null mice. Regardless, loss of PHEX or DMP1 result in the production or increased bioavailability of an unknown, local, bone-derived "FGF23 stimulating factor" or less likely an intrinsic cellular defect, leading to increased FGF23.

FGFR1 and possible paracrine/autocrine effects of other members of the FGF family may be a proximal signal mediating the effects of PHEX and DMP1 mutations. First, activating mutations of FGFR1 are associated with increased FGF23.[45] Second, microarray analysis of Hyp bone and Fgf23 promoter analysis have identified FGF-dependent regulation of Fgf23 transcription. Third, FGF have phosphaturic effects. For example, FGF2 is synthesized by osteoblasts, deposited in extracellular matrix and exerts autocrine or paracrine effects to inhibit osteoblastic differentiation,[106] also causes hypophosphatemia,[107] possibly through stimulation of FGF23. The phosphaturic effects of FGF7 might also be mediated via FGFR stimulation of FGF23.[108] Other putative phosphatonins, such as MEPE and secreted frizzled-related protein 4 (sFRP4) are also incorporated in this model. For example, MEPE actions may be mediated by binding of the ASARM peptide to PHEX and inhibition of its activity.[68,69] sFRP4 might inhibit BMP1-mediated processing of DMP1, leading to elevated FGF23 levels.

Evidence That Bone Turnover Regulates FGF23 Production

Several studies report elevation of FGF23 levels in association with primary alterations in bone remodelling. In case of low bone turnover, the hypothesis would predict that when bone resorption and formation activities are so reduced that there is almost no replacement of the existing mineralized matrix and therefore limited capacity for bone to buffer phosphate. Under these circumstances the continuous intake of phosphate creates a need for increasing phosphate excretion to avoid hyperphosphatemia that is met by increased FGF23 production by bone. A recent study supports the idea that the degree of bone turnover regulates FGF23 production by bone. In this study, low bone remodelling induced by antiresorptive agents, was associated with increased FGF23 circulating levels.[58] On the other hand, low bone turnover caused by low PTH levels in Gcm2 knockout mice are associated with low serum FGF23 levels. This might be explained by the dual effect of loss of PTH to inhibit bone remodeling and decrease $1,25(OH)_2D$ levels, the later which are necessary for maintaining FGF23 levels.[16] Idiopathic Phosphate Diabetes (IPD) leading to male osteoporosis have normal serum phosphate but low FGF23 serum levels.[109] Thus, the relationship between low bone remodelling and FGF23 levels is not always concordant.

In case of high bone turnover, the replacement rate of the existing matrix is such that no time is allowed to the resulting phosphate release to be buffered by bone and that existing reserves in phosphate are being released in excess.[110] Two different high bone turnover bone disease are associated with elevated FGF23 levels. The McCune-Albright syndrome (MIM 174800), also called Polyostotic Fibrous Dysplasia (PFD), is caused by a mosaicism for postzygotic activating mutations in the guanine nucleotide binding protein, alpha stimulating gene (*GNAS1*). This syndrome is characterized by the replacement of the medullary bone by fibrodysplastic tissue, together with hypophosphatemia and elevated circulating FGF23 levels in some patients.[110] Also, in Chronic Kidney Diseases (CKD), FGF23 levels are markedly increased and its actions to decrease CYP27B1 and increase CYP24 lead to diminish calcitriol levels and contributes to the development of secondary hyperparathyroidism.[111] In addition, parathyroidectomy is associated with reduced FGF23 levels in patients with renal failure, suggesting that PTH-mediated high bone turnover may lead to increased FGF23 prodution. The pattern of FGF23 expression in CKD, unlike the osteocytic predominance observed in hereditary hypophosphatemic disorders is located in peritrabeular osteoblasts. The common feature between McCune Albright syndrome and CKD is that both have excessive bone turnover mediated by activation of Gαs. FGF23 producing mesenchymal tumours causing Tumour-Induced Osteomalacia (TIO) may represent an benign neoplasia that produces FGF23 in a mesenchymal cell lineage.[112] Also, fracture healing is associated with an increase in FGF23 mRNA expression, particularly in osteoblasts and granulation tissue in the fracture callus. Whether increased FGF23 expression in these "high turnover" disorders is through a common pathway is not known. Osteoblasts express PTH receptors and addition of PTH or PTHrP to primary osteoblast cultures increases FGF23 expression.[113] In vivo, PTH(1-34) treatment increases FGF23 levels as well as bone turnover in some studies.[58] Other studies, however, found that administration of PTH to mice in doses resulting in increased bone remodeling resulted in a reduction of FGF23.[58] In addition, 1,25(OH)$_2$D deficient mice (1 alpha hydroxylase knock-out) display increased PTH, increased bone formation but undetectable FGF23 which revoques the hypothesis that PTH directly regulates FGF23 production.[58] Since the direct effects of PTH on FGF23 remain controversial, the origin of FGF23 in mesenchymal lineages and the diversity of clinical states leading to increased FGF23, suggests that special cells emerge during high turnover states that produce FGF23. Multiple myeloma which leads to increased bone resorption is associated with increased FGF23 levels may be an example of this mechanism. Chronic metabolic acidosis which causes increased bone resorption and impaired mineralization is also associated with increased circulating FGF23 levels and fractional excretion of phosphate.[65] FGF23 serum levels in senile or postmenopausal osteoporosis, in which bone resorption is also enhanced, has not yet been evaluated.

CONCLUSION

Bone metabolism is integrated with the gastrointestinal tract and kidney to maintain mineral homeostasis. Bone is both responsive to physiochemical and hormones, such as PTH and 1,25(OH)$_2$D, that regulate mineral metabolism and is also a source of FGF23, which targets the kidney and parathyroid gland to modulate phosphate 1,25(OH)$_2$D and PTH levels. Thus, in addition to its function as a storage compartment of minerals in equilibrium with systemic milieu, bone, through its hormonal actions of FGF23, actually regulates the systemic milieu to adjust serum phosphate and 1,25(OH)$_2$D levels in response to the bone remodelling/mineralization state and vitamin D status. In normal physiological conditions, a healthy skeleton with normal buffering capacity and adequate remodelling activity,

FGF23 AND BONE DISORDERS

basal physiological levels of FGF23 would contribute to the equilibrium between bone phosphate flux, renal phosphate reabsorption and $1,25(OH)_2D$-mediated gastrointestinal phosphate absorption. However, when there is a primary defect in bone mineralization, in which extracellular phosphate is not deposited into the collagenous matrix, FGF23 would be increased to increase the loss of phosphate in the urine to re-establish systemic phosphate homeostasis. Similarly, in states of primary reductions in bone remodelling status, in which bone resorption and formation activities are so reduced that the phosphate buffering capacity of bone is impaired, FGF23 would be increased to shunt the phosphate from bone deposition to kidney excretion.[114] The other extreme of bone remodelling, characterized by large release of phosphate from bone resorption in excess of that which can be redeposited into existing matrix, might also lead to increased FGF23-mediated phosphate wasting.[114] FGF23-mediated coordination of bone metabolism and systemic phosphate homeostasis provides an important physiological pathway to prevent the potential adverse effects of hyperphosphatemia and also to protect the organism from the potential toxic effects of excess vitamin D. Many questions remain regarding FGF23 regulation and function. Given the severity of the phenotypes of having too little or too much FGF23 in animal models, it is likely that alterations of FGF23 are clinical relevant in a variety of human diseases. If so, FGF23 levels may be an important biomarkers and agents that either stimulate or inhibit FGF23-mediated effects might be therapeutically useful in respectively managing disorders caused by FGF23 deficiency and excess.

REFERENCES

1. Quarles LD. Endocrine functions of bone in mineral metabolism regulation. J Clin Invest 2008; 118(12):3820-8.
2. Rizzoli R, Fleisch H, Bonjour JP. Role of 1,25-dihydroxyvitamin D3 on intestinal phosphate absorption in rats with a normal vitamin D supply. J Clin Invest 1977; 60(3):639-47.
3. Talmage RV, Doppelt SH, Fondren FB. An interpretation of acute changes in plasma 45Ca following parathyroid hormone administration to thyroparathyroidectomized rats. Calcif Tissue Res 1976; 22(2):117-28.
4. Ma YL, Cain RL, Halladay DL et al. Catabolic effects of continuous human PTH (1-38) in vivo is associated with sustained stimulation of RANKL and inhibition of osteoprotegerin and gene-associated bone formation. Endocrinology 2001; 142(9):4047-54.
5. Shiraki M, Gee MV, Baum BJ et al. Parathyroid hormone stimulates phosphate efflux through an apparently adenosine 3',5'-monophosphate-independent process in rat parotid cell aggregates. Endocrinology 1986; 118(5):2009-15.
6. Confavreux CB, Levine RL, Karsenty G. A paradigm of integrative physiology, the crosstalk between bone and energy metabolisms. Mol Cell Endocrinol 2009; 310(1-2):21-9.
7. Yamashita T, Yoshioka M, Itoh N. Identification of a novel fibroblast growth factor, FGF-23, preferentially expressed in the ventrolateral thalamic nucleus of the brain. Biochem Biophys Res Commun 2000; 277(2):494-8.
8. Itoh N, Ornitz DM. Evolution of the Fgf and Fgfr gene families. Trends Genet 2004; 20(11):563-9.
9. Itoh N, Ornitz DM. Functional evolutionary history of the mouse Fgf gene family. Dev Dyn 2008; 237(1):18-27.
10. Suzuki M, Uehara Y, Motomura-Matsuzaka K et al. betaKlotho is required for fibroblast growth factor (FGF) 21 signaling through FGF receptor (FGFR) 1c and FGFR3c. Mol Endocrinol 2008; 22(4):1006-14.
11. Kurosu H, Ogawa Y, Miyoshi M et al. Regulation of fibroblast growth factor-23 signaling by klotho. J Biol Chem 2006; 281(10):6120-3.
12. Urakawa I, Yamazaki Y, Shimada T et al. Klotho converts canonical FGF receptor into a specific receptor for FGF23. Nature 2006; 444(7120):770-4.
13. Yu X, Ibrahimi OA, Goetz R et al. Analysis of the biochemical mechanisms for the endocrine actions of fibroblast growth factor-23. Endocrinology 2005; 146(11):4647-56.
14. Li SA, Watanabe M, Yamada H et al. Immunohistochemical localization of Klotho protein in brain, kidney and reproductive organs of mice. Cell Struct Funct 2004; 29(4):91-9.
15. Yamazaki Y, Tamada T, Kasai N et al.. Anti-FGF23 neutralizing antibodies show the physiological role and structural features of FGF23. J Bone Miner Res 2008; 23(9):1509-18.

16. Liu S, Tang W, Zhou J et al. Fibroblast growth factor 23 is a counter-regulatory phosphaturic hormone for vitamin D. J Am Soc Nephrol 2006; 17(5):1305-15.
17. Bai X, Miao D, Li J et al. Transgenic mice overexpressing human fibroblast growth factor 23 (R176Q) delineate a putative role for parathyroid hormone in renal phosphate wasting disorders. Endocrinology 2004; 145(11):5269-79.
18. Larsson T, Marsell R, Schipani E et al. Transgenic mice expressing fibroblast growth factor 23 under the control of the alpha1(I) collagen promoter exhibit growth retardation, osteomalacia and disturbed phosphate homeostasis. Endocrinology 2004; 145(7):3087-94.
19. Shimada T, Mizutani S, Muto T et al. Cloning and characterization of FGF23 as a causative factor of tumor-induced osteomalacia. Proc Natl Acad Sci USA 2001; 98(11):6500-5.
20. Shimada T, Yamazaki Y, Takahashi M et al. Vitamin D receptor-independent FGF23 actions in regulating phosphate and vitamin D metabolism. Am J Physiol Renal Physiol 2005; 289(5):F1088-95.
21. Saito H, Kusano K, Kinosaki M. Human fibroblast growth factor-23 mutants suppress Na$^+$-dependent phosphate cotransport activity and 1alpha,25-dihydroxyvitamin D3 production. J Biol Chem 2003; 278(4):2206-11.
22. Zhang F, Zhai G, Kato BS et al. Association between KLOTHO gene and hand osteoarthritis in a female Caucasian population. Osteoarthritis Cartilage, 2007.
23. Kuro-o M, Matsumura Y, Aizawa H et al. Mutation of the mouse klotho gene leads to a syndrome resembling ageing. Nature 1997; 390(6655):45-51.
24. Ben-Dov IZ, Galitzer H, Lavi-Moshayoff V et al. The parathyroid is a target organ for FGF23 in rats. J Clin Invest 2007; 117(12):4003-8.
25. Wang H, Yoshiko Y, Yamamoto R et al. Overexpression of fibroblast growth factor 23 suppresses osteoblast differentiation and matrix mineralization in vitro. J Bone Miner Res 2008; 23(6):939-48.
26. Liu S, Zhou J, Tang W et al. Pathogenic role of Fgf23 in Hyp mice. Am J Physiol Endocrinol Metab 2006; 291(1):E38-49.
27. Shimada T, Kakitani M, Yamazaki Y et al. Targeted ablation of Fgf23 demonstrates an essential physiological role of FGF23 in phosphate and vitamin D metabolism. J Clin Invest 2004; 113(4):561-8.
28. Sitara D, Razzaque MS, Hesse M et al. Homozygous ablation of fibroblast growth factor-23 results in hyperphosphatemia and impaired skeletogenesis and reverses hypophosphatemia in Phex-deficient mice. Matrix Biol 2004; 23(7):421-32.
29. Hesse M, Fröhlich LF, Zeitz U et al. Ablation of vitamin D signaling rescues bone, mineral and glucose homeostasis in Fgf-23 deficient mice. Matrix Biol 2007; 26(2):75-84.
30. Sitara D, Razzaque MS, St-Arnaud R et al. Genetic ablation of vitamin d activation pathway reverses biochemical and skeletal anomalies in fgf-23-null animals. Am J Pathol 2006; 169(6):2161-70.
31. Stubbs J, Liu S, Tang W et al. Role of Hyperphosphatemia and 1,25(OH)$_2$D3 in Vascular Calcifications and Mortality in FGF23 Null Mice. J Am Soc Nephrol 2007; 17:689A.
32. Brownstein CA, Adler F, Nelson-Williams C et al. A translocation causing increased alpha-klotho level results in hypophosphatemic rickets and hyperparathyroidism. Proc Natl Acad Sci USA 2008; 105(9):3455-60.
33. Bai XY, Miao D, Goltzman D et al. The autosomal dominant hypophosphatemic rickets R176Q mutation in fibroblast growth factor 23 resists proteolytic cleavage and enhances in vivo biological potency. J Biol Chem 2003; 278(11):9843-9.
34. Benet-Pagès A, Lorenz-Depiereux B, Zischka H et al. FGF23 is processed by proprotein convertases but not by PHEX. Bone 2004; 35(2):455-62.
35. Feng JQ, Ward LM, Liu S et al. Loss of DMP1 causes rickets and osteomalacia and identifies a role for osteocytes in mineral metabolism. Nat Genet 2006; 38(11):1310-5.
36. Lorenz-Depiereux B, Bastepe M, Benet-Pagès A et al. DMP1 mutations in autosomal recessive hypophosphatemia implicate a bone matrix protein in the regulation of phosphate homeostasis. Nat Genet 2006; 38(11):1248-50.
37. Ling Y, Rios HF, Myers ER et al. DMP1 depletion decreases bone mineralization in vivo: an FTIR imaging analysis. J Bone Miner Res 2005; 20(12):2169-77.
38. Jonsson KB, Zahradnik R, Larsson T et al. Fibroblast growth factor 23 in oncogenic osteomalacia and X-linked hypophosphatemia. N Engl J Med 2003; 348(17):1656-63.
39. Weber TJ, Liu S, Indridason OS et al. Serum FGF23 levels in normal and disordered phosphorus homeostasis. J Bone Miner Res 2003; 18(7):1227-34.
40. Berndt T, Craig TA, Bowe AE et al. Secreted frizzled-related protein 4 is a potent tumor-derived phosphaturic agent. J Clin Invest 2003; 112(5):785-94.
41. Guo R, Quarles LD. Cloning and sequencing of human PEX from a bone cDNA library: evidence for its developmental stage-specific regulation in osteoblasts. J Bone Miner Res 1997; 12(7):1009-17.
42. The HYP Consortium, A gene (PEX) with homologies to endopeptidases is mutated in patients with X-linked hypophosphatemic rickets. Nat Genet 1995; 11(2):130-6.
43. Thompson DL, Sabbagh Y, Tenenhouse HS et al. Ontogeny of Phex/PHEX protein expression in mouse embryo and subcellular localization in osteoblasts. J Bone Miner Res 2002; 17(2):311-20.

FGF23 AND BONE DISORDERS　　　　　　　**81**

44. Liu S, Guo R, Simpson LG et al. Regulation of fibroblastic growth factor 23 expression but not degradation by PHEX. J Biol Chem 2003; 278(39):37419-26.
45. White KE, Cabral JM, Davis SI et al. Mutations that cause osteoglophonic dysplasia define novel roles for FGFR1 in bone elongation. Am J Hum Genet 2005; 76(2):361-7.
46. Benet-Pagès A, Orlik P, Strom TM et al. An FGF23 missense mutation causes familial tumoral calcinosis with hyperphosphatemia. Hum Mol Genet 2005; 14(3):385-90.
47. Larsson T, Davis SI, Garringer HJ et al. Fibroblast growth factor-23 mutants causing familial tumoral calcinosis are differentially processed. Endocrinology 2005; 146(9):3883-91.
48. Lyles KW, Burkes EJ, Ellis GJ et al. Genetic transmission of tumoral calcinosis: autosomal dominant with variable clinical expressivity. J Clin Endocrinol Metab 1985; 60(6):1093-6.
49. Ichikawa S, Imel EA, Kreiter ML et al. A homozygous missense mutation in human KLOTHO causes severe tumoral calcinosis. J Clin Invest 2007; 117(9):2684-2691.
50. Topaz O, Indelman M, Chefetz I et al. A deleterious mutation in SAMD9 causes normophosphatemic familial tumoral calcinosis. Am J Hum Genet 2006; 79(4):759-64.
51. Bennett EP, Hassan H, Clausen H. cDNA cloning and expression of a novel human UDP-N-acetyl-alpha-D-galactosamine. Polypeptide N-acetylgalactosaminyltransferase, GalNAc-t3. J Biol Chem 1996; 271(29):17006-12.
52. Segawa H, Yamanaka S, Ohno Y et al. Correlation between hyperphosphatemia and type II Na-Pi cotransporter activity in klotho mice. Am J Physiol Renal Physiol 2007; 292(2):F769-79.
53. Yoshida T, Fujimori T, Nabeshima Y. Mediation of unusually high concentrations of 1,25-dihydroxyvitamin D in homozygous klotho mutant mice by increased expression of renal 1alpha-hydroxylase gene. Endocrinology 2002; 143(2):683-9.
54. Razzaque MS, Sitara D, Taguchi T et al. Premature aging-like phenotype in fibroblast growth factor 23 null mice is a vitamin D-mediated process. FASEB J 2006; 20(6):720-2.
55. Amanzadeh J, Reilly RF, Jr. Hypophosphatemia: an evidence-based approach to its clinical consequences and management. Nat Clin Pract Nephrol 2006; 2(3):136-48.
56. Block GA, Klassen PS, Lazarus JM et al. Mineral metabolism, mortality and morbidity in maintenance hemodialysis. J Am Soc Nephrol 2004; 15(8):2208-18.
57. Masuyama R, Stockmans I, Torrekens S et al. Vitamin D receptor in chondrocytes promotes osteoclastogenesis and regulates FGF23 production in osteoblasts. J Clin Invest 2006; 116(12):3150-9.
58. Samadfam R, Richard C, Nguyen-Yamamoto L et al. Bone formation regulates circulating concentrations of fibroblast growth factor 23. Endocrinology 2009; 150(11):4835-45.
59. Tebben PJ, Singh RJ, Clarke BL et al. Fibroblast growth factor 23, parathyroid hormone and 1alpha,25-dihydroxyvitamin D in surgically treated primary hyperparathyroidism. Mayo Clin Proc 2004; 79(12):1508-13.
60. Kawata T, Imanishi Y, Kobayashi K et al. Parathyroid hormone regulates fibroblast growth factor-23 in a mouse model of primary hyperparathyroidism. J Am Soc Nephrol 2007; 18(10):2683-8.
61. Saji F, Shiizaki K, Shimada S et al. Regulation of fibroblast growth factor 23 production in bone in uremic rats. Nephron Physiol 2009; 111(4):p59-66.
62. Perwad F, Azam N, Zhang MY et al. Dietary and serum phosphorus regulate fibroblast growth factor 23 expression and 1,25-dihydroxyvitamin D metabolism in mice. Endocrinology 2005; 146(12):5358-64.
63. Ferrari SL, Bonjour JP, Rizzoli R. Fibroblast growth factor-23 relationship to dietary phosphate and renal phosphate handling in healthy young men. J Clin Endocrinol Metab 2005; 90(3):1519-24.
64. Nishida Y, Taketani Y, Yamanaka-Okumura H et al. Acute effect of oral phosphate loading on serum fibroblast growth factor 23 levels in healthy men. Kidney Int 2006; 70(12):2141-7.
65. Domrongkitchaiporn S, Disthabanchong S, Cheawchanthanakij R et al. Oral phosphate supplementation corrects hypophosphatemia and normalizes plasma FGF23 and 25-hydroxyvitamin D3 levels in women with chronic metabolic acidosis. Exp Clin Endocrinol Diabetes 2010; 118(2):105-12.
66. Prentice A, Ceesay M, Nigdikar S et al. FGF23 is elevated in Gambian children with rickets. Bone 2008; 42(4):788-97.
67. van Boekel G, Ruinemans-Koerts J, Joosten F et al. Tumor producing fibroblast growth factor 23 localized by two-staged venous sampling. Eur J Endocrinol 2008; 158(3):431-7.
68. Rowe PS, Kumagai Y, Gutierrez G et al. MEPE has the properties of an osteoblastic phosphatonin and minhibin. Bone 2004; 34(2):303-19.
69. Liu S, Rowe PS, Vierthaler L et al. Phosphorylated acidic serine-aspartate-rich MEPE-associated motif peptide from matrix extracellular phosphoglycoprotein inhibits phosphate regulating gene with homologies to endopeptidases on the X-chromosome enzyme activity. J Endocrinol 2007; 192(1):261-7.
70. Yuan B, Takaiwa M, Clemens TL et al. Aberrant Phex function in osteoblasts and osteocytes alone underlies murine X-linked hypophosphatemia. J Clin Invest 2008; 118(2):722-34.
71. Bowe AE, Finnegan R, Jan de Beur SM et al. FGF-23 inhibits renal tubular phosphate transport and is a PHEX substrate. Biochem Biophys Res Commun 2001; 284(4):977-81.

72. Guo R, Liu S, Spurney RF et al. Analysis of recombinant Phex: an endopeptidase in search of a substrate. Am J Physiol Endocrinol Metab 2001; 281(4):E837-47.
73. Campos M, Couture C, Hirata IY et al. Human recombinant endopeptidase PHEX has a strict S1' specificity for acidic residues and cleaves peptides derived from fibroblast growth factor-23 and matrix extracellular phosphoglycoprotein. Biochem J 2003; 373(Pt 1):271-9.
74. Addison WN, Nakano Y, Loisel T et al. MEPE-ASARM peptides control extracellular matrix mineralization by binding to hydroxyapatite: an inhibition regulated by PHEX cleavage of ASARM. J Bone Miner Res 2008; 23(10):1638-49.
75. Martin A, David V, Laurence JS et al. Degradation of MEPE, DMP1 and release of SIBLING ASARM-peptides (minhibins): ASARM-peptide(s) are directly responsible for defective mineralization in HYP. Endocrinology 2008; 149(4):1757-72.
76. Liu S, Brown TA, Zhou J et al. Role of matrix extracellular phosphoglycoprotein in the pathogenesis of X-linked hypophosphatemia. J Am Soc Nephrol 2005; 16(6):1645-53.
77. Bai X, Miao D, Panda D et al. Partial rescue of the Hyp phenotype by osteoblast-targeted PHEX (phosphate-regulating gene with homologies to endopeptidases on the X chromosome) expression. Mol Endocrinol 2002; 16(12):2913-25.
78. Erben RG, Mayer D, Weber K et al. Overexpression of human PHEX under the human beta-actin promoter does not fully rescue the Hyp mouse phenotype. J Bone Miner Res 2005; 20(7):1149-60.
79. Liu S, Guo R, Tu Q et al. Overexpression of Phex in osteoblasts fails to rescue the Hyp mouse phenotype. J Biol Chem 2002; 277(5):3686-97.
80. Feng JQ, Ward LM, Liu S et al. Loss of DMP1 causes rickets and osteomalacia and identifies a role for osteocytes in mineral metabolism. Nat Genet, 2006.
81. Narayanan K, Ramachandran A, Hao J et al. Dual functional roles of dentin matrix protein 1. Implications in biomineralization and gene transcription by activation of intracellular Ca^{2+} store. J Biol Chem 2003; 278(19):17500-8.
82. Ogbureke KU, Fisher LW. Expression of SIBLINGs and their partner MMPs in salivary glands. J Dent Res 2004; 83(9):664-70.
83. Goebel S, Lienau J, Rammoser U et al. FGF23 is a putative marker for bone healing and regeneration. J Orthop Res 2009.
84. Lu Y, Qin C, Xie Y et al. Studies of the DMP1 57-kDa functional domain both in vivo and in vitro. Cells Tissues Organs 2009; 189(1-4):175-85.
85. Liu S, Tang W, Zhou J et al. Distinct roles for intrinsic osteocyte abnormalities and systemic factors in regulation of FGF23 and bone mineralization in Hyp mice. Am J Physiol Endocrinol Metab 2008 Aug;295(2):E254-61.
86. Xiao ZS, Crenshaw M, Guo R et al. Intrinsic mineralization defect in Hyp mouse osteoblasts. Am J Physiol 1998; 275(4 Pt 1):E700-8.
87. Liu S, Zhou J, Tang W et al. Pathogenic role of Fgf23 in Dmp1-null mice. Am J Physiol Endocrinol Metab 2008; 295(2):E254-61.
88. Liu S, Zhou J, Tang W et al. Pathogenic Role of Fgf23 in Dmp1 Null Mice. Am J Physiol Endocrinol Metab, 2008.
89. Marie PJ, Travers R, Glorieux FH. Bone response to phosphate and vitamin D metabolites in the hypophosphatemic male mouse. Calcif Tissue Int 1982; 34(2):158-64.
90. Sitara D, Kim S, Razzaque MS et al. Genetic evidence of serum phosphate-independent functions of FGF-23 on bone. PLoS Genet 2008; 4(8):e1000154.
91. Liu S, Vierthaler L, Tang W et al. FGFR3 and 4 do not mediate renal effectso of FGF23 in vivo. J Am Soc Nephrol 2008; In Press.
92. He G, George A. Dentin matrix protein 1 immobilized on type I collagen fibrils facilitates apatite deposition in vitro. J Biol Chem 2004; 279(12):11649-56.
93. Rowe PS, Matsumoto N, Jo OD et al. Correction of the mineralization defect in hyp mice treated with protease inhibitors CA074 and pepstatin. Bone, 2006.
94. Laizé V, Martel P, Viegas CS et al. Evolution of matrix and bone gamma-carboxyglutamic acid proteins in vertebrates. J Biol Chem 2005; 280(29):26659-68.
95. Luo G, Ducy P, McKee MD et al. Spontaneous calcification of arteries and cartilage in mice lacking matrix GLA protein. Nature 1997; 386(6620):78-81.
96. Gopalakrishnan R, Suttamanatwong S, Carlson AE et al. Role of matrix Gla protein in parathyroid hormone inhibition of osteoblast mineralization. Cells Tissues Organs 2005; 181(3-4):166-75.
97. Narouz-Ott L, Maurer P, Nitshce DP et al. Thrombospondin-4 binds specifically to both collagenous and noncollagenous extracellular matrix proteins via its C-terminal domains. J Biol Chem 2000; 275(47):37110-7.
98. Posey KL, Hankenson K, Veerisetty AC et al. Skeletal abnormalities in mice lacking extracellular matrix proteins, thrombospondin-1, thrombospondin-3, thrombospondin-5 and type IX collagen. Am J Pathol 2008; 172(6):1664-74.

FGF23 AND BONE DISORDERS 83

99. Hankenson KD, Hormuzdi SG, Meganck JA et al. Mice with a disruption of the thrombospondin 3 gene differ in geometric and biomechanical properties of bone and have accelerated development of the femoral head. Mol Cell Biol 2005; 25(13):5599-606.

100. Ueno A, Miwa Y, Miyoshi K et al. Constitutive expression of thrombospondin 1 in MC3T3-E1 osteoblastic cells inhibits mineralization. J Cell Physiol 2006; 209(2):322-32.

101. Venta PJ, Welty RJ, Johnson TM et al. Carbonic anhydrase II deficiency syndrome in a Belgian family is caused by a point mutation at an invariant histidine residue (107 His——Tyr): complete structure of the normal human CA II gene. Am J Hum Genet 1991; 49(5):1082-90.

102. Mori S, Wu CY, Yamaji S et al. Direct binding of integrin alphavbeta3 to FGF1 plays a role in FGF1 signaling. J Biol Chem 2008; 283(26):18066-75.

103. Guo R, Rowe PS, Liu S et al. Inhibition of MEPE cleavage by Phex. Biochem Biophys Res Commun 2002; 297(1):38-45.

104. Rowe PS, Garrett IR, Schwarz PM et al. Surface Plasmon Resonance (SPR) confirms that MEPE binds to PHEX via the MEPE-ASARM motif: a model for impaired mineralization in X-linked rickets (HYP). Bone 2005; 36(1):33-46.

105. Lu Y, Ye L, Yu S et al. Rescue of odontogenesis in Dmp1-deficient mice by targeted re-expression of DMP1 reveals roles for DMP1 in early odontogenesis and dentin apposition in vivo. Dev Biol 2007; 303(1):191-201.

106. Fang MA, Glackin CA, Sadhu A et al. Transcriptional regulation of alpha 2(I) collagen gene expression by fibroblast growth factor-2 in MC3T3-E1 osteoblast-like cells. J Cell Biochem 2001; 80(4):550-9.

107. Nauman EA, Sakata T, Keaveny TM et al. bFGF administration lowers the phosphate threshold for mineralization in bone marrow stromal cells. Calcif Tissue Int 2003; 73(2):147-52.

108. Carpenter TO, Ellis BK, Insogna KL et al. FGF7—an inhibitor of phosphate transport derived from oncogenic osteomalacia-causing tumors. J Clin Endocrinol Metab, 2004.

109. Laroche M, Boyer JF, Jahafar H et al. Normal FGF23 levels in adult idiopathic phosphate diabetes. Calcif Tissue Int 2009; 84(2):112-7.

110. Riminucci M, Collins MT, Fedarko NS et al. FGF-23 in fibrous dysplasia of bone and its relationship to renal phosphate wasting. J Clin Invest 2003; 112(5):683-92.

111. Gutierrez O, Isakova T, Rhee E et al. Fibroblast growth factor-23 mitigates hyperphosphatemia but accentuates calcitriol deficiency in chronic kidney disease. J Am Soc Nephrol 2005; 16(7):2205-15.

112. Bahrami A, Weiss SW, Montgomery E et al. RT-PCR analysis for FGF23 using paraffin sections in the diagnosis of phosphaturic mesenchymal tumors with and without known tumor induced osteomalacia. Am J Surg Pathol 2009; 33(9):1348-54.

113. Rhee Y. FGF23 gene expression is upregulated by PTH receptor activation in osteocytes in vitro and in vivo: a parathyroid-bone link influencing the endocrine function of osteocytes. J Bone Miner Res 2009; 24(Suppl 1).

114. Seeman E. Bone modeling and remodeling. Crit Rev Eukaryot Gene Expr 2009; 19(3):219-33.

CHAPTER 5

FGF23, KLOTHO AND VITAMIN D INTERACTIONS:
What Have We Learned from In Vivo Mouse Genetics Studies?

M. Shawkat Razzaque

Department of Oral Medicine, Infection and Immunity, Harvard School of Dental Medicine, Boston, Massachusetts, USA; and Department of Pathology, Nagasaki University School of Biomedical Science, Nagasaki, Japan.
Email: mrazzaque@hms.harvard.edu; razzaque@med.nagasaki-u.ac.jp

Abstract: The molecular interactions of fibroblast growth factor 23 (FGF23), klotho and vitamin D coordinate to regulate the delicate phosphate levels of the body. Vitamin D can induce both FGF23 and klotho synthesis to influence renal phosphate balance. In the presence of klotho, FGF23 protein gains bioactivity to influence systemic phosphate homeostasis. Experimental studies have convincingly shown that in the absence of klotho, FGF23 is unable to regulate in vivo phosphate homeostasis. Furthermore, genetic inactivation of FGF23, klotho or both of the genes have resulted in markedly increased renal expression of 1-alpha hydroxylase [1α(OH)ase] and concomitant elevated serum levels of 1,25, dihydroxyvitamin D [1,25(OH)$_2$D] in the mutant mice. Vitamin D can induce the expression of both FGF23 and klotho while, FGF23 can suppress renal expression of 1α(OH)ase to reduce 1,25(OH)$_2$D activity. In this brief chapter, I will summarize the possible in vivo interactions of FGF23, klotho and vitamin D, in the light of recent mouse genetics studies.

INTRODUCTION

Phosphate is a widely distributed mineral ion and is an essential component of bone. Phosphate is also important for cell signaling, energy metabolism, nucleic acid synthesis and the maintenance of acid-base balance (urinary buffering).[1,2] The intestinal absorption, skeletal resorption and renal reabsorption coordinate to regulate physiologic phosphate balance.[3-5] Diseases affecting intestine, bone or kidney can lead to impaired phosphate balance.

Endocrine FGFs and Klothos, edited by Makoto Kuro-o.
©2012 Landes Bioscience and Springer Science+Business Media.

More than 70% of phosphate consumed through food is absorbed from the upper part of the intestine, while the excessive amount is mostly cleared out of the body through the kidneys. Phosphate transport in the enterocytes (intestine) and in the proximal epithelial cells (kidney) is mostly mediated by the sodium-phosphate (NaPi) cotransporter family proteins (NaPi-2a, NaPi-2b and NaPi-2c). Parathyroid hormone (PTH), vitamin D and FGF23 can directly or indirectly regulate NaPi transporter activities to control phosphate metabolism. PTH can reduce phosphate reabsorption in the proximal tubular epithelial cells by reducing NaPi-2a and NaPi-2c activities through facilitating internalization of NaPi transporter proteins from the apical side of the epithelial cells.[6] Phosphate transport across the proximal tubular epithelial cells is partly driven by a high extracellular sodium gradient, which is believed to be maintained by the membrane-associated Na,K-ATPase.[7-9] PTH can also mobilize phosphate from the bone into the bloodstream, possibly by enhancing osteoclastic bone resorption. Intestinal phosphate absorption, on the other hand, can be influenced by vitamin D through modulating the activities of NaPi-2b transporters. The active regulation of phosphate homeostasis is an evolving area of research and our understanding of such regulation is significantly enhanced by the identification of the potent phosphatonin: FGF23.[10,11]

FGF23

FGF23 is a ~30 kDa protein that is proteolytically processed to smaller N-terminal (~18 kDa) and C-terminal (~12 kDa) fragments. FGF23 can not only induce urinary phosphate excretion, but can also influence systemic vitamin D activity by suppressing the renal expression of $1\alpha(OH)$ase.[12] Animal studies using FGF23 transgenic mice have shown to reduce serum levels of $1,25(OH)_2D$.[13,14] FGF23 overproducing mice develop hypophosphatemia due to the increased renal excretion of phosphate. Similar to the animal studies, gain-of-function mutations of the *FGF23* gene, in patients with autosomal-dominant hypophosphatemic rickets (ADHR), are associated with excessive urinary phosphate wasting, leading to the development of rickets;[11] these mutations prevent the proteolytic cleavage of the FGF23 protein and the net effect enhances FGF23 activity in ADHR patients. In contrast to the increased activity of FGF23 in ADHR patients, decreased activity of FGF23 due to loss-of-function mutations in the *FGF23* gene in patients with familial tumoral calcinosis (FTC), usually develop hyperphosphatemia and ectopic calcification.[15] Of particular clinical importance, FGF23-mediated renal regulation of phosphate homeostasis is severely impaired in patients with chronic kidney disease (CKD), where despite increased serum levels of FGF23, patients develop hyperphosphatemia.

FGF23 AND CKD

Patients with advanced stages of CKD have elevated serum levels of FGF23, perhaps due to decreased renal clearance.[16] Despite increased serum levels of FGF23, CKD patients develop hyperphosphatemia and inability of FGF23 to restore the phosphate balance in CKD patients can trigger development of secondary hyperparathyroidism. Moreover, Calcitriol therapy in patients with CKD may also contribute to increased serum levels of FGF23.[17] Experimental studies have shown that both phosphate and $1,25(OH)_2D$ can independently increase circulating FGF23 levels.[18] Patients with CKD usually have low levels of 1,25(OH)$_2$D and since FGF23 can suppress vitamin D activity, the increased FGF23 levels in patients with CKD may, in turn reduce vitamin D activity to facilitate the development of compensatory secondary hyperparathyroidism.[19]

Increased serum levels of FGF23 have shown to be associated with an increased mortality of CKD patients undergoing hemodialysis.[20] Furthermore, a correlation between elevated serum FGF23 levels and an increased rate of left ventricular hypertrophy is also reported in CKD patients.[21,22] Additional experimental studies will determine whether such association is merely an epiphenomenon or a clinically important phenomenon. *Klotho* knockout mice mimic certain biochemical aspects of CKD patients.[23,24] For instance, *klotho* knockout mice are hyperphosphatemic despite significantly higher serum levels of FGF23 with shorter lifespan.[25,26] More importantly, the early, sudden death of *klotho* knockout mice is due to cardiac malfunction.[27] Determining whether FGF23 toxicity may induce the cardiac anomalies to contribute early mortality of *klotho* knockout mice may explain the pathologic role of increased serum FGF23 levels in patients with CKD.[24]

KLOTHO

Klotho is a Type 1 membrane protein[7] and can be released into the circulatory system as the "secreted form" when the short transmembrane domain is removed. A recent study has suggested that A Disintegrin And Metalloproteinases (ADAM)-10 and ADAM-17 are capable of cleavage klotho from the plasma membrane to induce klotho shedding.[28] Klotho is mostly expressed in the kidney (distal convoluted tubules), the parathyroid gland and the choroid plexus in the brain.[45] The *klotho* knockout mice have increased renal expression of Na-Pi-2a and Na-Pi-2c protein with concomitant hyperphosphatemia. Furthermore, *klotho* knockout mice have increased renal expression of 1α(OH)ase with elevated serum levels of 1,25(OH)$_2$D. Such serum biochemical changes in *klotho* knockout mice, along with other physical morphological and molecular changes are identical to those found in *Fgf23* knockout mice.[29] The identical phenotypes of these two mutant mice led to the discovery of klotho as an essential cofactor in FGF23 signaling pathways.

FGF23 SIGNALING

FGF23 can bind to multiple FGF receptors, including FGFR1c, FGFR3c and FGFR4 and activate downstream signaling events to exert diverse biological functions.[30-32] Subsequent studies have claimed that FGFR1 is the principal mediator of FGF23 effects in vivo.[33,34] FGF23 has much higher affinity towards FGF receptors in the presence of klotho. Klotho can increase the phosphorylation of FGF receptor substrate-2α and p44/42 mitogen-activated protein kinase (MAPK) and AKT in response to the exogenous treatment with recombinant FGF23 protein.[31,35] The binding of the FGF23-klotho-FGF receptor complex can activate downstream events by inducing phosphorylation of signal kinase proteins[31,32,35-38] to initiate the phosphate lowering actions of FGF23, which is partly mediated through the reduced activity of NaPi-2a and 1α(OH)ase in the proximal tubular epithelial cells. Recent studies, using mouse genetics as an in vivo tool, have convincingly shown that klotho is essential for FGF23-mediated regulation of systemic phosphate homeostasis.[39,40]

IN VIVO IMPORTANCE OF KLOTHO IN FGF23 FUNCTION

FGF23 transgenic mice develop hypophosphatemia due to severe urinary phosphate wasting. In similar lines of studies, genetic restoration of systemic actions of human FGF23 in *Fgf23* knockout mice can reverse hyperphosphatemia to hypophosphatemia.[14] Likewise, serum phosphate levels are significantly reduced following bioactive FGF23 injection into either wild-type or *Fgf23-/-* mice.[40] On the contrary, the injection of bioactive FGF23 protein into either *klotho-/-* mice or *Fgf23-/-/klotho-/-* double knockout mice did not change the serum phosphate levels,[40] implying that without klotho, FGF23 is unable to regulate phosphate homeostasis.

The PHEX, (a phosphate-regulating gene that is homologous to the endopeptidases of the X-chromosome) mutation leads to an increased serum accumulation of FGF23, causing severe hypophosphatemia due to excessive urinary phosphate loss in the mutant mouse [hypophosphatemic (*Hyp*) mouse] and in the human diseases [X-linked hypophosphatemia: (XLH) patients]. The inactivation of klotho in *Hyp* mice resulted in hyperphosphatemia in *Hyp/klotho* double mutant mice, even though double mutant mice have markedly increased serum levels of FGF23.[39] Therefore FGF23-mediated hypophosphatemic phenotypes of *Hyp* mice are a klotho-dependent process.[41,42] In a similar line of study, hypophosphatemic phenotypes of FGF23 transgenic mice can be reversed by genetic inactivation of klotho functions, reemphasizing the fact that klotho is indispensable in FGF23-mediated regulation of phosphate homeostasis.[43] In accord with the animal studies, a homozygous loss-of-function mutation in the human *Klotho* gene resulted in severe hyperphosphatemia even though the affected patient has elevated serum FGF23 levels.[44] Together, both human and animal studies provide compelling evidence of essential role of klotho in the FGF23-mediated regulation of systemic phosphate homeostasis and imply that klotho may be a potential therapeutic target to manipulate FGF23-mediated hypophosphatemic diseases.[24,41] Therapeutic manipulation of FGF23-klotho axis may, not only influence serum phosphate balance, but also can affect vitamin D homeostasis.

VITAMIN D AND FGF23-KLOTHO AXIS

$1,25(OH)_2D$ is the active metabolite of vitamin D, that is formed in the kidney through hydroxylation by the enzyme $1\alpha(OH)$ase.[45] Manipulating renal $1\alpha(OH)$ase enzyme activity, can therefore influence the $1,25(OH)_2D$ activity. $1,25(OH)_2D$ can induce expression of both FGF23 and klotho and increase renal phosphate wasting to reduce serum phosphate levels, while it can also influence increased intestinal phosphate absorption to increase serum phosphate levels. However, it is important to mention that in normal healthy conditions, coordinated interplay of different organ system help keep the phosphate balance within normal levels. In abnormal conditions, as in *klotho* knockout mice, a markedly increase renal expression of the *1α(OH)ase* gene results in elevated serum $1,25(OH)_2D$ levels.[46] Such increased serum levels of $1,25(OH)_2D$ are associated with abnormal soft tissue and vascular calcifications in *klotho-/-* mice, mostly due to increased serum levels of calcium and phosphate (Fig. 1).[47,48] Suppressing vitamin D activity from *klotho-/-* mice can reduce both serum calcium and phosphate levels in *klotho-/-/1α(OH)ase-/-* double knockout mice. Such changes in serum parameters in *klotho-/-/1α(OH)ase-/-* mice can completely eliminate extensive vascular and soft tissue calcifications that are consistently found in the *klotho-/-* mice (Fig. 1).[46]

Studies have found that both *klotho*[-/-] mice and *klotho*[-/-]/*NaPi2a*[-/-] double knockout mice have high serum calcium and 1,25(OH)₂D levels. As for serum phosphate, *klotho*[-/-] mice have hyperphosphatemia, while *klotho*[-/-]/*NaPi2a*[-/-] mice have hypophosphatemia. Despite significantly higher serum 1,25(OH)₂D levels, the soft tissue and vascular calcification were reduced or eliminated in *klotho*[-/-]/*NaPi2a*[-/-] double knockout mice.[49] This result suggests the presence of a 1,25(OH)₂D-independent calcification process that can be driven by elevated serum phosphate levels.[49] In the similar line of observation, it appears likely that the complete elimination of soft tissue and vascular calcifications in *klotho*[-/-]/*1α(OH)ase*[-/-] mice, may be the consequence of low serum levels of phosphate (Fig. 1).

CONCLUSION

Systemic phosphate homeostasis is tightly controlled by a limited number of essential molecules, including FGF23, klotho and vitamin D. Studies have shown that 1,25(OH)₂D

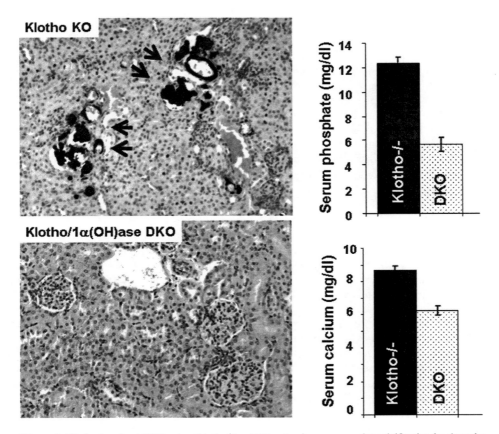

Figure 1. Klotho knockout (KO) mice (*klotho*[-/-]) exhibit extensive renovascular calcification by 6 weeks of age and such calcification is completely eliminated in *klotho*[-/-]/*1α(OH)ase*[-/-] double knockout (DKO) mice. Note that hyperphosphatemia, observed in *klotho*[-/-] mice, is reduced in *klotho*[-/-]/*1α(OH)ase*[-/-] DKO mice. Similar reduction of serum calcium level is also noted in DKO mice.

can induce the expression of FGF23 in osteocytes through a vitamin D response element (VDRE) that is present in the FGF23 promoter.[50] It is possible that 1,25(OH)$_2$D-mediated production of FGF23, in turn, can suppress 1α(OH)ase synthesis to reduce 1,25(OH)$_2$D production and can thereby form a feedback loop to maintain vitamin D homeostasis. As mentioned, complex in vivo genetic studies have shown that inactivation of FGF23-klotho axis in mice lead to the development of hypervitaminosis D, while elimination of vitamin D activity from mice can reduce serum FGF23 levels. Of relevance, serum FGF23 level is extremely high in *klotho$^{-/-}$* mice while almost undetectable in *klotho$^{-/-}$/1α(OH)ase$^{-/-}$* mice, again suggesting that vitamin D is a potent in vivo regulator of FGF23.[46] The overlapping phenotypes of *Fgf23*, *klotho* and *Fgf23/klotho* double knockout mice suggest that a limited number of molecules form the biological network to coordinately regulate mineral ion balance (Fig. 2). These mouse genetic studies provide the basis for the development of new therapeutic strategies to minimize the complications that arise from mineral ion imbalance. Further studies will determine whether monitoring the level of FGF23-klotho axis may help in defining the patients at risk for developing skeletal and vascular complications in patients with CKD and in assessing the response

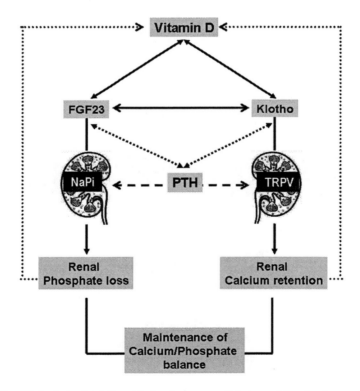

Figure 2. Simplified schematic outline showing how some of the essential molecules regulate mineral ion metabolism.[25] Note that vitamin D plays a crucial role in fine-tuning calcium and phosphate balance by interacting with both FGF23 and klotho. FGF23 can help in renal phosphate clearance by influencing NaPi transporter activities, while klotho can help in renal calcium retention by affecting the calcium-selective channel, TRPV5 activities. Please note that most of the FGF23-mediated functions are klotho dependent, while klotho has FGF23 independent functions.

to conventional and selective therapies. Fine-tuning the limitation of existing therapy, by manipulating FGF23-klotho axis, to reduce the disease burden will be a challenging yet clinically gratifying attempts to reduce the sufferings of patients.[41,51,52]

ACKNOWLEDGEMENTS

The author wishes to thank Drs. Teruyo Nakatani and Mutsuko Ohnishi of Harvard School of Dental Medicine, Boston for performing some of the original studies that are frequently discussed in this book chapter. Part of this book chapter is based on published review articles in the *Am J Physiol Renal Physiol* (2009) and *Nature Rev Endocrinol* (2009). Language editing: Rafiur Razzaque.

REFERENCES

1. Rastegar A. New concepts in pathogenesis of renal hypophosphatemic syndromes. Iran J Kidney Dis 2009;3:1-6.
2. Gaasbeek A, Meinders AE. Hypophosphatemia: an update on its etiology and treatment. Am J Med 2005;118:1094-101.
3. Berndt T, Kumar R. Phosphatonins and the regulation of phosphate homeostasis. Annu Rev Physiol 2007;69:341-59.
4. Imel EA, Econs MJ. Fibroblast growth factor 23: roles in health and disease. J Am Soc Nephrol 2005; 16:2565-75.
5. Quarles LD. Endocrine functions of bone in mineral metabolism regulation. J Clin Invest 2008; 118:3820-8.
6. Forster IC, Hernando N, Biber J, Murer H. Proximal tubular handling of phosphate: A molecular perspective. Kidney Int 2006; 70:1548-59.
7. Nabeshima Y, Imura H. alpha-Klotho: a regulator that integrates calcium homeostasis. Am J Nephrol 2008;28:455-64.
8. Razzaque MS. Klotho and Na$^+$,K$^+$-ATPase activity: solving the calcium metabolism dilemma? Nephrol Dial Transplant 2008; 23:459-61.
9. Imura A, Tsuji Y, Murata M et al. alpha-Klotho as a regulator of calcium homeostasis. Science 2007; 316:1615-8.
10. Yamashita T, Yoshioka M, Itoh N. Identification of a novel fibroblast growth factor, FGF-23, preferentially expressed in the ventrolateral thalamic nucleus of the brain. Biochem Biophys Res Commun 2000; 277:494-8.
11. ADHR_Consortium. Autosomal dominant hypophosphataemic rickets is associated with mutations in FGF23. The ADHR Consortium. Nat Genet 2000; 26:345-8.
12. Shimada T, Hasegawa H, Yamazaki Y et al. FGF-23 is a potent regulator of vitamin D metabolism and phosphate homeostasis. J Bone Miner Res 2004; 19:429-35.
13. Larsson T, Marsell R, Schipani E et al. Transgenic mice expressing fibroblast growth factor 23 under the control of the alpha1(I) collagen promoter exhibit growth retardation, osteomalacia and disturbed phosphate homeostasis. Endocrinology 2004; 145:3087-94.
14. DeLuca S, Sitara D, Kang K et al. Amelioration of the premature aging-like features of Fgf-23 knockout mice by genetically restoring the systemic actions of FGF-23. J Pathol 2008; 216:345-355.
15. Benet-Pages A, Orlik P, Strom TM et al. An FGF23 missense mutation causes familial tumoral calcinosis with hyperphosphatemia. Hum Mol Genet 2005; 14:385-90.
16. Imanishi Y, Inaba M, Nakatsuka K et al. FGF-23 in patients with end-stage renal disease on hemodialysis. Kidney Int 2004; 65:1943-6.
17. Nishi H, Nii-Kono T, Nakanishi S et al. Intravenous calcitriol therapy increases serum concentrations of fibroblast growth factor-23 in dialysis patients with secondary hyperparathyroidism. Nephron Clin Pract 2005; 101:c94-9.
18. Saito H, Maeda A, Ohtomo S et al. Circulating FGF-23 is regulated by 1alpha,25-dihydroxyvitamin D3 and phosphorus in vivo. J Biol Chem 2005; 280:2543-9.
19. Shigematsu T, Kazama JJ, Yamashita T et al. Possible involvement of circulating fibroblast growth factor 23 in the development of secondary hyperparathyroidism associated with renal insufficiency. Am J Kidney Dis 2004; 44:250-6.
20. Fliser D, Kollerits B, Neyer U et al. Fibroblast growth factor 23 (FGF23) predicts progression of chronic kidney disease: the Mild to Moderate Kidney Disease (MMKD) Study. J Am Soc Nephrol 2007; 18:2600-8.
21. Hsu HJ, Wu MS. Fibroblast growth factor 23: a possible cause of left ventricular hypertrophy in hemodialysis patients. Am J Med Sci 2009; 337:116-22.
22. Gutierrez OM, Januzzi JL, Isakova T et al. Fibroblast growth factor 23 and left ventricular hypertrophy in chronic kidney disease. Circulation 2009; 119:2545-52.
23. Kuro-o M. Klotho in chronic kidney disease—what's new? Nephrol Dial Transplant 2009; 24:1705-8.

FGF23-KLOTHO-VITAMIN D DURING PHOSPHATE TURNOVER 91

24. Razzaque MS. Does FGF23 toxicity influence the outcome of chronic kidney disease? Nephrol Dial Transplant 2009; 24:4-7.
25. Lanske B, Razzaque MS. Mineral metabolism and aging: the fibroblast growth factor 23 enigma. Curr Opin Nephrol Hypertens 2007; 16:311-8.
26. Lanske B, Razzaque MS. Premature aging in klotho mutant mice: cause or consequence? Ageing Res Rev 2007; 6:73-9.
27. Takeshita K, Fujimori T, Kurotaki Y et al. Sinoatrial node dysfunction and early unexpected death of mice with a defect of klotho gene expression. Circulation 2004; 109:1776-82.
28. Chen CD, Podvin S, Gillespie E et al. Insulin stimulates the cleavage and release of the extracellular domain of Klotho by ADAM10 and ADAM17. Proc Natl Acad Sci USA 2007; 104:19796-801.
29. Kuro-o M, Matsumura Y, Aizawa H et al. Mutation of the mouse klotho gene leads to a syndrome resembling ageing. Nature 1997; 390:45-51.
30. Mohammadi M, Olsen SK, Ibrahimi OA. Structural basis for fibroblast growth factor receptor activation. Cytokine Growth Factor Rev 2005; 16:107-37.
31. Kurosu H, Ogawa Y, Miyoshi M et al. Regulation of fibroblast growth factor-23 signaling by klotho. J Biol Chem 2006; 281:6120-3.
32. Urakawa I, Yamazaki Y, Shimada T et al. Klotho converts canonical FGF receptor into a specific receptor for FGF23. Nature 2006; 444:770-4.
33. Gattineni J, Bates C, Twombley K et al. FGF23 decreases renal NaPi-2a and NaPi-2c expression and induces hypophosphatemia in vivo predominantly via FGF receptor 1. Am J Physiol Renal Physiol 2009; 297(2):F282-91.
34. Liu S, Vierthaler L, Tang W et al. FGFR3 and FGFR4 do not mediate renal effects of FGF23. J Am Soc Nephrol 2008; 19:2342-50.
35. Medici D, Razzaque MS, Deluca S et al. FGF-23-Klotho signaling stimulates proliferation and prevents vitamin D-induced apoptosis. J Cell Biol 2008; 182:459-65.
36. Goetz R, Beenken A, Ibrahimi OA et al. Molecular insights into the klotho-dependent, endocrine mode of action of fibroblast growth factor 19 subfamily members. Mol Cell Biol 2007; 27:3417-28.
37. Yamazaki Y, Tamada T, Kasai N et al. Anti-FGF23 neutralizing antibodies show the physiological role and structural features of FGF23. J Bone Miner Res 2008; 23:1509-18.
38. Kuro-o M. Klotho as a regulator of fibroblast growth factor signaling and phosphate/calcium metabolism. Curr Opin Nephrol Hypertens 2006; 15:437-41.
39. Nakatani T, Ohnishi M, Razzaque MS. Inactivation of klotho function induces hyperphosphatemia even in presence of high serum fibroblast growth factor 23 levels in a genetically engineered hypophosphatemic (Hyp) mouse model. FASEB J 2009; 23:3702-11.
40. Nakatani T, Bara S, Ohnishi M et al. In vivo genetic evidence of klotho-dependent functions of FGF23 in regulation of systemic phosphate homeostasis. FASEB J 2009; 23:433-41.
41. Razzaque MS. FGF23-mediated regulation of systemic phosphate homeostasis: is Klotho an essential player? Am J Physiol Renal Physiol 2009; 296:F470-6.
42. Razzaque MS, Lanske B. The emerging role of the fibroblast growth factor-23-klotho axis in renal regulation of phosphate homeostasis. J Endocrinol 2007; 194:1-10.
43. Bai X, Dinghong Q, Miao D et al. Klotho ablation converts the biochemical and skeletal alterations in FGF23 (R176Q) transgenic mice to a Klotho-deficient phenotype. Am J Physiol Endocrinol Metab 2009; 296:E79-88.
44. Ichikawa S, Imel EA, Kreiter ML et al. A homozygous missense mutation in human KLOTHO causes severe tumoral calcinosis. J Clin Invest 2007; 117:2684-91.
45. Lanske B, Razzaque MS. Vitamin D and aging: old concepts and new insights. J Nutr Biochem 2007; 18:771-7.
46. Ohnishi M, Nakatani T, Lanske B et al. Reversal of mineral ion homeostasis and soft-tissue calcification of klotho knockout mice by deletion of vitamin D 1alpha-hydroxylase. Kidney Int 2009; 75:1166-72.
47. Tsujikawa H, Kurotaki Y, Fujimori T et al. Klotho, a gene related to a syndrome resembling human premature aging, functions in a negative regulatory circuit of vitamin D endocrine system. Mol Endocrinol 2003; 17:2393-403.
48. Memon F, El-Abbadi M, Nakatani T et al. Does Fgf23-klotho activity influence vascular and soft tissue calcification through regulating mineral ion metabolism? Kidney Int 2008; 74:566-70.
49. Ohnishi M, Nakatani T, Lanske B et al. In vivo genetic evidence for suppressing vascular and soft tissue calcification through the reduction of serum phosphate levels, even in the presence of high serum calcium and 1,25-dihydroxyvitamin-D levels. Circ Cardiovasc Genet 2009; 2:583-90.
50. Liu S, Tang W, Zhou J et al. Fibroblast growth factor 23 is a counter-regulatory phosphaturic hormone for vitamin D. J Am Soc Nephrol 2006; 17:1305-15.
51. Razzaque MS. Can fibroblast growth factor 23 fine-tune therapies for diseases of abnormal mineral ion metabolism? Nat Clin Pract Endocrinol Metab 2007; 3:788-9.
52. Razzaque MS. The FGF23 Klotho axis: endocrine regulation of phosphate homeostasis. Nat Rev Endocrinol 2009; 5:611-19.

CHAPTER 6

FGF23 AND THE PARATHYROID

Justin Silver and Tally Naveh-Many

Minerva Center for Calcium and Bone Metabolism, Nephrology Services, Hadassah Hebrew University Medical Center, Jerusalem, Israel.
Correspondence: Emails: silver@huji.ac.il and tally@cc.huji.ac.il

Abstract: Klotho and fibroblast growth factor 1 (FGFR1) are expressed not only in FGF23's classical target organ, the kidney, but also in other organs such as the parathyroid. FGF23 acts on the parathyroid to decrease PTH mRNA and serum PTH levels. It does this by activating the MAPK pathway. In chronic kidney disease there are very high levels of serum FGF23 together with increased serum PTH levels, implying resistance of the parathyroid to the action of FGF23. This has been shown in parathyroid tissue surgically removed from dialysis patients as well as in experimental models of uremia to be due to down-regulation of klotho-FGFR1 expression in the parathyroid. Moreover, the parathyroids of rats with advanced uremia do not respond to administered FGF23 by activation of the MAPK pathway or inhibition of PTH secretion. Therefore, there is down-regulation of parathyroid klotho-FGFR1 in CKD which correlates with the resistance of the parathyroid to FGF23. A further subject of great interest in this field is the effect of PTH to directly increase FGF23 expression by osteoblast like cells in culture and the observations that parathyroidectomy prevents and corrects the increased serum FGF23 level of experimental CKD as well as decreases FGF23 in patients with CKD. There is therefore a negative feedback loop between bone and the parathyroid.

INTRODUCTION

Administered FGF23 or excess endogenous FGF23, either due to mutations that render FGF23 catalytically stable or tumors that secrete FGF23, result in hypophosphatemia and a decrease in serum $1,25(OH)_2$ vitamin D_3.[1-3] Conversely, mice or humans with decreased levels of FGF23 or with genetic deletion of Klotho, the coreceptor essential for FGF23 function, have hyperphosphatemia and increased serum $1,25(OH)_2$ vitamin D_3 levels.[4,5] Congenital human FGF23 deficiency results in tumoral calcinosis and may be

---

Endocrine FGFs and Klothos, edited by Makoto Kuro-o.
©2012 Landes Bioscience and Springer Science+Business Media.

FGF23 AND THE PARATHYROID

either due to mutations in the gene coding for FGF23 itself or for a gene involved in the glycosylation of FGF23. In both instances the FGF23 is unstable and rapidly degraded. Mice with genetic deletion of FGF23 exhibit the same serum biochemistry as mice with deletion of Klotho.[6] In addition, klotho mRNA and protein were shown to be expressed at a far greater concentration in the kidney than in any other organ. It was therefore obvious that FGF23's action was determined by the presence of its receptor the Klotho-FGFR1c heterodimer, in the renal distal tubule and it remained to be explained how a distal tubule receptor transduces its signal to the proximal tubule.[7] However, Klotho was shown to be expressed not only in the kidney but also in other organs.

Takeshita et al generated *klotho*-null mice with a reporter gene system (*kl-geo*).[8] Homozygous *kl-geo* mice showed characteristic age-associated phenotypes that were almost identical to those of *kl/kl* mice. In the *kl-geo* mice, *klotho* expression was recognized exclusively in the kidney, in the sinoatrial node region of the heart, the parathyroid and choroid plexus. FGF23 binds to the Klotho-FGFR1c heterodimer and activates the MAPK signal transduction pathway.[4,9] In the kidney this results in decreased NaPi2a activity and resultant phosphaturia as well as inhibition of the 25(OH) vitamin D 1α(OH)ase, CYP27B1, leading to decreased levels of serum $1,25(OH)_2$ vitamin D. In the heart *klotho*[-/-] mice were shown to have sino-atrial conduction defects although no known physiological and pharmacological role of FGF23 on cardiac function has been shown. In the choroid plexus klotho has a role in calcium transfer mediated by Na^+/K^+-ATPase. The expression of klotho in the parathyroid was unexpected and therefore of particular interest.

EFFECT OF FGF23 ON THE PARATHYROID

Transgenic mice with FGF23 driven by the β-actin promoter showed all the biochemical and skeletal phenotypes seen in patients with autosomal dominant hypophosphatemic rickets (ADHR) or tumor induced osteomalacia, diseases that are due to marked increases in mutant or intact FGF23.[1,10] These mice had decreased serum calcium and $1,25(OH)_2$ vitamin D levels which would be expected to increase PTH secretion. However, PTH levels were decreased.[11] Three other studies with genetic manipulation of FGF23 showed increased serum PTH levels.[12-14] In two of these studies the increase in serum PTH was attributed to the decrease in serum $1,25(OH)_2$ vitamin D.[12,13]

EXPRESSION OF KLOTHO IN THE PARATHYROID[15]

Immunoblots with antiKlotho antibody showed Klotho expression in rat microdissected parathyroid.[15] There was no expression of Klotho protein in the thyroid and liver. Quantitative RT-PCR for Klotho mRNA showed high expression in parathyroid and kidney with negligible expression in thyroid, duodenum and liver and no expression at all in the spleen. Immunohistochemistry confirmed the localization of Klotho to the parathyroid and not the surrounding thyroid tissue and also showed the presence of FGFR1 in the parathyroid tissue. These results show that the parathyroid expresses the FGFR-Klotho receptor complex and suggest that the parathyroid is a target organ for FGF23.

FUNCTIONAL EFFECT OF FGF23 ON PTH EXPRESSION[15]

To study the effect of FGF23 on parathyroid function rats were injected intraperitoneally (ip) with full-length FGF23 harboring ADHR mutations (FGF23$^{R176Q/R179Q}$). The ADHR mutations were introduced to inhibit proteolytic inactivation of FGF23 and to increase its half-life. As a negative control we injected FGF23^{core}, an FGF23 variant lacking C-terminal residues past the ^{176}RXXR^{179} cleavage site for furin-like proteases.[16] This C-terminal truncation inactivates FGF23 by abrogating binding to Klotho.[16] FGF23$^{R176Q/R179Q}$ or FGF23^{core} were injected by daily ip injections for 5 days. Serum biochemistry showed the expected decrease in 1,25(OH)$_2$-vitamin D levels in the FGF23$^{R176Q/R179Q}$ treated rats. There was also a significant decrease in serum phosphorus with no significant changes in serum calcium. Importantly, serum PTH was decreased after FGF23$^{R176Q/R179Q}$ when compared to FGF23^{core} or saline. FGF23$^{R176Q/R179Q}$ also led to a marked decrease in PTH mRNA levels as measured by Northern blot and quantified by qRT PCR. FGF23 acts on the Klotho-FGFR through the MAPK pathway in cells transfected with Klotho as well as in the kidney as shown by an increased phosphorylation of ERK1/2 (4;17). FGF23$^{R176Q/R179Q}$ led to an increase in phospho-ERK1/2 in the parathyroid gland indicating activation of the MAPK pathway in the FGF23 treated rats.

Short-term experiments defined the FGF23 signal transduction in the parathyroid and its effect on PTH secretion. FGF23$^{R176Q/R179Q}$ decreased serum PTH at 10 and 30 min when given intravenously (iv) and after 40 min and 24 h when given ip. Wild type FGF23 without the R176Q/R179Q mutation also decreased serum PTH. FGF23$^{R176Q/R179Q}$ decreased PTH mRNA levels at 40 min as shown by qRT PCR. Therefore, FGF23 decreases PTH secretion and mRNA levels in short term experiments.

FGF23 ACTIVATES MAPK IN THE PARATHYROID[15]

To show that the MAPK pathway is required for the FGF23 effect on PTH secretion we used the MAPK inhibitor, U0126. We devised a novel method where FGF23 was given intravenously and the inhibitor U0126 was added topically. Rat parathyroid glands were exposed and submerged in PBS without or with the ERK1/2 inhibitor (U0126) immediately followed by iv injection of FGF23 or HEPES carrier. Serum PTH levels were increased by U0126 at 5 and 20 minutes. FGF23 decreased serum PTH levels at 20 min in the absence of the inhibitor, as before. Importantly, U0126 prevented the FGF23 induced decrease in serum PTH. To separate the direct effects of the MAPK inhibitor from other humoral factors we then performed in vitro studies.

In the absence of a parathyroid cell-line, we studied the direct action of FGF23 on in vitro organ cultures of rat parathyroids. Microdissected parathyroid glands were incubated in a medium with FGF23$^{R176Q/R179Q}$ or FGF23^{core}. Egr-1 mRNA levels at 10 min were increased in the parathyroids incubated with FGF23$^{R176Q/R179Q}$. FGF23$^{R176Q/R179Q}$ significantly decreased PTH secretion at 20, 40 and 60 minutes compared to FGF23^{core} control. We next studied the effect of the MAPK inhibitor U0126 on PTH secretion in response to FGF23. U0126 prevented the effect of FGF23 to decrease PTH secretion in parathyroid glands in vitro. Therefore, FGF23 acts directly on the parathyroid to activate the MAPK pathway and decrease PTH secretion and gene expression both in vivo and in vitro.[15]

FGF23 ACTS ON BOVINE PARATHYROID CELLS TO DECREASE PTH AND INCREASE 25(OH) VITAMIN D 1α HYDROXYLASE EXPRESSION

Krajisnik et al[18] incubated bovine parathyroid cells in primary culture with FGF23. They observed a rise in Egr-1 mRNA 1 h after FGF23(R176Q) treatment. In contrast, there was decrease in PTH mRNA 12 h after FGF23(R176Q) treatment, indicative of a late response gene. Their data show a direct effect of FGF23 on PTH expression. They also showed that FGF23 increased 25(OH) vitamin D 1α hydroxylase (CYP27B1) mRNA levels in the parathyroid cells and the resultant increased levels of parathyroid $1,25(OH)_2$ vitamin D may then act in an autocrine fashion to decrease PTH gene transcription.

FGF23 DECREASES SERUM PTH LEVELS IN TRANSGENIC MICE EXPRESSING THE HUMAN PTH GENE

Lavi-Moshayoff et al generated transgenic mice with the human gene (hPTH) expressed in the mouse parathyroid, using a bacterial artificial chromosome (BAC) containing the hPTH gene within its 144 kb chromosomal region (hPTH Tg mice).[19] The BAC construct maintains the native hPTH gene surrounding sequences and isolates it from positional effects. The transgenic mice had normal levels of serum mouse PTH (mPTH) in addition to both intact and bioactive hPTH. Recombinant ip FGF23 led to a decrease in serum mPTH in wt mice and in serum hPTH in the hPTH Tg mice, similar to the reported decrease in serum PTH in rats and mice after FGF23.[15,20] In the hPTH Tg mice there was a decrease in serum calcium and but change in serum Pi. The FGF23 signaling is conserved from mouse to man in the parathyroid.

FGF23 IN CHRONIC KIDNEY DISEASE (CKD)

In CKD there are markedly elevated levels of both FGF23 and PTH.[21] This finding implies resistance of the parathyroids to FGF23 in renal failure. The increase in serum FGF23 is found in early CKD[22] and at least in Stage 4 CKD[23] it precedes the decrease in serum $1,25(OH)_2D$ that was found in a large cohort of CKD patients to be the earliest change in mineral metabolism parameters.[24] The decrease in serum $1,25(OH)_2D$ is considered an important factor in the pathogenesis of secondary hyperparathyroidism[25] although in mice with selective deletion of the VDR in the parathyroid there were only moderate increases in serum PTH.[26] The increased FGF23 levels in renal failure patients correlates with the progression of the renal failure,[27,28] the prediction of the development of secondary hyperparathyroidism,[29] as well as the mortality in patients starting dialysis.[30] Therefore, it is important to understand the mechanisms that prevent the parathyroid from responding to the high levels of FGF23 in uremia.

THE MECHANISM OF THE RESISTANCE TO FGF23 IN THE PARATHYROID[31]

We have recently reported that a decrease in Klotho-FGFR1 expression and signal transduction may explain the resistance of the parathyroid to FGF23 in CKD.[31] Quantitative immunohistochemistry and RT-PCR using laser capture RNA retrieval showed that Klotho and FGFR1 protein and mRNA levels were decreased in parathyroid sections of rats with adenine high Pi diet induced advanced CKD. Moreover, recombinant FGF23 failed to decrease serum PTH or activate the MAPK pathway in parathyroids of rats with advanced CKD. In parathyroid organ culture, FGF23 decreased secreted PTH and PTH mRNA levels in control or early CKD rats but not in advanced CKD. Therefore, in advanced experimental CKD, there is a decrease in parathyroid Klotho and FGFR1 mRNA and protein levels. This decrease corresponds with the resistance of the parathyroid to FGF23 in vivo, which is sustained in parathyroid organ culture in vitro. We propose that the increased levels of FGF23 fail to decrease PTH levels in established CKD because of a down-regulation of its receptor heterodimer complex Klotho-FGFR1c.

In addition the laboratory of Rodriguez et al also studied the effect of FGF23 on parathyroid function in normal and uremic hyperplastic parathyroid glands.[37] Their results showed that in normal parathyroids, the effect of FGF23 was not constrained to a decrease in PTH production but also caused an increase in parathyroid calcium sensing receptor (CaR) and vitamin D receptor (VDR) expression. By contrast, in hyperplastic parathyroid glands, from rats with experimental CKD due to 5/6 nephrectomy, FGF23 did not reduce PTH production and did not affect CaR and VDR expression. Cell proliferation was reduced by FGF23 in normal but not in hyperplastic parathyroid glands. In normal parathyroids, FGF23 induced phosphorylation of ERK1/2 a key cell signal which mediated the action of FGF23. However, FGF23 failed to activate ERK1/2 pathway in hyperplastic parathyroid glands. There was also a low expression of FGF23 receptor (FGFR1) and Klotho protein in uremic hyperplastic parathyroid glands, which may explain the lack of response to FGF23. They concluded that in hyperparathyroidism secondary to renal failure, the parathyroid cells have low expression of FGFR1 and Klotho, which makes the parathyroid cells resistant to the inhibitory effect of FGF23.

In human parathyroid tissue Fukugama's laboratory has also shown a decrease in expression of klotho-FGFR1 in secondary hyperparathyroidism.[32] Komaba et al examined the expression of Klotho and FGFR1 in surgically excised parathyroid glands of uremic patients.[32] Compared with normal tissue, the expression of Klotho and FGFR1 decreased significantly in hyperplastic parathyroid glands, particularly in glands with nodular hyperplasia. This result suggests that the depressed expression of the Klotho–FGFR1c complex in hyperplastic glands may explain the resistance to extremely high FGF23 levels in uremic patients.[32,33]

Together, the results from these three laboratories demonstrate that the resistance of the parathyroid to FGF23 in experimental CKD is due to down-regulation of the Klotho-FGFR1 receptor.

THE MECHANISM OF THE INCREASED FGF23 LEVELS IN CKD

We have recently reported that PTH increases FGF23 expression in bone and causes the high FGF23 levels in CKD. Parathyroidectomy (PTX) both prevented and corrected the high FGF23 levels in adenine high Pi induced CKD rats.[38] The adenine diet led to an increase in FGF23 mRNA levels. Minipump infusion of PTH increased serum FGF23 and mRNA levels. Finally, we demonstrated a direct effect of PTH on bone by using osteoblast-like UMR106 cells. PTH increased FGF23 mRNA. Therefore, we showed that PTH increases FGF23 expression in vivo and in vitro. The effect of PTX to prevent and correct the increased FGF23 demonstrates the importance of PTH in the regulation of FGF23 levels. They also show that the high FGF23 levels in CKD are at least in part due to hyperparathyroidism and the direct action of PTH on bone to increase FGF23 expression. In a mouse model of primary hyperparathyroidism, PTH-cyclin D1 transgenic mice had higher serum FGF23 concentration than wild-type mice.[34] The serum FGF23 levels significantly and directly correlated with serum PTH and calcium levels and inversely correlated with phosphate levels. Serum FGF23 levels decreased in the transgenic mice after parathyroidectomy together with the decrease in serum calcium. Therefore, PTH and/or calcium correlated with changes in serum FGF23 levels in this model.[34] There is also a clinical correlate that illustrates the complexity of the relationship between PTH action, calcium and FGF23. In a patient with an activating mutation of the PTH/PTHrP receptor, namely Jansen's metaphyseal chondrodysplasia, serum FGF23 concentrations were found to be markedly and persistently elevated despite hypophosphatemia and normal 1,25D levels.[35] This observation suggested that serum FGF23 could be governed by activation of the PTH/PTHrP receptor in bone. However, the patient also had a high serum calcium which itself increases serum FGF23 levels.[36]

These observations point to a bone-parathyroid axis where the effect of FGF23 to suppress PTH expression, represents one arm of a novel endocrinologic feedback loop.[15] The second arm of this loop is the effect of PTH to increase serum FGF23 levels.

CONCLUSION

The klotho-FGFR1 complex is expressed in the parathyroid and FGF23 binds to this receptor complex to activate the MAPK pathway and thereby decrease PTH expression. In CKD there is a down-regulation of klotho and FGFR1 in the parathyroid and a decreased ability of FGF23 to activate the parathyroid MAPK pathway both

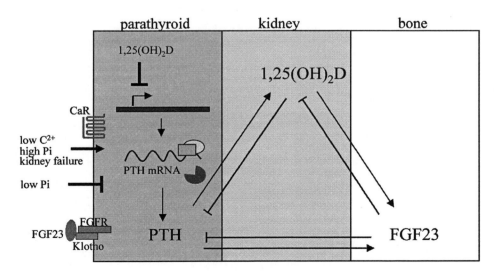

Figure 1. Interrelationships amongst calcium, phosphate, vitamin D, FGF23 and PTH.

in vivo and in vitro. In CKD there are very high serum levels of both FGF23 and PTH. In addition to the effect of FGF23 to decrease the renal synthesis of $1,25(OH)_2$ D_3 which itself would increase parathyroid gland activity and serum PTH, there is evidence that PTH itself acts on bone to increase FGF23 mRNA and serum levels. Therefore, FGF23 decreases serum PTH and PTH increases serum FGF23, a classic endocrine feedback loop (Fig. 1). In CKD the effect of PTH on bone to increase FGF23 expression is maintained but FGF23 is no longer able to suppress the synthesis and secretion of PTH.

REFERENCES

1. White KE, Evans WE, O'Riordan JLH et al. Autosomal dominant hypophosphataemic rickets is associated with mutations in FGF23. Nat Genet 2000; 26(3):345-348.
2. Shimada T, Mizutani S, Muto T et al. Cloning and characterization of FGF23 as a causative factor of tumor-induced osteomalacia. Proc Natl Acad Sci USA 2001; 98(11):6500-6505.
3. Shimada T, Hasegawa H, Yamazaki Y et al. FGF-23 is a potent regulator of vitamin D metabolism and phosphate homeostasis. J Bone Miner Res 2004; 19(3):429-435.
4. Kurosu H, Ogawa Y, Miyoshi M et al. Regulation of fibroblast growth factor-23 signaling by klotho. J Biol Chem 2006; 281(10):6120-6123.
5. Araya K, Fukumoto S, Backenroth R et al. A novel mutation in fibroblast growth factor 23 gene as a cause of tumoral calcinosis. J Clin Endocrinol Metab 2005; 90(10):5523-5527.
6. Kuro-o M. Klotho as a regulator of fibroblast growth factor signaling and phosphate/calcium metabolism. Curr Opin Nephrol Hypertens 2006; 15(4):437-441.
7. Farrow EG, Davis SI, Summers LJ et al. Initial FGF23-Mediated Signaling Occurs in the Distal Convoluted Tubule. J Am Soc Nephrol 2009; 20(5):955-960.
8. Takeshita K, Fujimori T, Kurotaki Y et al. Sinoatrial node dysfunction and early unexpected death of mice with a defect of klotho gene expression. Circulation 2004; 109(14):1776-1782.
9. Urakawa I, Yamazaki Y, Shimada T et al. Klotho converts canonical FGF receptor into a specific receptor for FGF23. Nature 2006; 444(7120):770-774.
10. White KE, Jonsson KB, Carn G et al. The autosomal dominant hypophosphatemic rickets (ADHR) gene is a secreted polypeptide overexpressed by tumors that cause phosphate wasting. J Clin Endocrinol Metab 2001; 86(2):497-500.
11. Shimada T, Urakawa I, Yamazaki Y et al. FGF-23 transgenic mice demonstrate hypophosphatemic rickets with reduced expression of sodium phosphate cotransporter type IIa. Biochem Biophys Res Commun 2004; 314(2):409-414.
12. Bai XY, Miao D, Goltzman D et al. The autosomal dominant hypophosphatemic rickets R176Q mutation in fibroblast growth factor 23 resists proteolytic cleavage and enhances in vivo biological potency. J Biol Chem 2003; 278(11):9843-9849.
13. Bai X, Miao D, Li J et al. Transgenic mice overexpressing human fibroblast growth factor 23 (R176Q) delineate a putative role for parathyroid hormone in renal phosphate wasting disorders. Endocrinol 2004; 145(11):5269-5279.
14. Larsson T, Marsell R, Schipani E et al. Transgenic mice expressing Fibroblast Growth Factor 23 under the control of the {alpha}1(I) collagen promoter exhibit growth retardation, osteomalacia and disturbed phosphate homeostasis. Endocrinol 2004; 145:3087-3094.
15. Ben Dov IZ, Galitzer H, Lavi-Moshayoff V et al. The parathyroid is a target organ for FGF23 in rats. J Clin Invest 2007; 117(12):4003-4008.
16. Goetz R, Beenken A, Ibrahimi OA et al. Molecular Insights into the Klotho-Dependent, Endocrine Mode of Action of FGF19 Subfamily Members. Mol Cell Biol 2007; 27(9):3417-3428.
17. Yamashita T, Konishi M, Miyake A. Fibroblast growth factor (FGF)-23 inhibits renal phosphate reabsorption by activation of the mitogen-activated protein kinase pathway. J Biol Chem 2002; 277(31):28265-28270.
18. Krajisnik T, Bjorklund P, Marsell R et al. Fibroblast growth factor-23 regulates parathyroid hormone and 1alpha-hydroxylase expression in cultured bovine parathyroid cells. J Endocrinol 2007; 195(1):125-131.
19. Lavi-Moshayoff V, Silver J, Naveh-Many T. Human PTH gene regulation in vivo using transgenic mice. Am J Physiol Renal Physiol 2009; 297(3):F713-F719.
20. Yamazaki Y, Tamada T, Kasai N et al. Anti-FGF23 neutralizing antibodies show the physiological role and structural features of FGF23. J Bone Miner Res 2008; 23(9):1509-1518.

FGF23 AND THE PARATHYROID 99

21. Imanishi Y, Inaba M, Nakatsuka K et al. FGF-23 in patients with end-stage renal disease on hemodialysis. Kidney Int 2004; 65(5):1943-1946.
22. Isakova T, Gutierrez O, Shah A et al. Postprandial mineral metabolism and secondary hyperparathyroidism in early CKD. J Am Soc Nephrol 2008; 19(3):615-623.
23. Westerberg PA, Linde T, Wikstrom B. Regulation of fibroblast growth factor-23 in chronic kidney disease. Nephrol Dial Transplant 2007; 22(11):3202-3207.
24. Levin A, Bakris GL, Molitch M et al. Prevalence of abnormal serum vitamin D, PTH, calcium and phosphorus in patients with chronic kidney disease: Results of the study to evaluate early kidney disease. Kidney Int 2007; 71(1):31-38.
25. Silver J, Kilav R, Naveh-Many T. Mechanisms of secondary hyperparathyroidism. Am J Physiol Renal Physiol 2002; 283(3):F367-F376.
26. Meir T, Levi R, Lieben L et al. Deletion of the vitamin D receptor specifically in the parathyroid demonstrates a limited role for the VDR in parathyroid physiology. Am J Physiol Renal Physiol 2009.
27. Gutierrez O, Isakova T, Rhee E et al. Fibroblast growth factor-23 mitigates hyperphosphatemia but accentuates calcitriol deficiency in chronic kidney disease. J Am Soc Nephrol 2005; 16(7):2205-2215.
28. Fliser D, Kollerits B, Neyer U et al. Fibroblast growth factor 23 (FGF23) predicts progression of chronic kidney disease: the mild to moderate kidney disease (MMKD) Study. J Am Soc Nephrol 2007; 18(9):2600-2608.
29. Nakanishi S, Kazama JJ, Nii-Kono T et al. Serum fibroblast growth factor-23 levels predict the future refractory hyperparathyroidism in dialysis patients. Kidney Int 2005; 67(3):1171-1178.
30. Gutierrez OM, Mannstadt M, Isakova T et al. Fibroblast growth factor 23 and mortality among patients undergoing hemodialysis. N Engl J Med 2008; 359(6):584-592.
31. Galitzer H, Ben-Dov IZ, Silver J, Naveh-Many T. Parathyroid cell resistance to fibroblast growth factor 23 in secondary hyperparathyroidism of chronic kidney disease. Kidney Int 2010; 77(3):211-218.
32. Komaba H, Goto S, Fujii H et al. Depressed expression of Klotho and FGF receptor 1 in hyperplastic parathyroid glands from uremic patients. Kidney Int 2009.
33. Komaba H, Fukagawa M. FGF23-parathyroid interaction: implications in chronic kidney disease. Kidney Int 2009.
34. Kawata T, Imanishi Y, Kobayashi K et al. Parathyroid hormone regulates fibroblast growth factor-23 in a mouse model of primary hyperparathyroidism. J Am Soc Nephrol 2007; 18(10):2683-2688.
35. Brown WW, Juppner H, Langman CB et al. Hypophosphatemia with elevations in serum FGF23 in a child with Jansen's Metaphyseal Chondrodysplasia (FGF23 in Jansen's Syndrome). J Clin Endocrinol Metab 2008.
36. Shimada T, Yamazaki Y, Takahashi M et al. Vitamin D receptor-independent FGF23 actions in regulating phosphate and vitamin D metabolism. Am J Physiol Renal Physiol 2005; 289(5):F1088-F1095.
37. Canalejo R, Canalejo A, Martinez-Moreno JM et al. FGF23 fails to inhibit uremic parathyroid glands. J Am Soc Nephrol 2010; 21(7):1125-1135.
38. Lavi-Moshayoff V, Wasserman G, Meir T et al. PTH increases FGF23 gene expression and mediates the high FGF23 levels of experimental kidney failure: a bone parathyroid feedback loop. AM J Physio Renal Physiol 2010; 299(4):F882-9.

CHAPTER 7

REGULATION OF ION CHANNELS
BY SECRETED KLOTHO

Chou-Long Huang

Department of Medicine, UT Southwestern Medical Center, Dallas, Texas, USA.
Email: Chou-Long.Huang@utsouthwestern.edu

Abstract: Klotho is an anti-aging protein predominantly expressed in the kidney, parathyroid glands and choroid plexus of the brain. Klotho exists in two forms, a membrane form and a soluble secreted form. Recent studies show that the secreted Klotho possess sialidase activity and regulates several ion channels via the activity. Removal of terminal sialic acids from N-glycan chains of the epithelial Ca^{2+} channel TRPV5 and the renal K^+ channel ROMK by secreted Klotho exposes the underlying disaccharide galactose-N-acetylglucosamine, a ligand for galectin-1. Binding to galectin-1 at the extracellular surface prevents internalization and leads to accumulation of the channels on the plasma membrane. Future studies will investigate whether secreted Klotho regulates cell-surface expression of other membrane glycoproteins via the same mechanism.

INTRODUCTION

Klotho is an anti-aging protein predominantly expressed in the distal tubules of kidney, parathyroid glands and choroid plexus of the brain. Klotho deficient mice exhibit multiple phenotypes resembling human aging.[1] In support of its function in aging suppression, overexpression of Klotho increases life span in mice.[2] Klotho protein is a Type-1 membrane polypeptide with a large amino-terminal extracellular domain, a membrane-spanning domain and a short intracellular carboxyl terminus.[1] The extracellular domain of Klotho is cleaved by the protease ADAM10 and secreted into blood, urine and cerebrospinal fluid.[2-4]

Endocrine FGFs and Klothos, edited by Makoto Kuro-o.
©2012 Landes Bioscience and Springer Science+Business Media.

HOMOLOGY WITH THE FAMILY 1 GLYCOSIDASES AND THE GLUCURONIDASE ACTIVITY OF KLOTHO

The extracellular domain consists of two internal repeats, each shares amino acid sequence homology to family 1 glycosidases.[1,5] Two highly conserved glutamate residues are critical for the enzymatic activities of family 1 glycosidases.[5,6] One acts as a nucleophile, the other as an acid-base catalyst. In Klotho, these two glutamate residues, however, are replaced by an asparagine at the acid-base catalyst position in the first internal repeat and an alanine or a serine at the nucleophile position in the second internal repeat, respectively.

Tohyama et al first examined the enzymatic activity of the secreted extracellular domain of Klotho.[7] Using artificial substrates, they found that secreted Klotho exhibits weak enzymatic activity toward β-glucuronic acids. The enzymatic activity is quite modest compared to the purified bovine liver β-glucuroniase (73 pmol/h/μg for Klotho vs 1940 pmol/h/μg for liver β-glucuronidase). Nonetheless, the enzymatic activity of Klotho toward artificial substrates is competitively inhibited by some natural steroid β-glucuronides. Tohyama et al concluded that secreted Klotho is a novel β-glucuronidase and that steroid β-glucuronides are potential candidate natural substrates of Klotho. Glucuronylation increases the water solubility of substances and enhances their excretion by the kidney. It was suggested that secreted Klotho may be involved in the clearance of steroids or related substrates.

SIALIDASE ACTIVITY AND REGULATION OF ION CHANNELS BY SECRETED KLOTHO

Chang et al reported that treatment with secreted Klotho increases cell-surface abundance of the epithelial Ca^{2+} channel TRPV5.[8] Mutation of the single asparagine residue responsible for N-linked glycosylation of the channel prevents the regulation by secreted Klotho. They found that purified bovine liver β-glucuronidase also regulates TRPV5 and a specific inhibitor of β-glucuronidase prevents klotho regulation of the channel. Chang et al concluded that the action of Klotho on TRPV5 is through hydrolysis of glucuronic acids from TRPV5. Glucuronic acids, though may be sugar moieties of cell-membrane protein in the form of heparin sulfate proteoglycans,[9] are extremely uncommon moieties of N-glycan chains of membrane glycoproteins such as TRPV5 channels.[10-12]

Cha et al recently provided compelling evidence that the effect of secreted Klotho on TRPV5 is mediated by a sialidase activity, rather than a glucuronidase activity.[13] The complex type N-glycan chains of glycoproteins consist of up to 4 branches initiated by the attachment of N-acetylglucosamine to underlying α3 or α6 mannose via β2, β4 or β6-linkage (Fig. 1).[10,11,12] Galactose is then added to N-acetylglucosamine forming the disaccharide galactose-N-acetylglucosamine, or N-acetyllactosame (LacNAc). Sialic acids often cap the terminal galactose residues of the N-glycan of glycoproteins via either α2,3- or α2,6-glycosidic linkage.[14] Enzymes responsible for synthesis of these glycosidic bonds are named α2,3- and α2,6-sialyltransferase, respectively.[15] Cha et al showed that secreted Klotho does not regulate TRPV5 expressed in Chinese hamster ovary (CHO) cells that do not contain endogenous α2,6-sialyltransferase. Forced expression of recombinant α2,6-sialyltransferase, but not α2,3-sialyltransferase, confers the regulation of TRPV5 by secreted Klotho. These results showed that secreted Klotho exhibits sialidase activity specific for α2,6-linked sialic acids. Cha et al also showed that treatment by purified liver

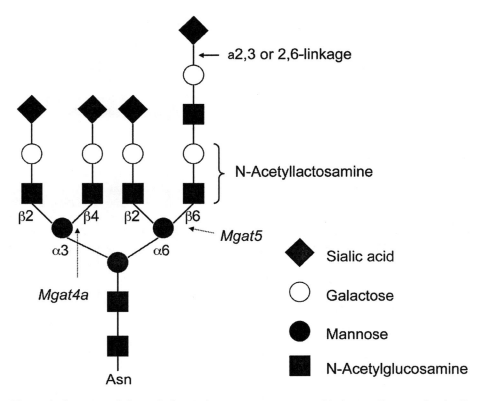

Figure 1. Structure of the typical complex type tetra-antennary N-glycans. See text for details. N-acetyllactosamine (LacNAc) is the ligand for galectins. Multiple LacNAc repeats may be present in the β6 branch (formed from underlying α6-mannose), thereby increasing the affinity of N-glycans for galectins. *Mgat5*, β1,6 N-acetylglucosaminyltransferase V, is responsible for synthesis of the β6 N-acetylglucosamine linkage on the α6-mannose; *Mgat4a*, β3,4 N-acetylglucosaminyltransferase IVa, is responsible for synthesis of the β4 N-acetylglucosamine linkage on the α3-mannose. Altered expression of these enzymes changes the number of branches of N-glycans and the number of LacNAc they carry, thus affecting binding of glycoprotein to galectin lattice on the cell surface. See text for specific examples of conditions/diseases that alter the expression of these enzymes.

β-glucuronidase increases TRPV5 expressed in a human embryonic kidney (HEK) cell line but not in CHO cells. Forced expression of recombinant α2,6-sialyltransferase in CHO cells also conferred the increase of TRPV5 by β-glucuronidase. Thus, the effect of purified liver β-glucuronidase on TRPV5 is due to off-target action toward sialic acids.

Cha et al further elucidated the mechanism of increase of cell surface abundance of TRPV5 by secreted Klotho.[13] Galectins are animal lectins that bind galactose-containing ligands.[16] Many galectins are secreted and present on the extracellular surface. Galectin-1 is ubiquitously expressed and binds LacNAc and α2,3-sialylated LacNAC, but not α2,6-sialylated LacNAc.[17] Cha et al found that knockdown of endogenous galectin-1 in HEK cells or extracellular application of antibody against galectin-1 prevents the increase of TRPV5 abundance by secreted Klotho. Moreover, secreted Klotho has no effect on TRPV5 when coexpressed with a dominant-negative dynamin mutant. Thus, removal of α2,6-sialic acids from N-glycan chains exposes underlying LacNAc for binding to galectin-1. This binding prevents TRPV5 from internalization via a dynamin-dependent process, leading to accumulation of channels on the cell surface (Fig. 2). It was further reported by Cha et al that secreted Klotho mediates removal of terminal sialic acids from N-glycans of the renal K+ channel ROMK and increases cell-surface abundance of the channel.[18] Interestingly, though internalization of ROMK and TRPV5 are both dynamin-dependent, different pathways are involved. Internalization of ROMK occurs via clathrin-coated vesicles whereas TRPV6 via caveolae.[19,20] Thus, the mechanism by which secreted Klotho increases cell-surface abundance of membrane proteins by enhancing galectin-N-glycan interaction and preventing internalization is independent of the mechanism of membrane retrieval.

CONTROL OF THE RESIDENT TIME OF MEMBRANE GLYCOPROTEINS VIA GALECTIN-N-GLYCAN INTERACTION

Binding to extracellular galectins via N-glycan chains has emerged as an important mechanism for regulation of cell-surface abundance of membrane glycoproteins, such as immune cells, surface receptors, cytokine receptors and glucose transporters.[10-12,21,22] Clustering of T-cell receptors (TCRs) by antigen presentation is critical for the activation of T-cells. Binding of TCRs via LacNAc to galectin-3 lattice at the cell surface restricts TCRs recruitment to the site of antigen presentation. It was reported that reduction of LacNAc in the N-glycans of TCRs caused by mutational deficiency

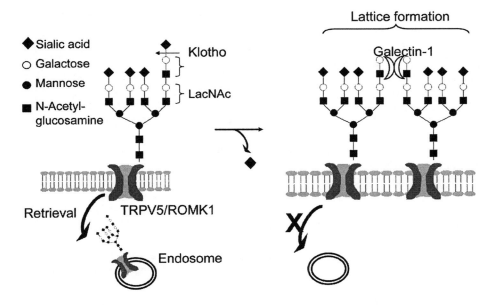

Figure 2. Mechanism for increase of cell surface abundance of ion channels by secreted Klotho. See text for details. LacNAc (N-acetyllactosamine) is a ligand for galectin-1. α2,6-linked sialic acids prevent binding of LacNAc to galectin-1. Removal of sialic acids from N-glycan of TRPV5 and ROMK channels by secreted Klotho allows LacNAc binding to galectin-1, preventing endocytosis of the channels.

of β1,6-*N*-acetylglucosaminyltransferase V (Mgat5), an enzyme that initiates β1,6 branching from α6-mannose (see Fig. 1), leads to increased T-cells activation in mice and likely increased susceptibility to autoimmune diseases.[10] Conversely, upregulation of Mgat5 increases the number of LacNAc in the N-glycans of cytokine receptors, such as epidermal growth factor and transforming growth factors receptors and enhances cell surface retention of the receptors by binding to galectin-3.[11] This modulation of cytokine receptors via Mgat5-mediated N-glycan processing underlies increased cell proliferation in some tumors.

Glut-2 glucose transporter in pancreatic β cells plays an important role in mediating glucose-induced insulin release by the pancreatic islets.[23] The density of Glut-2 in β cells is also modulated by the galectin-mediated retention mechanism. Ohtsubo et al reported that feeding high fat diet in mice downreguates the expression of glycosyltransferase Magt4a, which is one of two enzymes that synthesize the β4 N-acetylglucosamine linkage on the α3-mannose (see Fig. 1).[21] A decrease in Mgat4a reduces the half life of Glut-2 on the cell surface owing to the decrease in the number of LacNAc and in the binding of transporters to galectin.

In contrast to the above examples of regulation of synthesis and branching of N-glycans by affecting glycosyltransferases in the Golgi, secreted Klotho targets mature glycoproteins already at the cell surface and acting from the outside of cell. This action of Klotho represents a novel mechanism for regulation of cell surface glycoproteins.

REGULATION OF ION CHANNELS BY SECRETED KLOTHO VIA OTHER MECHANISMS

Secreted Klotho is a pleiotropic hormone.[24] Besides its ability to modify N-glycan, secreted Klotho interacts with fibroblast growth factor (FGF) receptors and modulates the affinity for ligands[25,26] and interferes with intracellular signaling of insulin and insulin-like growth factor.[2] More recently, Cha et al found that secreted klotho decreases cell-surface abundance of TRPC6,[27] a Ca^{2+}-permeant channel important in the regulation of cardiac and renal glomerular function.[28] This effect on TRPC6 is opposite to that on TRPV5 and ROMK channels and not mediated by the sialidase activity of Klotho.[27] The regulation of TRPC6 by secreted Klotho requires the presence of serum, suggesting that it may be mediated by an effect of Klotho to antagonize serum growth factors. Precise mechanisms by which secreted Klotho may regulate TRPC6 by antagonism of serum factors remain to be investigated.

ROLE OF REGULATION OF TRPV5 BY SECRETED KLOTHO IN THE OVERALL ANTI-AGING FUNCTION OF KLOTHO

Recent evidence indicates that premature aging of Klotho deficiency is related to deranged phosphate metabolism.[29] Mice lacking FGF-23, a bone-derived hormone that promotes renal phosphate excretion, have phenotypes identical to Klotho-deficient mice, including hyperphosphatemia, elevated 1,25-vitamin D_3 and premature aging.[30] Membrane Klotho interacts with FGF receptors to form high affinity coreceptors for FGF23.[25,26] Activation of FGFR-Klotho coreceptor complexes by FGF23 in the proximal tubule of the kidney suppresses the synthesis of 1,25-vitamin D_3.[31] 1,25-vitamin D_3 increases

REGULATION OF ION CHANNELS BY SECRETED KLOTHO

Ca^{2+} and phosphate absorption from gastrointestinal tract. The absolute requirement of Klotho for FGFR function explains why klotho deficiency and FGF23 deficiency cause the same phenotypes. Additionally, secreted Klotho regulates phosphate metabolism by inhibiting renal phosphate reabsorption via Na^+-coupled phosphate transporters in an FGF23-independent mechanism.[29] Thus, Klotho deficiency causes hyperphosphatemia by increasing intestinal phosphate absorption and increasing renal phosphate reabsorption. Hyperphosphatemia in Klotho deficient mice promotes vascular calcification and among other things leading to organ senescence. In support of this idea, dietary phosphate restriction prevents vascular calcification and rescues premature aging in Klotho-deficient mice.[29-31] Thus, the life-extending effect of Klotho may be, at least partly, related to its ability to promote renal phosphate excretion and inducing negative phosphate metabolism. Negative phosphate metabolism over time, however, leads to bone diseases. The stimulation of TRPV5 and renal Ca^{2+} reabsorption by secreted Klotho will promote positive Ca^{2+} balance and help preventing bone diseases from negative phosphate balance.

CONCLUSION AND FUTURE PERSPECTIVES

Emerging evidence indicates that secreted Klotho functions as a paracrine and endocrine hormone. As a paracrine hormone, Klotho secreted from distal tubules of the kidney working from the luminal side (= urinary space) increases cell surface abundance of the Ca^{2+} channel TRPV5 and K^+ channel ROMK in the distal renal tubules. These effects underlie an important function of Klotho in stimulation of Ca^{2+} reabsorption and K^+ secretion in the kidney.[18,32,33] As an endocrine hormone, secreted Klotho decreases cell surface expression of TRPC6 present in the heart and the kidney glomeruli. Upregulation of TRPC6 is implicated in the pathogenesis of pathological cardiac hypertrophy and glomerular proteinuria.[34,35] Future studies will investigate whether secreted Klotho is protective for pathological cardiac hypertrophy and glomerular proteinuria. Finally, it would be interesting to investigate whether secreted Klotho regulates cell-surface expression of other membrane glycoproteins and whether it occurs via the sialidase activity or via other pleiotropic effect of secreted Klotho.

ACKNOWLEDGEMENTS

Dr. Huang's research is supported by grants from National Institutes of Health and American Heart Association. Dr. Huang holds the Jacob Lemann Professorship in Calcium Transport of University of Texas Southwestern Medical Center.

REFERENCES

1. Kuro-o M, Matsumura Y, Aizawa H et al. Mutation of the mouse klotho gene leads to a syndrome resembling aging. Nature 1997; 390:45-51.
2. Kurosu H, Yamamoto M, Clark JD et al. Suppression of aging in mice by the hormone Klotho. Science 2005; 309:1829-1833.
3. Chen CD, Podvin S, Gillespie E et al. Insulin stimulates the cleavage and release of the extracellular domain of Klotho by ADAM10 and ADAM17. Proc Natl Acad Sci USA 2007; 104:19796-19801.

4. Imura A, Iwano A, Tohyama O et al. Secreted Klotho protein in sera and CSF: implication for posttranslational cleavage in release of Klotho protein from cell membrane. FEBS Lett 2004; 565:143-147.
5. Ito S, Fujimori T, Hayashizaki Y et al. Identification of a novel mouse membrane-bound family 1 glycosidase-like protein, which carries an atypical active site structure. Biochim Biophys Acta 2002; 1576:341-345.
6. Rye CS, Withers SG. Glycosidase mechanisms. Curr Opin Chem Biol 2000; 4:573-580.
7. Tohyama O, Imura A, Iwano A et al. Klotho is a novel β-glucuronidase capable of hydrolyzing steroid β-glucuronides. J Biol Chem 2004; 279:9777-9784.
8. Chang Q, Hoefs S, van der Kemp AW et al. The β-glucuronidase klotho hydrolyzes and activates the TRPV5 channel. Science 2005; 310:490-493.
9. Bishop JR, Schuksz M, Esko JD. Heparan sulphate proteoglycans fine-tune mammalian physiology. Nature 2007; 446:1030-1037.
10. Demetriou M, Granovsky M, Quaggin S et al. Negative regulation of T-cell activation and autoimmunity by Mgat5 N-glycosylation. Nature 2001; 409:733-739.
11. Partridge EA, Le Roy C, Di Guglielmo GM et al. Regulation of cytokine receptors by Golgi N-glycans processing and endocytosis. Science 2004; 306:120-124.
12. Stanley P. A method to the madness of N-glycan complexity? Cell 2007; 129:27-29.
13. Cha SK, Ortega B, Kurosu H et al. Removal of sialic acid involving Klotho causes cell-surface retention of TRPV5 channel via binding to galectin-1. Proc Natl Acad Sci USA 2008; 105:9805-9810.
14. Schauer R. Biosynthesis and function of N- and O-substituted sialic acids. Glycobiology 1991; 1:449-452.
15. Patel RY, Balaji PV. Identification of linkage-specific sequence motifs in sialyltransferases. Glycobiology 2006; 16:108-116.
16. Barondes SH, Cooper DN, Gitt MA et al. Galectins: Structure and function of a large family of animal lectins. J Biol Chem 1994; 269:20807-20810.
17. Leppanen A, Stowell S, Blixt O et al. Dimeric galectin-1 binds with high affinity to α2,3-sialylated and nonsialylated terminal N-acetyllactosamine units on surface-bound extended glycans. J Biol Chem 2005; 280:5549-5562.
18. Cha SK, Hu MC, Kurosu H et al. Reguation of ROMK1 channel and renal K+ excretion by Klotho. Mol Pharmacol 2009; 76:38-46.
19. Zeng WZ, Babich V, Ortega B et al. Evidence for endocytosis of ROMK potassium channels via clathrin-coated vesicles. Am J Physiol 2002; 283:F630-F639.
20. Cha SK, Wu T, Huang CL. Protein kinase C inhibits caveolae-mediated endocytosis of TRPV5. Am J Physiol Renal Physiol 2008; 294:F1212-F1221.
21. Ohtsubo K, Takamatsu S, Minowa MT et al. Dietary and genetic control of glucose transporter 2 glycosylation promotes insulin secretion in suppressing diabetes. Cell 2005;123:1307-1321.
22. Lau KS, Partridge EA, Grigorian A et al. Complex N-glycan number and degree of branching cooperate to regulate cell proliferation and differentiation. Cell 2007; 129:123-134.
23. Valera A, Solanes G, Fernández-Alvarez J et al. Expression of GLUT-2 antisense RNA in beta cells of transgenic mice leads to diabetes. J Biol Chem 1994; 269:28543-28546.
24. Nabeshima Y. Toward a better understanding of Klotho. Sci Aging Knowledge Environ 2006; 8:pe11.
25. Kurosu H, Ogawa Y, Miyoshi M et al. Regulation of fibroblast growth factor-23 signaling by Klotho. J Biol Chem 2006; 281:6129-6123.
26. Urakawa I, Yamazaki Y, Shimada T et al. Klotho converts canonical FGF receptor into a specific receptor for FGF23. Nature 2006; 444:770-774.
27. Cha SK, Kuroo M, Huang CL. The anti-aging hormone Klotho regulates cell surface abundance of TRPC6. J Am Soc Nephrol 2009; In press.
28. Dietrich A, Gudermann T. TRPC6. Handb Exp Pharmacol 2007; 179:125-141.
29. Kurosu H, Kuro-o M. The Klotho gene family and the endocrine fibroblast growth factors. Curr Opin Nephrol Hypertens 2008; 17:368-372.
30. Razzaque MS, Lanske B. The emerging role of the fibroblast growth factor-23-klotho axis in renal regulation of phosphate homeostasis. J Endocrinol 2007; 194:1-10.
31. Yoshida T, Fujimori T, Nabeshima Y. Mediation of unusually high concentrations of 1,25-dihydroxyvitamin D in homozygous klotho mutant mice by increased expression of renal 1alpha-hydroxylase gene. Endocrinology 2002; 143:683-689.
32. Tsuruoka S, Nishiki K, Ioka T et al. Defect in parathyroid-hormone-induced luminal calcium absorption in connecting tubules of Klotho mice. Nephrol Dial Transplant 2006; 21:2762-2767.
33. Alexander RT, Woudenberg-Vrenken TE, Buurman J et al. Klotho Prevents Renal Calcium Loss. J Am Soc Nephrol 2009; (Epub)
34. Winn MP, Conlon PJ, Lynn KL et al. A mutation in the TRPC6 cation channel causes familial focal segmental glomerulosclerosis. Science 2005; 308:1801-1804.
35. Kuwahara K, Wang Y, McAnally J et al. TRPC6 fulfills a calcineurin signaling circuit during pathologic cardiac remodeling. J Clin Invest 2006; 116:3114-3126.

CHAPTER 8

FGF23 IN CHRONIC KIDNEY DISEASE

Patricia Wahl and Myles Wolf*

Division of Nephrology and Hypertension, University of Miami Miller School of Medicine, Miami, Florida, USA.
*Corresponding Author: Myles Wolf—Email: mwolf2@med.miami.edu

Abstract: Chronic kidney disease (CKD) is a growing public health epidemic that is associated with a markedly increased risk of cardiovascular mortality. Disordered mineral metabolism and particularly, disordered phosphorus metabolism appears to be a contributing factor. Fibroblast growth factor 23 (FGF23) regulates phosphorus and vitamin D metabolism. Its levels increase progressively beginning in early CKD, presumably as a physiological adaptation to maintain normal serum phosphate levels or normal phosphorus balance. FGF23 promotes phosphaturia and decreases production of calcitriol. Recent studies suggest that increased FGF23 is associated with mortality, left ventricular hypertrophy, endothelial dysfunction and progression of CKD. These results were consistently independent of serum phosphate levels. At the very least, FGF23 is emerging as a novel biomarker that may help identify which CKD patients might benefit most from aggressive management of disordered phosphorus metabolism. It is also possible that markedly increased FGF23 levels in CKD could contribute directly to tissue injury in the heart, vessels and kidneys, an exciting question that is sure to be the topic of intense investigation in the near future.

INTRODUCTION: BURDEN OF CHRONIC KIDNEY DISEASE

Chronic kidney disease (CKD) is a growing public health epidemic that is estimated to affect 13% of the US adult population, or approximately 26 million Americans and far more patients worldwide.[1,2] The disease accounts for more than 24% of annual Medicare expenditures.[3] The growing burden of CKD reflects the impact of the rapidly increasing prevalence of diabetes and hypertension, which account for more than 50% of all adult cases of end stage renal disease requiring dialysis.[4] Indeed, the 30% increase in prevalence of CKD over the past decade has prompted the U.S. Renal Data System (USRDS) to issue for the first time a separate report documenting the magnitude of CKD in addition to its

---

Endocrine FGFs and Klothos, edited by Makoto Kuro-o.
©2012 Landes Bioscience and Springer Science+Business Media.

annual dialysis report: *The USRDS 2008 Annual Data Report: Atlas of Chronic Kidney Disease and End-Stage Renal Disease.*[4]

Although the overwhelming economic and medical impact of the growing dialysis population (Stage 5 CKD: >350,000) is widely recognized, >13 million people suffer from CKD Stages 3 or 4 which are risk factors for cardiovascular disease (CVD) mortality.[1,2,5] Indeed, besides progression of renal failure to the point that dialysis or kidney transplantation is required for survival, the most common and morbid adverse outcomes of CKD is CVD, including high rates of atherosclerosis, myocardial infarction, peripheral vascular disease, stroke, extensive arterial calcification with increased vascular stiffness, left ventricular hypertrophy and congestive heart failure.[6,7]

The tight relationship between CVD and CKD stems partly from their sharing a common set of risk factors such as diabetes and hypertension. New-onset CVD events are more common in all stages of CKD than in the general population and the outcomes related to these events are worse in CKD patients than in the general population. Furthermore, the presence of CVD is associated with more progressive CKD. In addition to these shared pathways, CKD appears to be an independent risk factor for CVD: the presence of CKD confers even greater risk than would be expected after taking into account the high frequency of traditional CVD risk factors.[8-10] This suggests the presence of additional, CKD-specific, CVD risk factors with disordered mineral metabolism among the most hotly investigated candidates.

Regardless of the mechanisms, based on the known disease-multiplying interrelationships between CKD and CVD, it is not surprising that patients with CKD are more likely to die than they are likely to develop end stage kidney failure requiring dialysis or a transplant.[11] The observation of dramatically increased risk of CVD in patients with CKD has fueled renewed emphasis on earlier diagnosis of CKD through widespread screening of at-risk populations, earlier intervention against traditional CVD risk factors in CKD patients and the search for new, CKD-specific risk factors to target for future interventions aimed at improving renal and cardiovascular outcomes.

Overview of Normal Kidney Function and Pathophysiology of CKD

CKD can target either of the two main structural components of the nephron, the glomeruli or tubules. Water, electrolytes and minerals are freely filtered through the glomerulus after which the tubules reabsorb desirable molecules and allow waste products to escape into the urine for excretion. The tubules can also secrete additional waste products to concentrate their excretion in the urine. The normal composition of a urinary component is that which maintains a normal amount of the substance in the blood stream. For example, in the setting of phosphorus loading, there is a marked increase in urinary phosphate excretion thereby maintaining normal phosphorus balance (also true for sodium, potassium and water loading). Conversely, when deprived of phosphorus, the fraction of filtered phosphate that is reabsorbed increases such that the kidney maintains normal body balance by conserving phosphate.

In the setting of CKD in which there is a progressive reduction in glomerular filtration, the kidney adapts by augmenting the filtration rate in the remaining nephrons, also known as hyperfiltration.[12] This helps maintain normal levels of serum creatinine despite reduced GFR, which can often obscure the diagnosis of early CKD. Additional homeostatic compensations in the tubules help maintain normal serum concentrations of sodium, potassium, calcium, phosphorus and normal water balance. However, over the long term hyperfiltration and its underlying mechanism of increased glomerular capillary pressure results in further injury to the remaining nephrons and thus progression of CKD if left untreated.[13,14]

The primary measurement of renal function is the glomerular filtration rate (GFR). Normal GFR is 120 ml/minute but normal values can vary by age, gender and race.[15] GFR can be calculated directly using iothalamate, iohexol or inulin clearance but these strategies are mostly used exclusively in the research setting. In clinical practice, 24 hour urine collections can be used to measure creatinine clearance, which is highly correlated with GFR. However, this approach is cumbersome and fraught with incorrect collections (too short or too long) that can under- or overestimate GFR.[16,17] For these reasons and given the need to screen large numbers of patients in an era when the time for clinical encounters is progressively shortening, 24 hour collections are now generally discouraged. The current recommendation is to screen for CKD using GFR estimating equations that incorporate spot measurements of serum creatinine and the patients' age, gender and race, most commonly, using the Modification of Diet in Renal Disease (MDRD) equation.[15] The estimated GFR (eGFR) has also replaced isolated assessments of serum creatinine levels, which dramatically underestimate the presence and severity of CKD, particularly early in its course when the opportunity for improved outcomes is greatest but most dependent on early diagnosis.[18]

Definition and Staging of CKD

A work group established by the National Kidney Foundation, the Kidney Disease Outcome Quality Initiative (KDOQI), was convened to standardize the nomenclature of CKD so that clinical care of CKD patients could be improved through stage-specific standardization of care.[15] The group defined CKD by the presence of either: (1) a sustained reduction in eGFR <60 mL/min/1.73 m^2 for more than 3 consecutive months; or (2) normal eGFR (>60) but evidence of chronic kidney injury (>3 months duration), either a structural or functional abnormality. For the latter, two examples (among many) include the presence of proteinuria, or the finding of multiple cysts underlying autosomal dominant polycystic kidney disease. Regardless of which definition is used to identify CKD in an individual patient, staging of CKD into 5 stages is defined solely on the basis of eGFR (Table 1): Stage 1, eGFR > 90; Stage 2, eGFR 60-89; Stage 3, eGFR 30-59; Stage 4, eGFR 15-29; and Stage 5, eGFR <15 mL/min/1.73 m^2.

Table 1. Stages of CKD[15]

Stage 1 CKD	Slightly diminished function; Kidney damage with normal or relatively high eGFR (>90 mL/min/1.73 m^2). Kidney damage is defined as pathologic abnormalities or markers of damage, including abnormalities in blood or urine test or imaging studies.
Stage 2 CKD	Mild reduction in eGFR (60-89 mL/min/1.73 m^2) with kidney damage. Kidney damage is defined as pathologic abnormalities or markers of damage, including abnormalities in blood or urine test or imaging studies.
Stage 3 CKD	Moderate reduction in eGFR (30-59 mL/min/1.73 m^2).
Stage 4 CKD	Severe reduction in eGFR (15-29 mL/min/1.73 m^2) Preparation for renal replacement therapy
Stage 5 CKD	Established kidney failure (eGFR <15 mL/min/1.73 m^2, or permanent renal replacement therapy (RRT)

Complications of CKD

CKD is a systemic disease that affects multiple organ systems during its course. With progressive loss of kidney function, patients with CKD develop multiple hemodynamic, hematologic and metabolic complications. The clinical impact of these consequences has been reviewed in detail recently elswewhere.[19] In the current chapter, focused attention is paid to the mineral metabolism abnormalities that are common in CKD, with specific emphasis to the role of FGF23 in CKD.

OVERVIEW OF NORMAL MINERAL METABOLISM AND ITS DISORDERS IN CKD

Disordered mineral metabolism is among the earliest and most common complications of CKD. In addition to its well-known effect on skeletal complications, the stature of disordered mineral metabolism in CKD has grown dramatically in recent years with the publication of a series of large epidemiological studies of dialysis and subsequently, earlier-stage CKD patients that demonstrated various components of disordered mineral metabolism to be independently associated with kidney disease progression, cardiovascular disease and mortality.[20-24] Understanding these observations requires a brief overview of mineral metabolism in health and in CKD. Under normal conditions, calcium and phosphorus balance, their serum levels and skeletal integrity and growth are tightly regulated by the interrelationships between parathyroid hormone (PTH), vitamin D and most recently, fibroblast growth factor 23 (FGF23). Collectively, these hormones contribute to the regulation of mineral absorption by the intestine, mineralization of bone and renal excretion of excess mineral.

Parathyroid Hormone

The primary role of PTH is to maintain a normal extracellular concentration of calcium, which is achieved by stimulating bone resorption and thus release of calcium into the circulation, increasing dietary calcium absorption by the gut that is stimulated by PTH-mediated increases in calcitriol production and increased tubular reabsorption of calcium which minimizes urinary calcium losses.[25-27] PTH is phosphaturic and as a result, when PTH levels rise in response to reduced calcium levels, its effects to normalize serum calcium are achieved with minimal impact on serum phosphate levels as bone release of phosphate is counterbalanced by increased phosphaturia.

Vitamin D

When exposed to ultraviolet B light, vitamin D is produced in the skin in the form of cholecalciferol (vitamin D_3). In addition, vitamin D_3 or vitamin D_2 (ergocalciferol, of plant origin) can each be ingested as a dietary supplement. Both vitamin D_3 and D_2 are converted in the liver to the vitamin D storage form, 25-hydroxyvitamin D (25D), which can then be converted to the biologically active 1,25 dihydroxyvitamin D (1,25D

FGF23 IN CHRONIC KIDNEY DISEASE

or calcitriol) by 25-hydroxyvitamin D 1-alpha-hydroxylase. The kidney is the primary source of 1-α hydroxylase and thus most of the circulating 1,25D, but additional extra renal tissues including monocytes, intestine, bone, parathyroid, beta cell and skin cells can also produce 1,25D for internal autocrine use.[28] Renal 1,25D production is controlled by modulating the activity of the renal 1-α hydroxylase with PTH the most potent stimulator and FGF23 the most important circulating inhibitor. Whether FGF23 has an effect on peripheral 1-α hydroxylase is not known. Vitamin D helps maintain calcium and phosphorus homeostasis by increasing dietary absorption of calcium and phosphorus in the small intestine and also contributes to bone formation and resorption.[28]

FGF23

FGF23 is a produced mainly by osteocytes.[29] Through binding to FGF receptor-klotho complexes (reviewed in detail elsewhere in this book) that are primarily expressed in the kidneys and the parathyroids, FGF23 exerts its physiological functions which include: inducing phosphaturia by down regulating renal sodium-phosphate cotransport in the proximal tubule (similar to the effect of PTH); inhibiting calcitriol production by down regulating the renal 1-α hydroxylase and stimulating the catabolic 24-hydroxylase; and inhibiting PTH secretion.[30-34] In healthy individuals, when phosphorus intake is high, FGF23 levels increase, which promote phosphaturia and inhibit calcitriol production thereby limiting the efficiency of dietary phosphorus absorption.[35-37] Conversely, on a low phosphorus diet, FGF23 levels fall, tubular conservation of filtered phosphate increases as does the renal production of 1,25D, which increases the efficiency of dietary phosphorus absorption.[35-37] The net result of these compensatory mechanisms is the maintenance of a normal serum phosphate level despite day-to-day fluctuation in dietary phosphorus intake. The effect of FGF23 to reduce PTH secretion can be understood as an additional mechanism for reducing calcitriol production. Indeed, PTH and FGF23 complete two separate but interrelated, counter-regulatory, negative endocrine feedback loops with vitamin D: FGF23 inhibits renal calcitriol production while calcitriol stimulates FGF23 secretion, whereas PTH stimulates renal calcitriol production while calcitriol inhibits PTH secretion.

As discussed elsewhere in this book, FGF23 was originally discovered in rare hereditary and acquired hypophosphatemic rachitic disorders in which primary FGF23 excess causes hypophosphatemia and inhibits the appropriate calcitriol response that would be expected in states of such severe hypophosphatemia.[38-42] In contrast, primary deficiency of biologically active FGF23 or resistance to its renal effects as in states of klotho deficiency result in a physiological syndrome characterized by hyperphosphatemia, increased calcitriol and extensive soft-tissue calcifications and premature death.[33,43,44] CKD can be viewed as the most common state of secondary FGF23 excess in which serum phosphate levels are normal to high, fractional excretion of phosphate is high, calcitriol levels low and FGF23 levels are markedly increased to a far greater extent that in the primary FGF23 syndromes (Fig. 1).

Disturbed Mineral Metabolism in Chronic Kidney Disease: Historical View

Patients with CKD develop several disturbances in mineral metabolism that begin early in the course of disease, often before the diagnosis of CKD is made. The historic perspective of disordered mineral metabolism and resulting secondary hyperparathyroidism in CKD was largely developed through animal models of end stage disease before FGF23

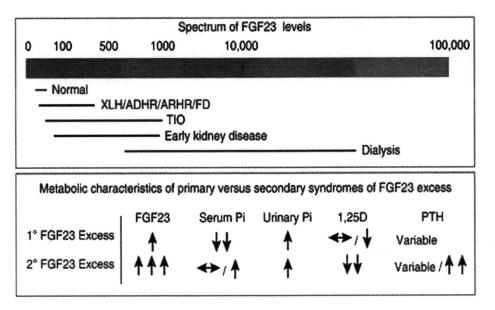

Figure 1. Spectrum of FGF23 level in CKD: The upper insert illustrates the spectrum of FGF23 levels that can be observed under normal conditions and in a variety of syndromes of FGF23 excess. Circulating FGF23 levels are 10- to 20-fold above normal (30-60 RU/mL using a C-terminal FGF23 assay) in patients with hereditary hypophosphatemic rickets syndromes, including X-linked hypophosphatemia (XLH), autosomal dominant hypophosphatemic (ADHR), autosomal recessive hypophosphatemic rickets (ARHR), and fibrous dysplasia (FD). Although FGF23 levels are often even higher in patients with tumor-induced osteomalacia (TIO), the highest levels are encountered in patients with kidney disease and especially in those on dialysis, in whom levels can reach concentrations more than 1000-fold above the normal range. The lower insert illustrates the differences in the metabolic characteristics of 'primary' syndromes of FGF23 excess, such as the hereditary diseases and TIO, versus 'secondary' syndromes of FGF23 excess, such as kidney disease. In addition to the severity of the FGF23 increase, the primary difference is normal to high serum phosphate (Pi) levels in patients with kidney disease compared to those with hypophosphatemia, which is the sin qua non of the hereditary syndromes. Although variable, 1,25D levels tend to be lower and the PTH levels higher in patients with kidney disease than in those with the hereditary syndromes. Urinary fractional excretion of phosphate is high in both predialysis kidney disease and genetic hypophosphatemic disorders. Originally published in Kidney Int. 2009;76(7):705-16.

was discovered and highlighted 3 main factors: hyperphosphatemia, hypocalcemia and 1,25D deficiency (Fig. 2).

It proposed that hyperphosphatemia develops when the ailing kidneys cannot sufficiently excrete phosphate loads while 1,25D deficiency was thought to develop when insufficient renal mass limited renal 1-α hydroxylase activity.[45-48] Hyperphosphatemia stimulates excessive PTH secretion, progressive parathyroid hyperplasia and exacerbates 1,25D deficiency by inhibiting 1-α hydroxylase.[48-51] Hyperphosphatemia and 1,25D deficiency contribute to hypocalcemia, which stimulates PTH further.[49] At the cellular level, parathyroid expression of the vitamin D and calcium sensing receptors declines progressively,[50,51] leading to parathyroid resistance to inhibition by 1,25D and calcium.[52-54] The net result is constitutive PTH secretion and progressive parathyroid hyperplasia.[54]

The historic perspective provided a strong explanation for mechanisms of secondary hyperparathyroidism in Stage 5 CKD but this model could not fully account for the findings observed in Stages 3-4. For example, in a small cross-sectional study, the

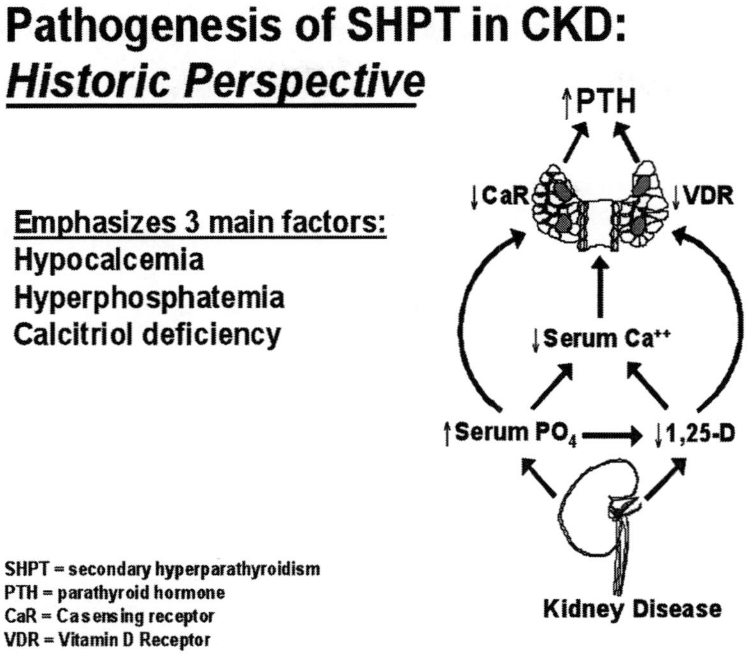

Pathogenesis of SHPT in CKD: *Historic Perspective*

Emphasizes 3 main factors:
Hypocalcemia
Hyperphosphatemia
Calcitriol deficiency

SHPT = secondary hyperparathyroidism
PTH = parathyroid hormone
CaR = Ca sensing receptor
VDR = Vitamin D Receptor

Figure 2. Pathogenesis of sHPT in CKD. The historic perspective of sHPT in CKD emphasizes hypocalcemia, hyperphosphatemia and calcitriol deficiency, contributing to a net result of increased PTH in CKD.

prevalence of 1,25D deficiency and secondary hyperparathyroidism were as high as 56% and 77%, respectively, in Stages 3-4 CKD when hyperphosphatemia (12%) and hypocalcemia (6%) were far less common.[55] In a subsequent study that is the largest of its type to date, 1,814 patients from primary care clinics across the United States with CKD Stages 2-5 were evaluated in a cross-sectional study to determine the timing of disturbances in mineral metabolism relative to eGFR.[56] The study found that increases in PTH occur early in the course of CKD, with levels >65 pg/mL observed in approximately 60% of patients with Stage 3 disease. The prevalence of secondary hyperparathyroidism grew as renal function deteriorated: 70% in Stage 4 and approximately 90% in Stage 5 disease. Vitamin D levels, both 25D and 1,25D, declined as eGFR decreased.

In contrast, hypocalcemia (<8.4 mg/dL) and hyperphosphatemia (>4.6 mg/dL) were relatively rare during the early stages of CKD with fewer than 10% of patients with an eGFR between 39 and 30 mL/minute/$1.73m^2$ affected. Even after the eGFR fell below 20 mL/minute/$1.73m^2$, the majority of patients still had normal serum calcium and phosphate levels.[56] Thus, unlike the experience on dialysis when target ranges for calcium and phosphorus were not achieved in 56% and 54% of patients,[57] respectively, calcium and phosphate levels are typically normal in most patients with predialysis CKD.[56] A follow-up analysis of the same SEEK cohort demonstrated that several of the abnormalities in mineral metabolism—secondary hyperparathyroidism, vitamin D deficiency and hyperphosphatemia—were more common and more severe among African-Americans compared with whites.[58]

Therefore, while it is clear that hyperphosphatemia and hypocalcemia independently exacerbate secondary hyperparathyroidism, among the three pathogenic mechanisms supported by the historic perspective, 1,25D deficiency develops earliest—when the eGFR decreases below -60 ml/min/1.73 m². Historically, 1,25D deficiency in CKD has been attributed to limited renal conversion of 25D to 1,25D due to "insufficient renal mass". However, 1,25D deficiency develops in early CKD when there is sufficient renal mass to support other proximal tubular functions such as erythropoietin production. Indeed, the prevalence of 1,25D deficiency exceeds the prevalence of anemia at all stages of CKD.[59] Although a higher than previously appreciated prevalence of 25D deficiency may contribute, CKD subjects with adequate 25D stores nonetheless develop 1,25D deficiency.[55,60] Thus, the historic perspective could not fully explain the etiology of 1,25D deficiency or the development of secondary hyperparathyroidism in early CKD.

Disturbed Mineral Metabolism in Chronic Kidney Disease: The Post-FGF23 View

Several of these discrepancies now appear to be explained by the excessive production of FGF23. Circulating levels of FGF23 rise progressively as renal function declines with levels already elevated as early as CKD Stages 2-3.[59,61,62] Importantly, the elevation of FGF23 levels is detectable long before hyperphosphatemia first appears. The latter highlights both the vital compensatory role of FGF23 to help maintain normal serum phosphate levels in CKD and the potential of FGF23 elevation as a more sensitive biomarker to identify disordered phosphorus metabolism in CKD before there is evidence of overt hyperphosphatemia. Current hypotheses suggest that CKD patients who maintain their usual phosphorus intake recruit the same physiological response as normal subjects fed a high phosphorus diet: increased FGF23 secretion which helps augment urinary phosphate excretion but in the process leads to decreased 1,25D levels.[63] While this FGF23 response appears to help maintain a normal serum phosphate levels it does so at the cost of an early reduction in 1,25D levels. By directly inhibiting calcitriol production and thereby releasing the parathyroids from its feedback inhibition, early FGF23 excess may be one of the key upstream steps in the pathogenesis of secondary hyperparathyroidism. This new perspective (Fig. 3) emphasizes the degree of phosphate intake relative to the degree of renal dysfunction and deemphasizes the need for overt hyperphosphatemia. Indeed, maintaining a "usual" phosphorus intake in the face of decreased renal function may be an initiating trigger in the pathogenesis of secondary hyperparathyroidism in CKD.

A number of observations support this view. Uremic animals fed a phosphorus restricted diet do not develop secondary hyperparathyroidism and secondary hyperparathyroidism in vitamin D receptor null mice can be rescued by a low phosphorus diet.[64-66] Human studies are corroborative.[67,68] Restricting phosphorus intake in CKD patients is associated with increased 1,25D and decreased PTH levels[69,70] although the mechanism was unknown. It now appears that differences in FGF23 levels could explain these findings. These studies support the pathogenic primacy of phosphorus intake in the development of 1,25D deficiency and secondary hyperparathyroidism, even in the absence of hyperphosphatemia. Indeed, total phosphorus intake seems to be more important than the serum phosphate levels as rats with experimental CKD develop progressively increased FGF23 and phosphaturia and decreased 1,25D levels before hyperphosphatemia appears,[71,72] consistent with our hypotheses. This is supported

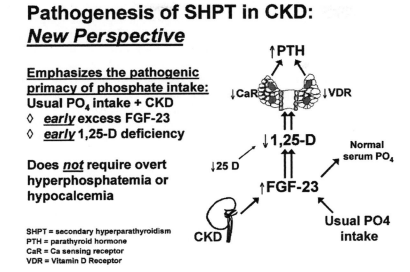

Figure 3. Pathogenesis of sHPT in CKD. The new perspective of sHPT in CKD emphasizes the degree of phosphate intake relative to the degree of renal dysfunction and de emphasizes the need for overt hyperphosphatemia. Early FGF23 excess may be a key upstream event of increased PTH in CKD.

by cross-sectional studies of CKD patients in which FGF23 levels increased markedly as eGFR declined and increased FGF23 was the strongest independent predictor of decreased 1,25D levels.[59] Indeed, the univariate association between decreased GFR and decreased 1,25D was completely extinguished when FGF23 was added to the model, suggesting that FGF23 rather than reduced GFR may be the key factor that underlies 1,25D deficiency in early CKD. To date, a handful of small pilot studies have suggested that dietary phosphorus binders can effectively reduce FGF23 levels in CKD which mirror similar results in animal models of CKD and in healthy volunteers.[37,73-75]

Clinical Implications for Screening

Our updated understanding of the pathogenesis of disordered mineral metabolism and the temporal aspects of when individual abnormalities appear in patients' laboratory tests has important clinical implications. Whereas in the past, clinicians would test patients for the presence of secondary hyperparathyroidism typically only after they manifested an abnormal calcium or phosphate levels, current data suggest and thus clinical guidelines recommend that PTH screening occur solely on the basis of the degree of renal dysfunction and should be performed even if calcium and phosphate levels are normal as they typically are. Currently, it is recommended that such screening begin once the eGFR reaches 60 ml/min/1.73m^2, however, these guidelines did not consider that secondary hyperparathyroidism begins even earlier in a large proportion of minority patients who also happen to bear a disproportionate burden of CKD.[76,77] Thus, it seems prudent to screen minority patients once CKD is diagnosed, regardless of eGFR. Regardless of the optimal timing to initiate screening in specific CKD subpopulations, the emphasis is clearly on earlier screening than in the past.

Obviously, screening for a disease or complication is only justified if it can lead to a change in clinical management that ultimately improves the patients' clinical outcome. The growing interest in identifying disordered mineral metabolism in its earliest stages is fueled largely by the observation that its therapies—primarily, active vitamin D and dietary phosphorus binders—may improve survival. Although the studies that underlie these findings were not randomized, the consistent results from observational studies along with the consistent finding of increased risk of adverse outcomes associated with disordered mineral metabolism has driven the growing trend to screen and treat these disorders in earlier stages of CKD.[78,79] Indeed, based on the results summarized below, disordered mineral metabolism and most prominently phosphorus metabolism, is now being considered as a novel risk factor for adverse cardiovascular and renal outcomes. Whether abnormal phosphorus metabolism in CKD proves to be a *causal* factor in the development of adverse outcomes will require definitive proof that can only be achieved by randomized trials that test the effects of phosphorus reduction strategies on outcomes.

DISORDERED PHOSPHORUS METABOLISM AND ADVERSE CLINICAL OUTCOMES

Phosphorus Excess and Adverse Clinical Outcomes in CKD

In addition to its well-known adverse effects on bone and the parathyroids, hyperphosphatemia has also emerged as a novel risk factor for kidney disease progression and cardiovascular disease in CKD. When mildly uremic rats were administered diets with high phosphorus content they experienced accelerated kidney disease progression, and epidemiological studies suggest that higher serum phosphate levels predict more rapid CKD progression. In support of these findings, animal and human studies found that phosphorus-restricted diets slow the decline in kidney function.[66-68]

Arterial calcification causing arterial stiffening is an important phenotype of vascular injury in CKD that is independently associated with mortality.[80,81] Hyperphosphatemia is independently associated with greater burden of arterial calcification in dialysis and predialysis CKD and even in patients with normal renal function.[82-86] A potential mechanism for arterial calcification induced by phosphorus-excess is phosphorus-dependent transformation of vascular smooth muscle cells into bone forming osteoblast-like cells in the arterial media leading to ectopic ossification in the walls of the large conduit arteries.[87] Another cardiovascular phenotype that is a risk factor for mortality and also may be associated with phosphorus excess is left ventricular hypertrophy. Left ventricular hypertrophy is common in CKD patients and contributes to high rates of congestive heart failure and sudden cardiac death, which are leading causes for CVD-related mortality in CKD. In animal studies, excess dietary phosphorus loading led to hyperphosphatemia and left ventricular hypertrophy, whereas intensive daily hemodialysis that can rapidly reverse hyperphosphatemia is associated with regression of LVH,[88-90] suggesting a potential effect of phosphate.

Finally, many studies have identified hyperphosphatemia as an independent risk factor for mortality among dialysis patients.[20,91,92] Indeed, it was these dialysis studies that initially highlighted the potential toxicity of disordered phosphorus metabolism. As described above, however, hyperphosphatemia is relatively uncommon in predialysis CKD. Thus, it is important to note that subsequent studies of predialysis and non-CKD populations confirmed an increased future risk of mortality in association with subtle increases in serum phosphate levels that were well within the normal range.[21,23,93] These findings in healthy population indirectly support the analogous results in CKD, further strengthening the body of data that promote phosphorus excess as a novel risk factor for cardiovascular disease and mortality in CKD.

Limitations of Serum Phosphate for Clinical Management of Early CKD

Despite the considerable excitement surrounding hyperphosphatemia as a risk factor for mortality and novel candidate for therapeutic intervention in CKD, there are major limitations that will likely preclude its use as a biomarker for clinical practice in the vast majority of predialysis patients. For example, the absolute differences in serum phosphate levels that conferred increased risk of mortality are too small to be used reliably in the clinical management of individual patients given that these differences are actually smaller than the diurnal and postprandial variability in serum levels across the hours of the day.[94,95] Thus, the discovery of FGF23 has not only revolutionized our understanding of mineral metabolism physiology, it has also presented a novel potential biomarker to stratify phosphorus-related risk with far greater resolution than the serum phosphate itself, especially in patients with early CKD in whom phosphate levels are routinely normal. Preliminary studies support the role of FGF23 as novel biomarker of adverse outcomes in CKD.

FGF23 Excess and Adverse Clinical Outcomes in CKD

Like phosphorus excess, increased FGF23 levels have been shown to be independently associated with a variety of adverse renal and cardiovascular outcomes. However, in most cases, the results were independent of serum phosphate levels and the reported associations were stronger for FGF23 compared to serum phosphate, highlighting the potential superiority of FGF23 as a biomarker compared to phosphate.

In the early 1980s, a link between dietary phosphate intake in CKD patients and the rate of progression to renal failure was established and more recent evidence suggests increased levels of FGF23—potentially reflecting chronically increased dietary phosphate loads—may also predict a more rapid deterioration of renal function.[96] Consistent with previous reports, in a cross-sectional evaluation of 227 nondiabetic adult patients with CKD Stages 1-4, higher serum values of the Ca x P product, PTH and FGF23 were observed with progressive CKD stages.[59,62] In a subsequent 53 month prospective analysis of 177 patients from this cohort, older age, higher protein excretion rates, lower glomerular filtration rates, higher serum phosphate, PTH and FGF23 levels were all associated with an increased rate of CKD progression, defined as a doubling of serum creatinine or the need for renal replacement therapy during the follow-up period. However, in multivariable analyses, only baseline eGFR and FGF23 were independent predictors of progression.[97] The etiology of the association between increased FGF23 and renal failure progression is not understood but, if confirmed, could reflect detrimental effects of phosphate itself, or an uncharacterized toxic effect of FGF23 on the renal parenchyma. Importantly, the effect of FGF23 was independent of serum phosphate levels suggesting that FGF23 is superior to isolated serum phosphate measurements as a biomarker of phosphorous-related toxicity.

Several recent studies examined FGF23 and common cardiovascular phenotypes in CKD.[98-102] While the results for vascular calcification are conflicting,[102,103] several reports linked higher FGF23 levels to increased left ventricular mass index and left ventricular hypertrophy in dialysis patients and earlier-stage CKD.[98,101] Additional confirmation of an FGF23-vascular disease link came from studies of healthy non-CKD patients, in which higher FGF23 levels were independently associated with impaired vascular reactivity and with increased arterial stiffness.[100] The association between increased FGF23 and left ventricular hypertrophy in predialysis CKD patients was independent of traditional risk factors and serum phosphate levels, which were not associated with left ventricular mass index.[98] In these studies, the effect of FGF23 was independent of serum phosphate levels suggesting again that FGF23 is superior to isolated serum phosphate measurements as a biomarker of phosphorous-related toxicity.

FGF23 Excess and Mortality on Dialysis

One of the largest studies to date of FGF23 and hard clinical outcomes was a prospective study of FGF23 levels at the initiation of dialysis and risk of subsequent mortality on dialysis (Fig. 4).[104] By helping to prevent or attenuate the severity of hyperphosphatemia, increased FGF23 levels would appear to be a protective compensation in CKD. However, the increased FGF23 also inhibits renal production of 1,25D leading to severe, prolonged 1,25D deficiency, which itself is a risk factor for mortality. Whereas hyperphosphatemia and low 1,25D levels are independent risk factors for mortality in CKD, whether increased FGF23 is protective or harmful in terms of mortality was unknown.

In a prospective study of 10,044 incident hemodialysis patients, the association between FGF23 and phosphate with mortality was assessed using measurements of these analytes from samples that were collected at the first outpatient hemodialysis session among patients new to dialysis.[104] All patients were included in the analysis of phosphate and mortality while the FGF23 analyses used a nested, prospective case-control sampling. Cases were patients who died during the first year on dialysis and controls were those who survived the first year. To minimize confounding by hyperphosphatemia, frequency matching was used to randomly select 50 cases and 50 controls within each quartile of baseline serum phosphate so that a final sample of 200 cases and 200 controls with balanced serum phosphate levels were selected. FGF23 was assayed in two ways and parallel analyses of each were performed: using a C-terminal (cFGF23) assay and an intact FGF23 (iFGF23) assay.

One-year mortality was 178 deaths/1000 patient-years at risk. Only a serum phosphate level in the highest quartile (>5.5 mg/dl) was independently associated with mortality compared to levels <3.5 mg/dl and the effect size was modest (HR 1.2; 95% CI 1.1, 1.4). In the nested, case-control sample (n = 400), there was strong correlation (r = 0.74, P < 0.01) between cFGF23 (median 1752; interquartile range [IQR] 1089, 4019 RU/ml) and iFGF23 (713; IQR 579, 951 pg/ml) levels. The median cFGF23 was significantly higher in patients who died versus those who survived in the overall sample (2260; IQR 1196, 5296 vs 1406; IQR 989, 2741 RU/ml; P < 0.01) and within each phosphate quartile except the highest. Increased log-cFGF23 levels were independently associated with increased risk of mortality in all models. When examined in quartiles, there was a monotonic increase in risk of mortality with increasing cFGF23 in univariate, case-mix adjusted and fully adjusted models with odds ratio of death approaching 6 in the highest vs lowest quartile.[104] Median FGF23 levels were significantly lower among blacks and Hispanics compared to whites (blacks: 1579; IQR 966, 2959; P < 0.01; Hispanics: 1336; IQR 1094, 2262; P < 0.01; vs whites: 2016; IQR

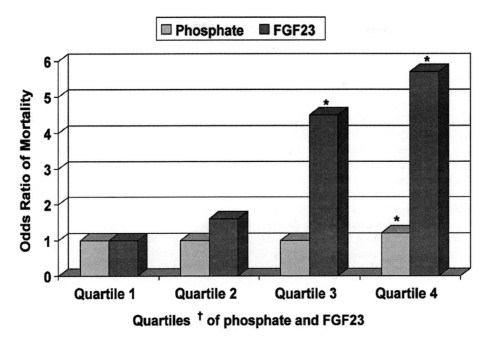

Figure 4. Increased FGF23 levels are associated with mortality in hemodialysis.

1132, 4865) and blacks with FGF23 levels below the median had significantly lower risk of mortality than comparable whites.

Several important observations emerged from this study. Increased FGF23 levels were for the first time associated with mortality and the results were independent of serum phosphate levels and other known risk factors. There was a strong "dose-response" relationship and virtually no confounding. Furthermore, the magnitude of risk associated with FGF23 was dramatically larger than the analogous results for serum phosphate, which were comparable to previous reports. These results indicate that serum phosphate levels provide only a partial assessment of risk associated with disordered phosphorus metabolism, especially when serum levels are relatively normal. In contrast, FGF23 was most informative when serum phosphate was normal. This is of great relevance to early CKD. Stage 3 CKD is the most common cause of disordered phosphorus metabolism but since the vast majority of the estimated 15 million affected in the US has normal serum phosphate levels, few are treated with phosphorus reduction strategies such as dietary phosphorus restriction and phosphorus binders that appear capable of lowering FGF23 levels.[74] These data suggest that FGF23 may be a more sensitive biomarker to help identify which normophosphatemic patients might benefit from phosphorus reduction strategies and how to titrate these therapies.

Another important aspect of this study was the tight correlation between FGF23 levels measured using either assay strategy.[104] This is important because it has been assumed that intact FGF23 assays must be preferable because C-terminal fragments could theoretically accumulate in CKD as is the case for PTH. This study and several others demonstrate highly correlated and virtually identical results using either assay.[37,97] A recent report suggests that the assays yielded qualitatively identical results because FGF23 fragments do not accumulate to a significant extent in CKD and thus, both assays measured the same dramatically increased levels of biologically active FGF23.[105]

This was also the first report of racial and ethnic differences in FGF23 levels.[104] While further validation is needed, these results are plausible as they appear to explain other known racial differences in mineral metabolism: compared to whites, blacks demonstrate decreased urinary phosphate excretion and increased serum phosphate levels despite increased PTH and significantly higher 1,25D despite significantly decreased 25D substrate. Through its effects on phosphaturia and 1α-hydroxylase, decreased FGF23 levels could account for these discrepancies.

Finally, the strong association between FGF23 and adverse outcomes raises an important question: is FGF23 excess a biomarker of another true risk factor or is it a direct uremic toxin? For example, it is possible that any impact of FGF23 on mortality may reflect effects of 1,25D deficiency, toxicity due to high phosphorus load, or klotho deficiency. Alternatively, it is possible that at markedly elevated levels, as observed in CKD, FGF23 could induce direct tissue injury that contributes to mortality, presumably through klotho-independent mechanisms. Further research is needed to dissect these possibilities.

Therapeutic Implications of FGF23 Research

The growing understanding of FGF23 physiology and its potential role as a biomarker in CKD has potentially important clinical implications: earlier institution of phosphorus-related therapies in predialysis CKD patients with normal serum phosphate but elevated FGF23 levels. Thus, it could be envisioned that FGF23 screening of early CKD patients with normal serum phosphate levels could be used to identify candidate patients for early dietary phosphorus restriction and phosphate binder therapy, just as PTH screening is justified in patients with normal calcium levels to identify those who may nevertheless benefit from initiation of active vitamin D therapy. This physiological-based approach would replace the current standard of delaying therapy for disordered phosphorus metabolism until the serum level is abnormally elevated. However, before such a strategy could be brought to clinical practice, it must be demonstrated that FGF23 levels can be safely lowered in early CKD patients with elevated levels. A handful of recent studies examined the effect of phosphate binders on FGF23 levels.[73,74] A short-term 6-wk dose titration study evaluated the effect of calcium acetate versus sevelamer hydrochloride on PTH and FGF23 levels in forty patients with CKD Stages 3-4.[74] Treatment was associated with improved control of secondary hyperparathyroidism without corresponding changes in serum phosphate levels. Sevelamer but not calcium acetate also lowered FGF23 levels. In another study, forty-six patients undergoing maintenance hemodialysis therapy were randomly treated with 3 g sevelamer hydrochloride and 3 g of calcium bicarbonate ($CaCO_3$), or 3 g of $CaCO_3$ alone.[73] Although the serum phosphate levels were comparable before treatment, the levels were significantly lower in the patients treated with sevelamer hydrochloride + $CaCO_3$ than those with $CaCO_3$ alone after 4 weeks of treatment. FGF23 levels significantly decreased after 4 weeks of the treatment with sevelamer hydrochloride + $CaCO_3$ from the pretreatment levels, while no changes were observed in the patients treated with $CaCO_3$ alone. Thus, more aggressive treatment with binders

FGF23 IN CHRONIC KIDNEY DISEASE

reduced FGF23 levels in dialysis patients, presumably by blocking intestinal phosphorus absorption. These results need to be confirmed and extended in larger and more long-term prospective studies. In predialysis patients, placebo controlled studies are justified and should be performed. Studies that focus on the effects of dietary phosphorus restriction should also be performed, in comparison to or in combination with binders. If FGF23 can be successfully lowered in this population, a potential landmark randomized trial of binders and diet versus placebo with mortality as the outcome could be envisioned on the not too distant horizon.[106]

Other therapies commonly used to treat secondary hyperparathyroidism in CKD also impact FGF23 levels, although the data are limited. Vitamin D sterol therapy, which is commonly used to treat elevated PTH levels in CKD patients, would also be expected to increase FGF23 levels, but in the two studies that examined this, the effect was modest, albeit in relatively short-term studies.[107,108] In contrast, cinacalcet hydrochloride, an allosteric activator of the calcium-sensing receptor that is also used to lower PTH levels in dialysis patients, had a modest FGF23 reducing effect in one study.[108] Whether these differences are clinically relevant or will have a meaningful impact on outcomes is unknown at present. Additional study is required.

CONCLUSION

FGF23 was originally discovered as a cause of rare rachitic disorders but its greatest clinical impact may ultimately lie in the management of disordered mineral metabolism in CKD patients. While there are many unanswered questions about the biology of FGF23 in health and its pathophysiology in CKD, gradually, a clinical role appears to be emerging and this will guide additional research. At the very least, FGF23 appears to represent a most promising biomarker for when to initiate therapy with dietary restriction or phosphate binders and could pave the way for earlier treatment of greater numbers of patients, which is the current emphasis in most aspects of CKD management. Alternatively, it is possible that FGF23 could represent much more: potentially an elusive uremic toxin with direct adverse effects on the heart, vessels, kidneys, and, perhaps, elsewhere. If that could be proven, then FGF23 itself, rather than serum phosphate or phosphorus loads, could become a direct target for intervention. More in depth knowledge into the regulation and function of FGF23 in CKD will likely improve our understanding of the pathogenesis of disordered mineral metabolism, lead to changes in its medical management and establish additional biological links between disorders of mineral metabolism and associated comorbidities in patients with progressive renal failure.

REFERENCES

1. Coresh J, Selvin E, Stevens LA et al. Prevalence of chronic kidney disease in the United States. JAMA 2007; 298:2038-2047.
2. Snyder JJ, Foley RN, Collins AJ. Prevalence of CKD in the United States: a sensitivity analysis using the National Health and Nutrition Examination Survey (NHANES) 1999-2004. Am J Kidney Dis 2009; 53:218-228.
3. USRDS: the United States Renal Data System. Am J Kidney Dis 2003; 42:1-230.
4. U S Renal Data System, USRDS 2010 Annual Data Report: Atlas of Chronic Kidney Disease and End-Stage Renal Disease in the United States, National Institutes of Health, National Institute of Diabetes and Digestive and Kidney Diseases, Bethesda, MD, 2010
5. Muntner P, He J, Hamm L et al. Renal insufficiency and subsequent death resulting from cardiovascular disease in the United States. J Am Soc Nephrol 2002; 13:745-753.

122   ENDOCRINE FGFs AND KLOTHOS

6. Levin A. Clinical epidemiology of cardiovascular disease in chronic kidney disease prior to dialysis. Semin Dial 2003; 16:101-105.
7. Schiffrin EL, Lipman ML, Mann JF. Chronic kidney disease: effects on the cardiovascular system. Circulation 2007; 116:85-97.
8. Weiner DE, Tighiouart H, Stark PC et al. Kidney disease as a risk factor for recurrent cardiovascular disease and mortality. Am J Kidney Dis 2004; 44:198-206.
9. Elsayed EF, Tighiouart H, Griffith J et al. Cardiovascular disease and subsequent kidney disease. Arch Intern Med 2007; 167:1130-1136.
10. Go AS, Chertow GM, Fan D et al. Chronic kidney disease and the risks of death, cardiovascular events and hospitalization. N Engl J Med 2004; 351:1296-1305.
11. Keith DS, Nichols GA, Gullion CM et al. Longitudinal follow-up and outcomes among a population with chronic kidney disease in a large managed care organization. Arch Intern Med 2004; 164:659-663.
12. Yu HT. Progression of chronic renal failure. Arch Intern Med 2003; 163:1417-1429.
13. Brenner BM. Remission of renal disease: recounting the challenge, acquiring the goal. J Clin Invest 2002; 110:1753-1758.
14. Schieppati A, Remuzzi G. The June 2003 Barry M. Brenner Comgan lecture. The future of renoprotection: frustration and promises. Kidney Int 2003; 64:1947-1955.
15. K/DOQI clinical practice guidelines for evaluation of chronic kidney disease: evaluation, classification and stratification. Am J Kidney Dis 2002; 39:S46-S76.
16. Stevens LA, Coresh J, Greene T et al. Assessing kidney function—measured and estimated glomerular filtration rate. N Engl J Med 2006; 354:2473-2483.
17. Stevens LA, Levey AS. Measured GFR as a confirmatory test for estimated GFR. J Am Soc Nephrol 2009; 20:2305-2313.
18. Perrone RD, Madias NE, Levey AS. Serum creatinine as an index of renal function: new insights into old concepts. Clin Chem 1992; 38:1933-1953.
19. Abboud H, Henrich WL. Stage IV chronic kidney disease. N Engl J Med 2009; 362:56-65.
20. Block GA, Klassen PS, Lazarus JM et al. Mineral metabolism, mortality and morbidity in maintenance hemodialysis. J Am Soc Nephrol 2004; 15:2208-2218.
21. Kestenbaum B, Sampson JN, Rudser KD et al. Serum phosphate levels and mortality risk among people with chronic kidney disease. J Am Soc Nephrol 2005; 16:520-528.
22. Bhuriya R, Li S, Chen SC et al. Plasma parathyroid hormone level and prevalent cardiovascular disease in CKD stages 3 and 4: an analysis from the Kidney Early Evaluation Program (KEEP). Am J Kidney Dis 2009; 53:S3-10.
23. Voormolen N, Noordzij M, Grootendorst DC et al. High plasma phosphate as a risk factor for decline in renal function and mortality in predialysis patients. Nephrol Dial Transplant 2007; 22:2909-2916.
24. Kovesdy CP, Ahmadzadeh S, Anderson JE et al. Secondary hyperparathyroidism is associated with higher mortality in men with moderate to severe chronic kidney disease. Kidney Int 2008; 73:1296-1302.
25. Slatopolsky E. The role of calcium, phosphorus and vitamin D metabolism in the development of secondary hyperparathyroidism. Nephrol Dial Transplant 1998; 13 Suppl 3:3-8.
26. Juppner H, Potts JT Jr. Immunoassays for the detection of parathyroid hormone. J Bone Miner Res 2002; 17 Suppl 2:N81-86.
27. Goodman WG, Quarles LD. Development and progression of secondary hyperparathyroidism in chronic kidney disease: Lessons from molecular genetics. Kidney Int 2007.
28. Brown AJ, Dusso A, Slatopolsky E. Vitamin D. Am J Physiol 1999; 277:F157-175.
29. Quarles LD. Endocrine functions of bone in mineral metabolism regulation. J Clin Invest 2008; 118:3820-3828.
30. Saito H, Maeda A, Ohtomo S et al. Circulating FGF-23 is regulated by 1alpha,25-dihydroxyvitamin D3 and phosphorus in vivo. J Biol Chem 2005; 280:2543-2549.
31. Shimada T, Hasegawa H, Yamazaki Y et al. FGF-23 is a potent regulator of vitamin D metabolism and phosphate homeostasis. J Bone Miner Res 2004; 19:429-435.
32. Saito H, Kusano K, Kinosaki M et al. Human fibroblast growth factor-23 mutants suppress Na+-dependent phosphate cotransport activity and 1alpha,25-dihydroxyvitamin D3 production. J Biol Chem 2003; 278:2206-2211.
33. Shimada T, Kakitani M, Yamazaki Y et al. Targeted ablation of Fgf23 demonstrates an essential physiological role of FGF23 in phosphate and vitamin D metabolism. J Clin Invest 2004; 113:561-568.
34. Ben-Dov IZ, Galitzer H, Lavi-Moshayoff V et al. The parathyroid is a target organ for FGF23 in rats. J Clin Invest 2007; 117:4003-4008.
35. Ferrari SL, Bonjour JP, Rizzoli R. Fibroblast growth factor-23 relationship to dietary phosphate and renal phosphate handling in healthy young men. J Clin Endocrinol Metab 2005; 90:1519-1524.
36. Antoniucci DM, Yamashita T, Portale AA. Dietary phosphorus regulates serum fibroblast growth factor-23 concentrations in healthy men. J Clin Endocrinol Metab 2006; 91:3144-3149.
37. Burnett SM, Gunawardene SC, Bringhurst FR et al. Regulation of C-terminal and intact FGF-23 by dietary phosphate in men and women. J Bone Miner Res 2006; 21:1187-1196.

FGF23 IN CHRONIC KIDNEY DISEASE

38. Jonsson KB, Zahradnik R, Larsson T et al. Fibroblast growth factor 23 in oncogenic osteomalacia and X-linked hypophosphatemia. N Engl J Med 2003; 348:1656-1663.
39. White KE, Jonsson KB, Carn G et al. The autosomal dominant hypophosphatemic rickets (ADHR) gene is a secreted polypeptide overexpressed by tumors that cause phosphate wasting. J Clin Endocrinol Metab 2001; 86:497-500.
40. Autosomal dominant hypophosphataemic rickets is associated with mutations in FGF23. Nat Genet 2000; 26:345-348.
41. Shimada T, Mizutani S, Muto T et al. Cloning and characterization of FGF23 as a causative factor of tumor-induced osteomalacia. Proc Natl Acad Sci USA 2001; 98:6500-6505.
42. De Beur SM, Finnegan RB, Vassiliadis J et al. Tumors associated with oncogenic osteomalacia express genes important in bone and mineral metabolism. J Bone Miner Res 2002; 17:1102-1110.
43. Benet-Pages A, Orlik P, Strom TM et al. An FGF23 missense mutation causes familial tumoral calcinosis with hyperphosphatemia. Hum Mol Genet 2005; 14:385-390.
44. Stubbs JR, Liu S, Tang W et al. Role of hyperphosphatemia and 1,25-dihydroxyvitamin D in vascular calcification and mortality in fibroblastic growth factor 23 null mice. J Am Soc Nephrol 2007; 18:2116-2124.
45. Slatopolsky E, Delmez JA. Pathogenesis of secondary hyperparathyroidism. Nephrol Dial Transplant 1996; 11 Suppl 3:130-135.
46. Slatopolsky E, Finch J, Denda M et al. Phosphorus restriction prevents parathyroid gland growth. High phosphorus directly stimulates PTH secretion in vitro. J Clin Invest 1996; 97:2534-2540.
47. Portale AA, Halloran BP, Morris RC, Jr. Physiologic regulation of the serum concentration of 1,25-dihydroxyvitamin D by phosphorus in normal men. J Clin Invest 1989; 83:1494-1499.
48. Portale AA, Halloran BP, Murphy MM et al. Oral intake of phosphorus can determine the serum concentration of 1,25-dihydroxyvitamin D by determining its production rate in humans. J Clin Invest 1986; 77:7-12.
49. Llach F, Massry SG. On the mechanism of secondary hyperparathyroidism in moderate renal insufficiency. J Clin Endocrinol Metab 1985; 61:601-606.
50. Brown AJ, Zhong M, Ritter C et al. Loss of calcium responsiveness in cultured bovine parathyroid cells is associated with decreased calcium receptor expression. Biochem Biophys Res Commun 1995; 212:861-867.
51. Fukuda N, Tanaka H, Tominaga Y et al. Decreased 1,25-dihydroxyvitamin D3 receptor density is associated with a more severe form of parathyroid hyperplasia in chronic uremic patients. J Clin Invest 1993; 92:1436-1443.
52. Mitlak BH, Alpert M, Lo C et al. Parathyroid function in normocalcemic renal transplant recipients: evaluation by calcium infusion. J Clin Endocrinol Metab 1991; 72:350-355.
53. Fukagawa M. Cell biology of parathyroid hyperplasia in uremia. Am J Med Sci 1999; 317:377-382.
54. Rodriguez M, Nemeth E, Martin D. The calcium-sensing receptor: a key factor in the pathogenesis of secondary hyperparathyroidism. Am J Physiol Renal Physiol 2005; 288:F253-264.
55. LaClair RE, Hellman RN, Karp SL et al. Prevalence of calcidiol deficiency in CKD: a cross-sectional study across latitudes in the United States. Am J Kidney Dis 2005; 45:1026-1033.
56. Levin A, Bakris GL, Molitch M et al. Prevalence of abnormal serum vitamin D, PTH, calcium and phosphorus in patients with chronic kidney disease: results of the study to evaluate early kidney disease. Kidney Int 2007; 71:31-38.
57. Young EW, Akiba T, Albert JM et al. Magnitude and impact of abnormal mineral metabolism in hemodialysis patients in the Dialysis Outcomes and Practice Patterns Study (DOPPS). Am J Kidney Dis 2004; 44:34-38.
58. Gutierrez OM, Isakova T, Andress DL et al. Prevalence and severity of disordered mineral metabolism in Blacks with chronic kidney disease. Kidney Int 2008; 73:956-962.
59. Gutierrez O, Isakova T, Rhee E et al. Fibroblast growth factor-23 mitigates hyperphosphatemia but accentuates calcitriol deficiency in chronic kidney disease. J Am Soc Nephrol 2005; 16:2205-2215.
60. Gonzalez EA, Sachdeva A, Oliver DA et al. Vitamin D insufficiency and deficiency in chronic kidney disease. A single center observational study. Am J Nephrol 2004; 24:503-510.
61. Larsson T, Nisbeth U, Ljunggren O et al. Circulating concentration of FGF-23 increases as renal function declines in patients with chronic kidney disease, but does not change in response to variation in phosphate intake in healthy volunteers. Kidney Int 2003; 64:2272-2279.
62. Shigematsu T, Kazama JJ, Yamashita T et al. Possible involvement of circulating fibroblast growth factor 23 in the development of secondary hyperparathyroidism associated with renal insufficiency. Am J Kidney Dis 2004; 44:250-256.
63. Komaba H, Fukagawa M. FGF23-parathyroid interaction: implications in chronic kidney disease. Kidney Int 2009.
64. Denda M, Finch J, Slatopolsky E. Phosphorus accelerates the development of parathyroid hyperplasia and secondary hyperparathyroidism in rats with renal failure. Am J Kidney Dis 1996; 28:596-602.
65. Lopez-Hilker S, Dusso AS, Rapp NS et al. Phosphorus restriction reverses hyperparathyroidism in uremia independent of changes in calcium and calcitriol. Am J Physiol 1990; 259:F432-437.
66. Kusano K, Segawa H, Ohnishi R et al. Role of low protein and low phosphorus diet in the progression of chronic kidney disease in uremic rats. J Nutr Sci Vitaminol (Tokyo) 2008; 54:237-243.
67. Barsotti G, Giannoni A, Morelli E et al. The decline of renal function slowed by very low phosphorus intake in chronic renal patients following a low nitrogen diet. Clin Nephrol 1984; 21:54-59.

68. Koizumi T, Murakami K, Nakayama H et al. Role of dietary phosphorus in the progression of renal failure. Biochem Biophys Res Commun 2002; 295:917-921.
69. Martinez I, Saracho R, Montenegro J et al. The importance of dietary calcium and phosphorous in the secondary hyperparathyroidism of patients with early renal failure. Am J Kidney Dis 1997; 29:496-502.
70. Portale AA, Booth BE, Halloran BP et al. Effect of dietary phosphorus on circulating concentrations of 1,25-dihydroxyvitamin D and immunoreactive parathyroid hormone in children with moderate renal insufficiency. J Clin Invest 1984; 73:1580-1589.
71. Hasegawa H, Iijima K, Shimada T et al. FGF-23 plays a critical role in the development of reduced serum 1,25-dihydroxyvitamin D levels associated with renal insufficiency (Abstract). J Am Soc Nephrol 2003; 14:40A.
72. Yamashita T. Involvement of FGF-23 in abnormal vitamin D and mineral metabolism associated with renal insufficiency (Abstract). J Am Soc Nephrol 2002; 13:577A.
73. Koiwa F, Kazama JJ, Tokumoto A et al. Sevelamer hydrochloride and calcium bicarbonate reduce serum fibroblast growth factor 23 levels in dialysis patients. Ther Apher Dial 2005; 9:336-339.
74. Oliveira RB, Cancela ALE, Graciolli FG et al. Early control of PTH and FGF23 in normophosphatemic CKD patients: a new target in CKD-MBD therapy? CJASN 2009; In Press.
75. Nagano N, Miyata S, Abe M et al. Effect of manipulating serum phosphorus with phosphate binder on circulating PTH and FGF23 in renal failure rats. Kidney Int 2006; 69:531-537.
76. McClellan W, Warnock DG, McClure L et al. Racial differences in the prevalence of chronic kidney disease among participants in the Reasons for Geographic and Racial Differences in Stroke (REGARDS) Cohort Study. J Am Soc Nephrol 2006; 17:1710-1715.
77. Xue JL, Eggers PW, Agodoa LY et al. Longitudinal study of racial and ethnic differences in developing end-stage renal disease among aged medicare beneficiaries. J Am Soc Nephrol 2007; 18:1299-1306.
78. K/DOQI clinical practice guidelines for bone metabolism and disease in chronic kidney disease. Am J Kidney Dis 2003; 42:S1-201.
79. (KDIGO) CKD-MBD Work Group. KDIGO clinical practice guideline for the diagnosis, evaluation, prevention and treatment of Chronic Kidney Disease-Mineral and Bone Disorder (CKD-MBD). Kidney Int Suppl 2009:S1-130.
80. Blacher J, Asmar R, Djane S et al. Aortic pulse wave velocity as a marker of cardiovascular risk in hypertensive patients. Hypertension 1999; 33:1111-1117.
81. Klassen PS, Lowrie EG, Reddan DN et al. Association between pulse pressure and mortality in patients undergoing maintenance hemodialysis. JAMA 2002; 287:1548-1555.
82. Cozzolino M, Brancaccio D, Gallieni M et al. Pathogenesis of vascular calcification in chronic kidney disease. Kidney Int 2005; 68:429-436.
83. Blacher J, Guerin AP, Pannier B et al. Arterial calcifications, arterial stiffness and cardiovascular risk in end-stage renal disease. Hypertension 2001; 38:938-942.
84. Goodman WG, Goldin J, Kuizon BD et al. Coronary-artery calcification in young adults with end-stage renal disease who are undergoing dialysis. N Engl J Med 2000; 342:1478-1483.
85. Moe SM, O'Neill KD, Reslerova M et al. Natural history of vascular calcification in dialysis and transplant patients. Nephrol Dial Transplant 2004; 19:2387-2393.
86. Adeney KL, Siscovick DS, Ix JH et al. Association of serum phosphate with vascular and valvular calcification in moderate CKD. J Am Soc Nephrol 2009; 20:381-387.
87. Jono S, McKee MD, Murry CE et al. Phosphate regulation of vascular smooth muscle cell calcification. Circ Res 2000; 87:E10-17.
88. Achinger SG, Ayus JC. Left ventricular hypertrophy: is hyperphosphatemia among dialysis patients a risk factor? J Am Soc Nephrol 2006; 17:S255-261.
89. Ayus JC, Mizani MR, Achinger SG et al. Effects of short daily versus conventional hemodialysis on left ventricular hypertrophy and inflammatory markers: a prospective, controlled study. J Am Soc Nephrol 2005; 16:2778-2788.
90. Culleton BF, Walsh M, Klarenbach SW et al. Effect of frequent nocturnal hemodialysis vs conventional hemodialysis on left ventricular mass and quality of life: a randomized controlled trial. JAMA 2007; 298:1291-1299.
91. Young EW, Albert JM, Satayathum S et al. Predictors and consequences of altered mineral metabolism: the Dialysis Outcomes and Practice Patterns Study. Kidney Int 2005; 67:1179-1187.
92. Kalantar-Zadeh K, Kuwae N, Regidor DL et al. Survival predictability of time-varying indicators of bone disease in maintenance hemodialysis patients. Kidney Int 2006; 70:771-780.
93. Menon V, Greene T, Pereira AA et al. Relationship of phosphorus and calcium-phosphorus product with mortality in CKD. Am J Kidney Dis 2005; 46:455-463.
94. Isakova T, Gutierrez O, Shah A et al. Postprandial mineral metabolism and secondary hyperparathyroidism in early CKD. J Am Soc Nephrol 2008; 19:615-623.
95. Markowitz M, Rotkin L, Rosen JF. Circadian rhythms of blood minerals in humans. Science 1981; 213:672-674.
96. Haut LL, Alfrey AC, Guggenheim S et al. Renal toxicity of phosphate in rats. Kidney Int 1980; 17:722-731.

FGF23 IN CHRONIC KIDNEY DISEASE

97. Fliser D, Kollerits B, Neyer U et al. Fibroblast growth factor 23 (FGF23) predicts progression of chronic kidney disease: the Mild to Moderate Kidney Disease (MMKD) Study. J Am Soc Nephrol 2007; 18:2600-2608.
98. Gutierrez OM, Januzzi JL, Isakova T et al. Fibroblast growth factor 23 and left ventricular hypertrophy in chronic kidney disease. Circulation 2009; 119:2545-2552.
99. Mirza MA, Larsson A, Melhus H et al. Serum intact FGF23 associate with left ventricular mass, hypertrophy and geometry in an elderly population. Atherosclerosis 2009; 207:546-551.
100. Mirza MA, Larsson A, Lind L et al. Circulating fibroblast growth factor-23 is associated with vascular dysfunction in the community. Atherosclerosis 2009; 205:385-390.
101. Hsu HJ, Wu MS. Fibroblast growth factor 23: a possible cause of left ventricular hypertrophy in hemodialysis patients. Am J Med Sci 2009; 337:116-122.
102. Jean G, Terrat JC, Vanel T et al. High levels of serum fibroblast growth factor (FGF)-23 are associated with increased mortality in long haemodialysis patients. Nephrol Dial Transplant 2009; 24:2792-2796.
103. Kojima F, Uchida K, Ogawa T et al. Plasma levels of fibroblast growth factor-23 and mineral metabolism in diabetic and nondiabetic patients on chronic hemodialysis. Int Urol Nephrol 2008; 40:1067-1074.
104. Gutierrez OM, Mannstadt M, Isakova T et al. Fibroblast growth factor 23 and mortality among patients undergoing hemodialysis. N Engl J Med 2008; 359:584-592.
105. Shimada T, Urakawa I, Isakova T et al. Circulating Fibroblast Growth Factor 23 in Patients with End-Stage Renal Disease Treated by Peritoneal Dialysis Is Intact and Biologically Active. J Clin Endocrinol Metab 2009.
106. Isakova T, Gutierrez OM, Chang Y et al. Phosphorus binders and survival on hemodialysis. J Am Soc Nephrol 2009; 20:388-396.
107. Nishi H, Nii-Kono T, Nakanishi S et al. Intravenous calcitriol therapy increases serum concentrations of fibroblast growth factor-23 in dialysis patients with secondary hyperparathyroidism. Nephron Clin Pract 2005; 101:c94-99.
108. Wetmore JB, Liu S, Krebill R et al. Effects of cinacalcet and concurrent low-dose vitamin D on FGF23 levels in ESRD. Clin J Am Soc Nephrol 2010;5(1):110-6.

CHAPTER 9

SECRETED KLOTHO AND CHRONIC KIDNEY DISEASE

Ming Chang Hu, Makoto Kuro-o and Orson W. Moe*

Charles and Jane Pak Center for Mineral Metabolism and Clinical Research, Departments of Internal Medicine, Physiology, Pediatrics, and Pathology, University of Texas Southwestern Medical Center, Dallas, Texas, USA.
*Corresponding Author: Orson W. Moe—Email: orson.moe@utsouthwestern.edu

Abstract: Soluble Klotho (sKl) in the circulation can be generated directly by alterative splicing of the Klotho transcript or the extracellular domain of membrane Klotho can be released from membrane-anchored Klotho on the cell surface. Unlike membrane Klotho which functions as a coreceptor for fibroblast growth factor-23 (FGF23), sKl, acts as hormonal factor and plays important roles in anti-aging, anti-oxidation, modulation of ion transport, and Wnt signaling. Emerging evidence reveals that Klotho deficiency is an early biomarker for chronic kidney diseases as well as a pathogenic factor. Klotho deficiency is associated with progression and chronic complications in chronic kidney disease including vascular calcification, cardiac hypertrophy, and secondary hyperparathyroidism. In multiple experimental models, replacement of sKl, or manipulated up-regulation of endogenous Klotho protect the kidney from renal insults, preserve kidney function, and suppress renal fibrosis, in chronic kidney disease. Klotho is a highly promising candidate on the horizon as an early biomarker, and as a novel therapeutic agent for chronic kidney disease.

INTRODUCTION

As one approach almost one and one-half decade of research since its discovery in 1997, Klotho biology is eventually entering into the public health domain where its dysregulation is implicated in human pathophysiology, and its diagnostic and therapeutic potential are beginning to enter the realm of reality. We will take a preclinical viewpoint of the status of Klotho in this aspect using kidney disease as a vehicle. Klotho is a single transmembrane 130 KDa protein encoded by the *Klotho* gene. A secreted form of Klotho of 70 kD is a product of alternative splicing[4-7] and the extracellular domain

---

Endocrine FGFs and Klothos, edited by Makoto Kuro-o.
©2012 Landes Bioscience and Springer Science+Business Media.

SECRETED KLOTHO AND CHRONIC KIDNEY DISEASE

of membrane Klotho can be released into blood[8,9] thus functioning as a circulating substance to exert multiple systemic biological actions on distant organs.[10-13] This cleaved extracellular domain of membrane Klotho is referred as soluble Klotho (sKl) in this chapter. sKl could function as a β-glucuronidase[7,14] and sialidase.[15,16] Klotho is principally synthesized in kidney and brain although it is expressed in multiple organs.[6,17] Klotho-deficient mice ($Kl^{-/-}$) manifest multi-organ premature aging while over-expression of Klotho through virus-mediated delivery or genetic overexpression can rescue the Klotho-deficient phenotype at baseline[6,19,20] and enhance mice's resistance to oxidant stress and ischemic damage.[6,21-23]

One important biological function of Klotho is to maintain mineral homeostasis, which is both fibroblast growth factor-23 (FGF23)-dependent and FGF23-independent. From a renal point of view, sKl regulates urinary calcium,[16,24] potassium[15] and phosphorus excretion.[14] In addition, Klotho suppresses 1α-hydroxylase in the kidney to regulate calcium metabolism[25,26] and participates in the regulation of PTH synthesis in parathyroid gland by FGF23.[27,28]

The study of Klotho in human biology has been limited to association studies of clinical features with Klotho polymorphisms with potential but yet to be determined significance.[29-36] The association of genetic variations in the *KLOTHO* gene with high mortality in hemodialysed patients was shown, but this link could be modified by activated vitamin D supplementation.[38] This modification of association of nucleotide polymorphisms in the *KLOTHO* gene may result from epigenetic upregulation of Klotho expression by Vitamin D[26] and subsequently affects the outcome of CKD patients. Interestingly, unlike the myriads of phenotypic features described in the rodent models of Klotho deficiency or excess, both Klotho deficiency caused by loss-of-function missense mutation[45] and presumed Klotho overexpression due to translocation in human *Klotho* gene[47] display disturbances in mineral metabolism but surprisingly both deficient and excess states have high FGF23 and hyperparathyroidism. These findings support the important role of Klotho in mineral homeostasis and suggest that there is still higher level of complexity that is beyond our comprehension at the moment.

In this book, there are several comprehensive reviews addressing Klotho's effect on aging, renal ion channels and transporter, and Pi homeostasis. This chapter will be mainly focused on the role of sKl in chronic kidney diseases and potential clinical implications.

KIDNEY: THE MAJOR SITE FOR KLOTHO SYNTHESIS

The association of kidney disease with Klotho can be view as intuitive considering the expression pattern of Klotho. The distribution of *Klotho* mRNA is restricted to few tissues with the strongest expression in the kidney and weaker ones in the brain, heart, parathyroid gland and testis.[6,48] RT-PCR detects *Klotho* expression in more tissues including aorta, colon, pituitary gland, thyroid gland, pancreas, and gonads but the kidney still has the strongest Klotho expression. From the onset, the findings suggest that the kidney might be a major source of endocrine Klotho. The higher Klotho protein in the suprarenal vein than that in infrarenal vein in rodents (personal observation) and the low circulating plasma levels observed in rodent with renal failure[2] strongly suggests that the kidney is a main source for Klotho in the blood circulation (Fig. 1A).

In mammalian kidney including mouse, rat and human, Klotho is prominently expressed in distal convoluted tubules (DCT),[17] but Klotho is also unequivocally found

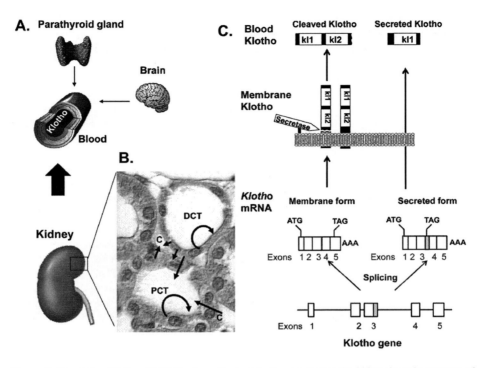

Figure 1. Circulating Klotho. A) Klotho is synthesized in few organs but the kidney is main resource of circulating blood Klotho. Whether parathyroid gland and brain contribute significantly to circulating Klotho is not clear. B) Both renal proximal convoluted tubules (PCT) and distal convoluted tubules (DCT) produce Klotho. Klotho is released into the capillaries (C) and systemic circulation. The relative contribution of these two segments to circulating Klotho is unknown. Conversely, Klotho from the blood may enter the urine potentially via transcytosis of the tubular epithelium. Luminal Klotho from the PCT can travel axially down the nephron to the DCT lumen. Klotho from PCT cells and/or DCT cells may also exert paracrine actions on tubular epithelium. C) Membrane form of Klotho transcript arises from Klotho gene. Secreted Klotho be derived from alternative RNA splicing. The internal splice donor site is in exon 3 of Klotho gene. The resultant alternatively spliced transcript contains 50 bp insertion after exon 3 (grey) with an in-frame translation stop codon introduced. The product contains only Kl1 and is released into blood circulation. On the other hand, the Klotho protein encoded by membrane form of Klotho transcript is a plasma membrane-anchored protein in Klotho producing cells. Extracellular domain of membrane Klotho containing Kl1 and Kl2 repeats is shed and cleaved by α and β-secretases, and released into blood stream. Thus in blood circulation, there are two forms of Klotho, one is derived from cleavage of the extracellular domain of membrane Klotho. Another one is secreted membrane derived from an alternatively spliced Klotho transcript.

in the proximal convoluted tubule (PCT) although in lower levels compared to DCT.[14] The presence of *Klotho* mRNA is confirmed by RT-PCR in microdissected rat nephron segments, and in OK cells, a renal proximal tubule-like cell line. The PCT mRNA rules out Klotho protein uptake as an explanation for Klotho antigen in the PCT (Fig. 1B).[14]

Klotho protein is present in cerebrospinal fluid (CSF),[49] blood[2,49,50] and urine of mammals.[2,14,50] This soluble Klotho protein of 130 kDa by SDS-PAGE[49] is different from the secreted Klotho encoded by a spliced transcript of Klotho with a predicted size of 70 kDa.[4,5] The extracellular domain of Klotho is shed from full length membrane Klotho protein by α-secretase, ADAM (a disintegrin and metalloprotease) 10 and 17 and β-secretase, β-APP cleaving enzyme 1, and released into blood circulation

SECRETED KLOTHO AND CHRONIC KIDNEY DISEASE 129

(Fig. 1C).[9] The cleavage and release of the extracellular domain of Klotho by ADAM10 and ADAM17 is stimulated by insulin and inhibited by metalloproteinase inhibitors.[8] ADAM 10 is expressed in the DCT and is potentially perfectly poised to cleave Klotho.[51] To date, whether other molecules are also able to regulate membrane Klotho shedding is not known. The substrates for ADAM 10/17 are massively broad and what regulates ADAM10/17-mediated ectodoamin shedding requires much more studies. This is a serious void in the database as the control of release from the parental pre-molecule is a crucial step for any endocrine substance.

sKl plays important roles in a variety of physiological and pathological processes including modulation of Wnt signal transduction,[11] anti-oxidation[52] and renal ion channels[15,16,24] and transporters.[14] Another possible role of the cleaved extracellular domain of Klotho may act to modify the signal transduction role played by membrane Klotho as a coreceptor for FGF23 but there is no data to date to support Klotho's ability to quench soluble FGF23 (as a decoy) or to block the Klotho-FGF receptor complex (as a competitive inhibitor).

In addition to the final urine, Klotho protein is detected in the proximal tubule urine.[14] This location of Klotho is absolutely necessary for it to function as a phosphaturic substance.[14] Currently, the origin of this proximal luminal Klotho is unclear. It may be directly cleaved from proximal tubular cell or derived from the pertibular blood circulation via transcytosis (Fig. 1B). The only route that cannot be true is glomerular filtration as Klotho is too large for the glomerular basement membrane selectivity filter. Indeed, there is no detectable Klotho antigen in the glomerular Bowman space proto-urine (personal observation).

The kidney is a principal organ in calcium[16,24,53] and phosphate[14] homeostasis and directly modulates blood 1,25-$(OH)_2$ Vitamin D_3, and indirectly PTH, and FGF23 levels.[25,27,28,54-56] In addition to being commanded by calciotropic hormones as an operative, the kidney is endowed with the calcium sensing receptor[57] which renders it a self-sufficient afferent and efferent organ for calcium homeostasis. Although the identity of the phosphate sensor is elusive, the proximal tubule is clearly responding to ambient phosphate concentrations[58] and exerts it effect by regulating Na-phosphate cotransporters directly[59] and 1α-hydroxylase expression and activity.[60] Although 1α-hydroxylation is present in many organs,[60] the proximal tubule activity is the most important and regulated.[61] Klotho protein is poised in the central place of this network of mineral metabolism homeostasis where it is in direct apposition with the sensors and effectors. The physical proximity and functional coupling of the glomerulus, proximal, and distal tubule renders this ensemble a perfect locale for a paracrine, autocrine, and endocrine hormone.

When normal rats were subjected to ischemia-reperfusion injury, adenovirus-mediated Klotho gene delivery results in significantly improved serum creatinine, dramatically ameliorated renal histological changes, and remarkably diminished apoptotic cells in the kidneys.[23] It is important to note that the gene transfer occurred only in the liver and not in kidney. This experimental finding strongly suggests that Klotho functions as a circulating substance to exert renoprotection.

CHRONIC KIDNEY DISEASE: A STATE OF KLOTHO DEFICIENCY

As a general principle, if the organ of origin of an endocrine substance is diseased, it is logical to suspect that endocrine deficiency of that substance ensues. There are many similar features between the clinical manifestations of CKD with the phenotypes of $Kl^{-/-}$ mice (Table 1), which is the first suggestion that CKD might be a state of Klotho deficiency. Experimental data and clinical findings have thus far supported this view. There is a significant reduction of renal Klotho transcript and protein in CKD with varied etiology including ischemic perfusion injury, subtotal nephrectomy, oxidant stress, and exposure to angiotensin II (Ang II) and calcineurin inhibitors, and also in human CKD including chronic glomerulonephritis, obstructive nephropathy, diabetic nephropathy, and chronic graft rejection (Table 2). Hyperphosphatemia is a prominent abnormality in CKD, and is a major contributor to cardiovascular disease, which accounts for significant mortality in this population. Severe hyperphosphatemia is also found in Klotho-deficient mice.[6,14] High FGF23 level in the blood is a feature in patients with CKD,[62,63] and could contribute to decline of the 1α-hydroxylase activity, hyperparathyroidism, bone disease and cardiac hypertrophy in CKD.[63,64] This can be viewed as a revisited version of the "trade-off hypothesis" in CKD. High plasma FGF23 is also observed in Klotho-deficient mice.[65] In patients with CKD, ectopic calcification is frequently encountered.[66,67] Interestingly, $Kl^{-/-}$ mice have very severe and wide-spread ectopic calcifications including vascular calcification.[2,6]

A hypothesis can be raised that the Klotho deficiency may be one of the factors triggering complications in CKD and correction of Klotho deficiency may be a therapeutic possibility for treatment of kidney diseases.

Renal Klotho Deficiency

Aging Kidney

Aging is a complicated and chronic degenerative biologic process that affects kidney function and morphology. Aging is associated with decline in kidney function[68] coincident with a progressive decrease in the number of functioning nephrons and increase in glomerular and tubulointerstitial scarring.[69] These changes may cause secondary abnormal systemic hemodynamics, and disturbed mineral and hormonal homeostasis.[70]

The aged kidney is more prone to ischemic injury and nephrotoxin.[71,72] Epidemiological data clearly shows that CKD is common in the elderly and associated with higher morbidity and mortality. More than 40% Canadian residents at 65 years or older has

Table 1. Similar clinical characteristics between *Klotho*$^{-/-}$ animals and CKD subjects

	Klotho$^{-/-}$	CKD
Blood creatinine	↑	↑↑↑
Renal Klotho expression	↓↓↓	↓↓
Serum Pi	↑↑↑	↑↑
Serum FGF23	↑↑↑	↑↑↑
Ectopic calcification	Present	Present
Artherosclerosis	Present	Present
Growth	Severe retardation	Retardation in children
Anemia	No or mild	Severe
Life span	↓↓↓	↓↓

Table 2. Endocrine and Renal Klotho deficiency in chronic kidney disease in humans and in rodent models

Category of Kidney Disease	Cause or Animal Model	Renal Klotho Protein		Renal Klotho mRNA		Blood Klotho Assay	Urine Klotho Assay	Human or Rodent			Citation
		IB	IHC	NB	qPCR	Assay	Assay	Mouse	Rat	Human	
CKD	CGN	✓	✓	✓						✓	1
CKD	DN	✓	✓	✓						✓	1
Graft rejection	CGR	✓	✓	✓						✓	1
CKD	Npx+IRI	✓	✓	✓	✓	IP + IB	IB	✓	✓		2
CKD	5/6 Npx	✓		✓					✓		3
CKD	ICGN				✓						18
CKD	Npx in apo-E$^{-/-}$	✓		✓	✓	ELISA		✓			37
Hypertension	SHR			✓					✓		3
Hypertension	DOCA			✓					✓		3
Hypertension	Ang II	✓	✓	✓					✓		21
Hypertension	Ang II	✓		✓					✓		39
DM	OLETF			✓					✓		3
DM	Streptozotocin	✓						✓			40
Nephrotoxicity	CsA	✓	✓		✓			✓			41
Nephrotoxicity	FK	✓	✓		✓			✓			41
Nephrotoxicity	SRL+CsA or FK	✓	✓		✓			✓			41
Nephrotoxicity	CsA	✓	✓		✓						42
Childrn/adults	Normal					ELISA				✓	43
Age/aged kidney	Normal		✓						✓		44
Age/aged kidney	Normal	✓						✓			46

CGN: Chronic glomerulonephritis; CsA: cyclosporine A; DM: Diabetes mellitus; DN: Diabetic nephropathy; DOCA-salt: Deoxycorticosterone acetate and high salt intake (DOCA-Salt); FK: FK-506, or called Tacrolimus; IB: Immunoblot; ICGN: ICR-derived spontaneous glomerulonephritis; IHC: Immunohistochemistry; IP: immunoprecipitation; NB: Northern blot; Npx: nephrectomy; OLETF: the Otsuka Long-Evans Tokushima Fatty Rat; RT: Reverse transcription; SHR: Spontaneous hypertension; SRL: Sirolimus; qPCR: quantitative or real time PCR.

moderate CKD (estimated GFR < 60 ml/min/1.73 m²). Among American CKD patients, about 30% are from age 75 to 84 and 50% of patients of age 85 to 100 have at least moderate CKD.[73]

Aged mice (29 months) have low renal Klotho protein expression compared to young mice (4 weeks).[46] Furthermore, aged rats (male, 27 months) have significantly higher serum creatinine than that of young rats (12-months). Notably, aged rats have significantly decreased renal Klotho protein levels along with increase in oxidative stress, over production of proinflammatory cytokine and activation of endothelin signal transduction.[44]

In one report using a newly developed human plasma Klotho assay, an inverse relationship was found between the plasma Klotho levels and age.[43] One possibility is that the physiologic decrease in renal function with age reduces Klotho expression and shedding into circulation, which consequently promote multi-organ senescence. On the other hand, age-related renal Klotho decline renders the kidneys more susceptible to insults such as hypertension, ischemia, and nephrotoxins. However regardless of how Klotho is down-regulated with aging, Klotho supplementation has potential beneficial impact in renoprotection and slowing of aging.

Rodent CKD from Ablation of Renal Mass

CKD rats with subtotal (5/6) nephrectomy have hypertension, proteinuria, azotemia, anemia, lower urinary concentrating ability[74] and vascular calcification.[75] 5/6th nephrectomized rats have low *Klotho* mRNA expression in the kidney at 8 weeks after surgery.[3] Low renal Klotho protein and mRNA expression is also demonstrated in another rodent model generated by unilateral nephrectomy plus contralateral ischemic reperfusion injury followed by high dietary phosphate intake.[2] Renal histology reveals the same changes as those in the 5/6th subtotal ablation model. In addition, animals have high plasma creatinine concentration, hyperphosphatemia, anemia, and ectopic calcification in soft tissues including the kidney, aorta, heart and stomach.

Similar findings are observed in apolipoprotein E-deficient (apo-E$^{-/-}$) uremic mice induced by nephrectomy plus electrocautery.[37] Furthermore, apo-E$^{-/-}$ uremic mice have lower blood Klotho than apo-E$^{-/-}$ non-uremic mice.[37]

Immune-Mediated Chronic Glomerulonephritis

In addition to CKD model by ablation of renal mass, immune-mediated chronic glomerulonephritis, which is a major cause of CKD in humans, also have reduced renal Klotho mRNA.[18] Imprinting Control Region (ICR) strain-derived mice with spontaneous chronic glomerulonephritis due to mutation in *Tensin2* have shorter life span, which is reversed by Klotho overexpression. Klotho overexpression improves renal function, and ameliorates renal histology. The improvements are associated with less superoxide anion generation and lipid peroxidation, and with decreased levels markers of cell senescence, mitochondrial DNA fragmentation, and apoptosis. Thus Klotho protein might serve as a renoprotective factor by diminishing oxidative stress, cell senescence and apoptosis.[18]

Metabolic Syndrome and Diabetes

The association of kidney disease with the metabolic syndrome is known in humans and rodents.[76-78] OLETF rat is a rodent model for the metabolic syndrome.[79,80] OLETF rats have lower level of *Klotho* mRNA in the kidneys than age-matched control rats.[3,80]

The administration of thiazolidinedione to OLETF rats significantly increases renal *Klotho* mRNA expression, and attenuates abnormal lipid and glucose metabolism and reduces systolic blood pressure.[80] The effects of thiazolidinedione on OLETF rats is also observed in OLETF rats overexpressing the *Klotho* gene, suggesting that the capacity of thiazolidinedione to restore the phenotypes in OLETF rats may results from an increase in renal Klotho expression.[80]

In rats with streptozotocin-induced diabetes,[40] Klotho protein in the kidneys is notably decreased along with kidney destruction.[40] Both insulin and phloridzin corrects hyperglycemia, reverses the reduced renal Klotho expression, and improves kidney function and histology of diabetic rats. Klotho protein in Madin-Darby Canine Kidney cell (MDCK) is reduced in vitro by incubation in high glucose medium. Insulin has been shown to participate in the shedding of extracellular domain of Klotho,[8] which may increase the blood Klotho concentrations.

Hemodynamic Effect in Animal Models

Emerging data suggest a role of *Klotho* in hypertension. In spontaneous hypertensive rats and in rats administered deoxycorticosterone acetate plus high salt intake, which is a volume-dependent type of hypertension, renal *Klotho* mRNA expression is significantly reduced compared to normotensive control rats.[3] Hypotension *per se* does not seem to affect renal Klotho, because *Klotho* gene and protein are not decreased in the kidneys of rats with low systolic blood pressure due to myocardial infarction or phlebotomy.[3,81] However, decreased renal Klotho expression is observed in rats with hypotension induced by lipopolysaccharide (LPS) injection.[81] Thus the upstream regulator is more likely inflammatory mediators including TNF-α and IFN-γ released during LPS injection than hypotension (Table 3).[82]

Calcineurin Inhibitor Related Nephrotoxicity

Calcineurin inhibitors (CNI's) such as cyclosporine A (CsA) and tacrolimus (also called FK-506, FK) were introduced for immunosuppression in organ transplantation.[83,84] CNI's are also widely used for treatment of immune-mediated nephrotic syndrome and glomerulonephritis.[85,86] But CNI-induced nephrotoxicity is frequently a cumbersome limitation for its clinical utilization.[87,88] Mice treated with CsA or FK have reduced renal *Klotho* mRNA, and protein, and increased urinary 8-hydroxy-2'-deoxyguanosine (8-OHdG), an oxidative stress marker, excretion compared with vehicle-treated mice. In addition, there is a strong inverse correlation between Klotho protein levels and urinary 8-OHdG excretion. The CNIs-induced down-regulation of renal Klotho expression is correlated with increased renal angiotensinogen and renin expression, tubulointerstitial fibrosis and urinary 8-OHdG excretion, which are reversed by angiotensin II Type I receptor blocker, Losartan.[42] These results suggest that Ang II-induced Klotho down regulation in the kidneys may be involved in CsA nephropathy.

Moreover, combination of CNI's with sirolimus (SRL), commonly used to enhance immunosuppression for renal transplantation[89] accelerates CNI-induced oxidative stress and down-regulates renal Klotho expression in the kidney.[41] Those results may alarm the higher nephrotoxicity induced by mTOR and CNI's combination.

Table 3. Factors that downregulate renal Klotho in CKD setting

---

Loss of functional kidney mass
Abnormal cytokine production
        TNF-α ↑
        IFN-γ ↑
Oxidative stress
        H_2O_2 ↑
        Lipid peroxidation ↑
Over activation of renal RAS
        Ang II ↑
Abnormal hormone
        1,25-VD₃ ↓
Disturbed mineral metabolism
        Pi overload
        Blood Pi ↑
Small uremic toxin
        Indoxyl sulphate ↑

Human CKD

Thus far, measurements of blood Klotho in human CKD have been limited. Renal expression of the *Klotho* gene is reported to be markedly decreased in patients with CKD (Table 2). The levels of Klotho mRNA and protein are reduced in nephrectomy samples from patients with CKD or end stage renal disease (ESRD), obstructive nephropathy, rejected transplanted kidneys, diabetic nephropathy and chronic glomerulonephritis.[1] Klotho transcript and protein expression in kidneys from human and animal is clearly decreased. However, the relationship between the renal Klotho expression and the plasma Klotho levels remains to be studied and the mechanisms underlying the relationship between renal Klotho down-regulation and kidney diseases in humans remain unclear. These are studies of the highest priority at this juncture.

Endocrine Klotho Deficiency in CKD

The apo-E⁻/⁻ 5/6th nephrectomized mice have decreased plasma Klotho and this decline increases with age. Furthermore, the reduction of plasma Klotho was more dramatic compared to non-uremic apo-E⁻/⁻ mice.[37] In another CKD model of uninephrectomy plus contralateral ischemia reperfusion in mice and rats, plasma Klotho concentration was remarkably decreased in CKD compared to Sham animals. The degree of decreased plasma Klotho is similar in magnitude to that of decreased Klotho protein in the kidneys and in the urine.[2] Transgenic mice with Klotho overexpression also have reduction in their renal, blood and urine Klotho levels with CKD but still maintain levels comparable to wild type mice without kidney disease (Fig. 2) and have better kidney function and less vascular calcification.[2] Soft tissue calcium content is inversely related to Klotho levels. Klotho overexpressing mice have lower, and haploinsufficient $Kl^{+/-}$ mice have higher calcium contents than WT mice do at baseline and after induction of CKD.[2] Thus

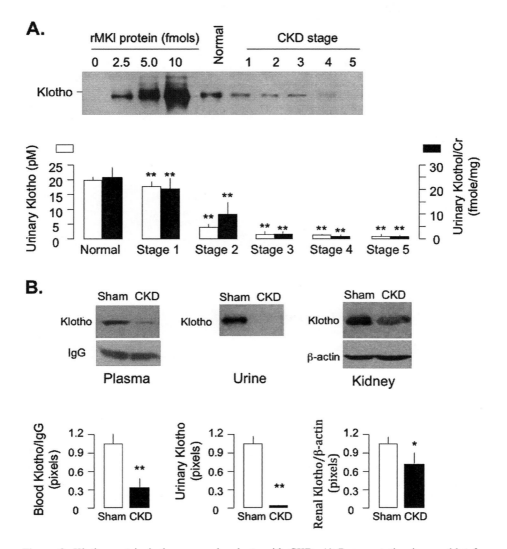

Figure 2. Klotho protein in humans and rodents with CKD. A) Representative immunoblot for Klotho protein with serial dilutions of known concentration of recombinant mouse Klotho (rMKl) as standard and concentrated human urine samples of identical amount of creatinine. Lower panel shows summary of urinary Klotho protein concentration (open bars) and of Klotho normalized by creatinine (black bars) of normal subjects and CKD patients. Significant difference when $P < 0.05$ between groups analyzed by one-way ANOVA followed by Student-Newman-Keuls test. *: $P < 0.05$, **: $P < 0.01$ vs. Normal. B) Klotho levels in plasma, urine, and kidney in a murine model of CKD. Representative blots of Klotho protein in plasma, urine and kidney in CKD mice. Co-IP of Klotho in 100 μl of mouse serum was followed by immunoblot. IgG heavy chain = loading control. Urine Klotho was examined by directly immunoblotting ~40 μl of animal fresh urine with identical amount of creatinine. Klotho protein in the kidney was analyzed by immunoblotting 30 μg of the total kidney lysate. Lower panel shows summary of blood, urinary and renal Klotho protein of Sham (open bars) and CKD (black bars) mice. Quantitative data represented as means ± SEM and analyzed by the unpaired t-test. *$P < 0.05$, **$P < 0.01$ between Sham and CKD mice.

in rodents, CKD is a state of endocrine Klotho deficiency in addition to renal Klotho deficiency, the systemic Klotho levels efficiently affect the renal function preservation as well as calcification development in CKD animals.[2]

Equivalent data in human plasma Klotho are yet to be acquired. Urinary Klotho levels of CKD patients are significantly decreased and this decrease starts at very early stage and sustainably reduced with progression of CKD (Fig. 2).[2] Thus far, rodent plasma, kidney and urine Klotho appears to covary rather tightly.[2] To date, one ELISA kit for plasma Klotho measurement is available and using this kit, plasma Klotho levels are found to be negatively related to the plasma creatinine concentration in children and adults without kidney disease.[43] Further confirmation is required.

ROLE OF KLOTHO DEFICIENCY IN PROGRESSION AND COMPLICATIONS OF CKD

In rodents, endocrine and renal Klotho deficiency is striking in CKD. The most important question is whether this is a mere biomarker of the presence of CKD or whether the Klotho deficiency is a pathogenetic factor for CKD development, progression, and complications.[2]

Potential Mechanism of Klotho Deficiency

Renal, plasma, and urine Klotho are decreased very acutely upon ischemic-reperfusion injury[50] and in CKD, it is one of the earliest abnormality.[2] This begs the question as to how Klotho is decreased with kidney injury.

Phosphotoxicity

Klotho is a phosphaturic hormone. Klotho deficiency impairs phosphaturia,[14] and consequently accelerates Pi accumulation in CKD. The higher level of serum Pi, the greater the degree of soft tissue calcification, and the greater risk of mortality.[90] Control of serum Pi significantly improves survival in CKD and ESRD patients, decreases vascular calcification, and suppresses proliferation of parathyroid glands.[90-92] There is little doubt that the prevailing thought is that Pi is one of "uremic toxins" but the concept of phosphotoxicity extends beyond just CKD.[93]

Phosphate overload suppresses Klotho expression in the kidney. Normal mice fed high Pi diet have dramatically decreased Klotho protein and mRNA in the kidney, while Klotho hypomorph mice fed low Pi diet regain part of their Klotho expression.[94] Whether the blood Klotho concentration is regulated by physiologic changes in dietary Pi is yet to be determined.

Vitamin D Deficiency

Low $1,25\text{-}(OH)_2\text{-vitamin } D_3$ is a major component of disturbance of mineral metabolism, and is conventionally attributed to cause bone disease and secondary hyperparathryoidism in CKD.[95] The increase in plasma FGF23 in CKD plays an important

SECRETED KLOTHO AND CHRONIC KIDNEY DISEASE 137

role in suppression of 1α-hydroxylase in the kidney and initiates or accelerates vitamin D deficiency.[96,97] Importantly, administration of 1,25-(OH)$_2$-vitamin D$_3$ induces klotho expression in the kidney.[26] It is plausible that low vitamin D level in CKD further worsens renal Klotho deficiency (Table 3).

Angiotensin II

Long-term infusion of Ang II down-regulates renal Klotho mRNA and protein. Administration of norepinephrine,[21,39] or L-NAME, an inhibitor of NO synthesis,[81] cause a comparable hypertensive effect as that of Ang II, but does not affect renal Klotho expression. Continuous infusion of low dose of Ang II in rats that does not affect systolic blood pressure, also down-regulates Klotho mRNA in the kidney[21,39] suggesting that the Ang II-induced Klotho down-regulation is not dependent on blood pressure. In addition, both losartan, an angiotensin Type 2 receptor antagonist, and hydralazine block the rise in blood pressure by Ang II, but only losartan completely blocks the Ang II-induced decrease in Klotho indicating that the Ang II effect on Klotho is via an Ang II Type 2 receptor.[21] This suggests a significant role of Ang II in the regulation of Klotho expression in the kidney and infers important therapeutic possibilities.

Oxidative Stress

Oxidative stress in the kidney originates not only from infiltrating inflammatory cells, but also from damaged kidney cells. Oxidative stress does not only trigger acute kidney injury but also accelerates chronic progression of kidney disease.[98,99] H$_2$O$_2$ suppresses Klotho mRNA and protein expression in cultured mIMCD$_3$ cells.[22] Adenovirus-mediated overexpression of the *klotho* gene reduces apoptosis in mIMCD$_3$ cells, suggesting that klotho may be protective against oxidative stress injury and apoptosis at least in this cell line.[22]

Indoxyl Sulfate Toxicity

Indoxyl sulfate is one of uremic toxins, and significantly increased in the blood of CKD.[100,101] It is not only a biomarker for kidney damage,[101] but also seems to promote progression of CKD and cardiovascular disease in CKD.[101] Indoxyl sulfate reduces renal Klotho expression, and contributes to cell senescence in the kidneys and enhances renal fibrosis of hypertensive rats.[102] The mechanism of Klotho down-regulation in the kidney of CKD is complicated and is summarized in Table 3.

Potential Pathogenetic Roles of Klotho Deficiency in Kidney Disease

The fact that Klotho-deficient mice have more severe kidney damage and fibrosis, and Klotho overexpressors have milder kidney dysfunction and fibrosis after CKD induction surgery, strongly supports the notion that Klotho might be a pathogenic intermediate for CKD development.[2]

Reduced Ability of Regeneration

Klotho deficiency is associated with stem cell dysfunction and depletion, which is part of normal aging.[11] The decrease in stem cell number is associated with an increase in progenitor cell senescence. Secreted Klotho binds to various Wnt family members and inhibits their biological activity. Wnt signaling activity is significantly increased in tissues from $Kl^{-/-}$ mice, while suppressed by genetic Klotho overexpression.[11] Administration of exogenous Wnt could stimulate Wnt signal transduction, and trigger or accelerate cell senescence both in vitro and in vivo. Thus, Klotho appears to be a secreted Wnt antagonist and antagonizes mammalian aging.

Cell senescence is a complicated process present not only in normal aging but also in several pathophysiological states.[18,103-108] Secondary Klotho deficiency in kidney diseases could enhance cell senescence accompanying oxidative stress.[18,102-109] Cell culture studies confirm that Klotho deficiency directly promotes senescence of renal epithelial cells.[110] Klotho supplementation could block senescence induced by oxidative stress.[52,111] Excessive senescence or apoptosis and secondary stem cell deletion might decrease the ability of kidney to defend against renal insults and impair regeneration.

Abnormal Endothelial Function and Angiogenesis

Abnormal endothelial function and impairment of angiogenesis and vasculogenesis can delay kidney regeneration post injury and contribute to progression of CKD and aging of the kidney.[112-116] There are two abnormal aspects in the vasculature of the $Kl^{-/-}$ mice. One is abnormal vasodilatation due to abnormal endothelial function;[117] another is impaired angiogenesis and vasculogenesis.[118] The aortas of $Kl^{+/-}$ mice has exaggerated contractile response to norepinephrine and attenuated responses to acetylcholine and superoxide dismutase. $Kl^{+/-}$ mice have low urinary NO metabolites (NO_2^-/NO_3^-) and cGMP concentrations, but normal the prostaglandin.[117] Low NO which is as result of low expression of NO synthase[117] may be one of causes of hypertension in $Kl^{+/-}$ mice (personal observation).

The defect in angiogenesis and vasculogenesis is well known with aging, but its mechanism is not completely understood. $Kl^{+/-}$ mice have delayed angiogenesis and vasculogenesis and low blood-flow after hindlimb ischemia.[119] The HMG co-enzyme A reductase inhibitor cerivastatin increases Klotho levels and restores the impaired neovascularization.[118] The low vasculogenesis and angiogenesis in $Kl^{+/-}$ mice might be attributable to down-regulation of vascular endothelial growth factor (VEGF) in the aorta.[117]

Klotho incubation mitigates increased cell senescence and apoptosis in endothelial cells triggered by oxidative stress or by Klotho deficiency.[111] In addition, Klotho suppresses TNF-α-induced expression of intracellular adhesion molecule-1 and vascular cell adhesion molecule-1, attenuates NF-kappaB activation, and reverses the inhibition of eNOS phosphorylation by TNF-α. Thus Klotho protein may have another role in protection of vasculature by inhibition of endothelial inflammation.[120]

Renal Fibrosis

Renal fibrosis is one striking histological characteristic in CKD, and is generally attributed to epithelial-mesenchymal transition (EMT).[121-126] TGF-β1, a major fibrotic

SECRETED KLOTHO AND CHRONIC KIDNEY DISEASE

progression protein, plays a key role in EMT in the kidneys.[125,127-129] $Kl^{-/-}$ mice have more appreciable renal tubulointerstitial fibrosis,[130] which is associated with up-regulation of TGF-β1 in the kidneys. The renal fibrosis in unilateral ureteral obstruction (UUO) is accompanied by up-regulation of TGF-β1 and fibronectin, and down-regulation of Klotho mRNA and protein. These alterations are more exaggerated in $Kl^{+/-}$ UUO mice than in WT UUO mice. HK2 cells incubated with TGF-β1 have down-regulation of Klotho expression which is attenuated with TGF-β1 receptor inhibitor (ALK5 inhibitor).[130] Secreted Klotho alleviates renal fibrosis induced by UUO and suppresses expression of fibrosis markers and TGF-β1 target genes (Snail, Twist), but does not reduce TGF-β1 expression in UUO kidney.[131] Furthermore, Klotho suppresses TGF-β1 induced Smad2 phosphorylation in a rat tubular epithelial cell line (NRK52E), suggesting that Klotho protein suppresses renal fibrosis primarily through inhibiting TGF-β1 signaling.[131]

Plasminogen activator inhibitor-1 (PAI-1) is a member of the serine protease inhibitor family and regulates fibrinolysis and proteolysis through inhibiting plasminogen activation.[132,133] PAI-1 levels are increased in the kidneys of human with glomerulonephritis, hypertensive nephrosclerosis, diabetic nephropathy, and chronic allograft nephropathy, CysA nephrotoxicity[134] and lupus nephritis in female MRL *lpr/lpr* mice.[132,133] PAI-1 mRNA and activity are strikingly elevated in multiple tissues in $Kl^{-/-}$ mice compared with WT mice. This increase in PAI-1 is age-dependent and linked to the development of ectopic calcification, and glomerular fibrin deposition in the kidneys of $Kl^{-/-}$ mice.[135] Thus, the lack of Klotho increases PAI-1 mRNA and activity.

Potential Contribution of Klotho Deficiency to Complications of CKD

The emerging view is that Klotho deficiency not only worsens renal disease but also exacerbates extrarenal complications in CKD.

Ectopic Calcification

Cardiovascular calcificaion is a heterogeneous disorder with overlapping yet distinct mechanisms of initiation and progression.[67,136,137] Vascular calcification is a dynamic process resulting from the imbalance between promoters and inhibitors.[136,138] Abnormal FGF23 and Klotho can contribute to ectopic calcification in soft tissue.[2,6,65]

The fact that $Kl^{-/-}$ mice have extensive ectopic calcification in soft tissue, which is also observed in CKD subjects strongly suggests a pathogenetic association between Klotho deficiency and calcification. CKD animals and human patients have significantly low Klotho protein and mRNA in the kidneys (Table 2) and CKD animals have low plasma and urinary Klotho protein.[2] Administration of exogenous Klotho protein (personal observation) or increasing Klotho level via genetically engineered manipulation[2] significantly inhibits vascular calcification in CKD animals suggesting that Klotho deficiency is associated with vascular calcification in CKD.

Elevated plasma Pi is associated with vascular calcification in experimental animals and in CKD patients and cellular Pi influx is believed to be mediated by the NaPi-3 group of Na^+ -coupled transporters (also named Pit1 and Pit2) in vascular smooth muscle cells (VMSC).[139] $Kl^{-/-}$ mice have high levels of Pit1 and Pit2 mRNA and higher tissue calcium contents compared to WT litter mates. In addition, $Kl^{+/-}$ CKD mice have higher and Tg-Kl CKD mice have lower Pit1/2 mRNA in the aortas than WT CKD mice. Upregulation of expression and activity of Pit1/2 in the aortas which may result from high ambient Pi and/or unknown uremic toxin (s) could increase intracellular Pi concentration, induce VSMC de-differentiation, and increase or trigger vascular calcification.[2]

FGF23-deficient mice have ectopic calcification in soft tissues including vasculature, raising the possibility that FGF23 may be protective for the vasculature. However, the high plasma FGF23 in CKD patients challenges this notion. One possible explanation might be that the target organ loses the response to FGF23 in CKD. Klotho and FGFR1/3 expression are down-regulated when human aorta derived-SMCs are incubated with high phosphate and calcium. These cells exhibit concomitant osteo/chondrocytic transformation and loss of contractile phenotype. De-differentiation of SMC and loss of the ability to respond to FGF23 may result from down-regulationof Klotho-FGFR1/3 expression in the arteries.[140]

Cardiac Hypertrophy

Cardiac hypertrophy in CKD, also referred to as "uremic cardiomyopathy" by some, is another pathological feature of cardiovascular complication. Clinically, it is characterized by cardiac arrest or sudden death, left ventricular hypertrophy, and congestive heart failure, which may be distinct from hypertensive and ischemic cardiomyopathy.[141,142] Left ventricular hypertrophy is more frequent in CKD patients than the general population when age and gender are matched, and has a negative impact on cardiovascular prognosis.[143] In addition to traditional risk factors such as smoking, hypertension, dyslipidemia, diabetes, anemia, AV fistula, volume overload, hypoalbuminemia, oxidative stress, chronic inflammation, and secondary hyperparathyroidism,[144] additional risk factors include inappropriate activation of the RAS (renin-angiotensin-aldosterone system),[145,146] vitamin D deficiency,[147] high FGF23 in blood,[148] and more recently, low Klotho in blood.[2] We have observed left ventricular hypertrophy in Klotho-deficient animals without kidney disease (personal observation).

In the heart, Klotho is expressed solely at the sinoatrial node.[48] The high rate of sudden death of $Kl^{-/-}$ mice under restraint is likely caused by sinoatrial node dysfunction.[48] Intrinsic heart rate after pharmacological blockade of autonomic nerves in $Kl^{-/-}$ mice was significantly lower than that in WT mice. The sinus node recovery time after overdrive pacing is significantly longer in $Kl^{-/-}$ mice than in WT mice; but there is no degenerative structural change in the sinoatrial node, suggesting that normal Klotho gene expression is essential for the sinoatrial node to function as a dependable pacemaker under conditions of stress.[48] Whether sKl also functions as regulator of the pacemaker is still unclear. Recently, we have observed cardiac fibrosis in $Kl^{-/-}$ mice, and this change is progressive with aging (personal observation). These findings suggest that Klotho may suppress intrinsic fibrogenesis in heart, but whether the mechanism of fibrosis is same between the heart and the kidney is not known. Since there is normally no Klotho expression in the ventricles, the cardiac phenotype in $Kl^{-/-}$ mice is mostly likely due to deficiency in circulating Klotho or other factors. There is no data on Klotho expression in the heart of CKD patients. Thus the direct role of Klotho deficiency on the high cardiovascular mortality and morbidity in CKD patients remains to be confirmed.

Secondary Hyperparathyroidism

Secondary hyperparathyroidism is part of the CKD-Metabolic bone disease (MBD) spectrum.[149,150] The role of PTH in CKD-MBD is well known and the role of FGF23 and

SECRETED KLOTHO AND CHRONIC KIDNEY DISEASE

Klotho in CKD-MBD is attracting attention. CKD subjects have high blood level and over activity of FGF23[62] and low Klotho and FGFR(s) in parathyroid gland[151,152] although some studies have argued for up-regulation of Klotho protein in uremic parathyroid gland.[153] In normal subjects with normal kidney function, FGF23 plays a crucial role, both as a phosphaturic factor[12,64,154] and as a suppressor of active vitamin D (1,25D) production in the kidney,[97] and of PTH production in the parathyroid gland.[28] In contrast, FGF23 fails to inhibit PTH production probably due to down-regulation of Klotho and FGFR(s) in parathyroid gland in the uremic setting.[151,152]

Bone Disease

Renal bone disease is the skeletal component of CKD-MBD (Mineral and Bone Disorder), and is characterized by either high or low turn-over and either presence or absence of mineralization defect.[155,156] Secondary hyperparathyroidism, poor calcium intake, Pi retention, deficiency of 1,25-dihydroxy vitamin D, high FGF23 and low Klotho may all potentially contribute to CKD-MBD.[157-159] $Kl^{-/-}$ mice have high blood FGF23, and low turn-over form of osteopenia[160,161] associated with reduced number of osteoblast progenitors and low activity of osteoblastic cells in ex vivo bone marrow cultures derived from $Kl^{-/-}$ mice. This suggests independent impairment of both osteoblast and osteoclast differentiation leading to low-turnover osteopenia observed in $Kl^{-/-}$ mice.[160,161] The bone disease in $Kl^{-/-}$ mice is not the same as those in CKD patients. The direct effect of Klotho on CKD-MBD remains to be clarified.

MECHANISMS OF HOW SECRETED KLOTHO PRESERVES KIDNEY FUNCTION AND AMELIORATES COMPLICATIONS IN CKD

Induction of Phosphaturia

Hyperphosphatemia results mainly from renal phosphorus retention,[14,26] and is a prominent feature in the $Kl^{-/-}$ mice.[6,14] The restoration of Klotho levels via genetic manipulation,[20] or viral-based delivery[162] successfully normalizes blood phosphate level. $Kl^{-/-}$ mice display increased activity of Na-coupled phosphate (NaPi) cotransport and elevation of NaPi-2a and NaPi-2c cotransporter proteins compared with wild-type (WT) mice.[163] This suggests that the hyperphosphatemia at least in part is of renal origin. Moreover, NaPi-2b protein and mRNA in gut are higher in $Kl^{-/-}$ mice than in Wt mice[163] indicating that increased intestinal phosphorus absorption may exacerbate augmentation of blood Pi concentration in $Kl^{-/-}$ mice. Transgenic Klotho overexpressing mice ($Tg-Kl$) have lower blood Pi, while renal fractional excretion of phosphorus (FE_{phos}) is increased indicating a renal leak of Pi.[14] Injection of sKl significantly increases FE_{phos} and decreases blood Pi in the normal rat.[14]

Klotho protein inhibits NaPi cotransport activity in renal brush border membrane (BBM) and in OK cells.[14] Klotho dramatically reduces NaPi-2a abundance in apical NaPi-2a protein in kidney and OK cells after 4 or more hours in vivo and in vitro respectively, indicating that the more sustained effects of Klotho on NaPi involves the canonical pathway of NaPi-2a internalization.[14]

The extracellular domain of Klotho contains two tandem repeats with 20%-40% amino acid identity with members of the glycosidase family including β-glucosidase, and has β-glucuronidase-like enzymatic activity.[6,7] NaPi-2a is a glycosylated protein.[164] The direct and acute inhibition of NaPi transport by Klotho can be mimicked by recombinant β-glucuronidase but not by sialidase. The substrate of this glycuronidase activity in the BBM is not known. While protease inhibitors abolish the proteolysis, they do not reverse the Klotho-induced inhibition of transport, indicating that Klotho-induced deglycosylation is sufficient and that subsequent proteolysis is not required to suppress Na-dependent Pi transport. Klotho modulates NaPi-2a in a biphasic fashion with dual distinct mechanisms. It acutely (<4 hrs) decreases its intrinsic transport activity via removal of glucuronate from some yet to be identified substrate, followed by proteolytic cleavage, and in a second phase (>4 hrs) induces changes in cell surface NaPi-2a.[14]

Hyperphosphatemia is universally observed in CKD patients, and is a potent predictor of cardiovascular morbidity and mortality.[90,165] Controlling blood phosphate by restriction of intake,[166] phosphate binder,[92] and more efficient dialysis[167] all improve clinical outcome in CKD patients. Undoubtedly, lack of the phosphaturic action of Klotho protein is an important pathogenic factor in CKD and means of restoring Klotho is of potential benefit.

Antivascular Calcification: Inhibition of Type III Na-Phosphate Cotransporter

Klotho can decrease soft tissue calcification by lowering plasma phosphate levels through promotion of phosphaturia. In addition, overexpression of Klotho improves kidney function after CKD induction surgery. Both of these actions can reduce the calcium content in soft tissues. The third most important and direct effect of Klotho on inhibition of calcification is the modulation of the NaPi-3's-Pit1 and Pit2, key modulators for Pi influx into vascular smooth muscle cells (VSMC).[2]

While there is abundant calcification in the multiple organs of both *WT* and *Kl*[+/-] CKD mice, *Tg-Kl-CKD* animals have very few or no calcification.[2] Calcium content is higher in the aortas and the kidneys of CKD than Sham in both the *Wt* and *Kl*[+/-] mice. This increase is ameliorated by overexpression of Klotho.[82] *Pit1, Pit2,* and *Runx2* (a marker of osteoblast-like phenotype) mRNA are increased and *SM22* (a marker of contractile smooth muscle cell) is decreased in *Kl*[-/-], while overexpression of Klotho has the opposite effect. Klotho may control the balance between differentiation and de-differentiation of VMSC.[2] CKD induces a similar pattern as Klotho deficiency and Klotho overexpression completely blocked the changes induced by CKD. When rat VMSC are grown in vitro, Klotho inhibits Na-dependent Pi influx and minimizes the mineralization induced by high ambient Pi.[2] The up-regulation of *Runx2,* and down-regulation of *SM22* by high Pi is reversed by recombinant Klotho protein, suggesting that Klotho directly blocks Pi-induced de-differentiation of rat VMSC.[2]

Anti-Oxidation

The pathogenic effect of reactive oxygen species in initiation or exacerbation of kidney damage is well-known.[22,168-171] Part of the anti-aging effect of Klotho is thought to be mediated by anti-oxidation or increased resistance to oxidant stress.[172]

Kl[-/-] mice have high level of oxidative stress and damage, while *Tg-Kl* mice have lower levels as compared with *WT* mice.[172] Klotho-transfected mIMCD$_3$ cells exposed to H_2O_2, show fewer apoptotic cells.[22] Paraquat-induced increase in lipid peroxidation is suppressed in HeLa cells treated with Klotho protein.[172] These results suggest that the Klotho may be involved in prevention of oxidative injury and apoptosis possibly related to activation of FOXO transcription signal and stimulation of MnSOD.[172]

Anticell Senescence

The permanent and irreversible growth arrest of cell and cell senescence are central paradigms of aging[173] and diseases secondary to ischemia, toxin, inflammation, and so on.[18,103,104,106,108,109,174] Senescent cells may secrete altered levels of growth factors, increase susceptibility to apoptosis, and delay the repair and regeneration in the aging kidney. Thus prevention of cell senescence may provide an important beneficial impact on aging as well as kidney disease.[106] Klotho functions as a secreted Wnt antagonist by directly binding to Wnt3/5 proteins klotho and inhibits cell senescence.[11]

Cell senescence and oxidative stress are closely associated and implicated in acute and chronic kidney disease. Mice with spontaneous chronic glomerular disease carrying a mutation in *Tensin2* have low renal Klotho, high level of lipid peroxidation, superoxide anion production, mitochondrial oxidative stress, and severe cell senescence in the kidney.[18] Genetic Klotho overexpression ameliorates renal injury associated with a dramatic improvement in mitochondria damage, reduction in senescent cells, decreased oxidant stress, and reduced apoptosis in the kidney.[18]

Suppression of Angiotensin II Effects

It is known that the angiotensin II (AngII) activity and production are significantly upregulated in a variety of kidney diseases. Ang II is a proinflammatory mediator and oxidant.[175-179] Animal experiments clearly revealed that Ang II decreases Klotho mRNA and protein expression in the kidney which could be blocked by Angiotensin Type I receptor antagonist.[21] Moreover, Klotho gene transfer also blocks Ang II-induced kidney damage, suggesting that the Ang II may act upstream of Klotho and produce pathogenic effects by reducing Klotho.[21]

In addition, Ang II receptor antagonists do not only block the downregulation of Klotho, but also abrogate the induction of TGF-β, p-p38 and p53 expression in cultured NRK cells by Ang II. Therefore, blocking Ang II action causes upregulation of Klotho and exerts a cytoprotective role.[180] Similarly HMG-CoA reductase inhibitors effectively blunt Ang II-induced reduction of Klotho expression in mIMCD$_3$ cell line.[181] Upregulation of Klotho may be one of therapeutic mechanisms of HMG-CoA reductase inhibitor action.

VALUE OF MEASURING KLOTHO IN CKD

CKD is one of major diseases affecting human life span and quality of life. Early diagnosis is paramount for one to initiate early and effective treatment. Extensive effort has been devoted to search for early biomarker for kidney diseases focusing mostly on acute kidney disease and less on CKD by transcriptomics, proteomics, metabolomics, lipidomics, and gene arrays.[182-184] In addition to traditional clinical parameters such as eGFR, proteinuria, albuminuria, N-acetyl-beta-D-glucosaminidase, and cystatin C,[185-190] novel proteins as adiponectin,[191,192] γ-Glutamyltransferase,[193,194] endothelin-1,[195,196] uric acid,[197-199] and FGF23[200] have been proposed as biomarkers but remained to be validated in population studies.

Klotho has two potentials in terms of its role as a biomarker. First is that fall in plasma or urinary Klotho can be one of the earliest abnormality in CKD from a variety of causes. The most promising data thus far is that urinary Klotho is reduced at a very early stage CKD (Stage 1 and 2) and is progressively lowered with declining eGFR (Table 2).[2] Hopefully, urinary Klotho protein could be an ideal early biomarker for CKD; as it declines much earlier than other parameters (Fig. 3). Second, in the presence of CKD, Klotho levels may bear prognostic value in predicting progression and complications of CKD. Judging from the rodent model, for the same degree of renal insufficiency and hyperphosphatemia, the lower the Klotho level, the more severe the soft tissue calcification.[2] A longitude study is required to reveal the prognostic value of Klotho in CKD patients.

To date, there are limited assays available to determine the blood Klotho in humans. Immunoprecipitation from human plasma have proven to be not readily reproducible.[47] Recently, one sandwich ELISA became available to detect soluble Klotho in human blood.[43] This first report states that human serum Klotho ranges from 239 to 1266 pg/ml with a mean of 562 ± 146 pg/ml in normal adults. It is inversely related to blood creatinine and age. Normal children (7.1 ± 4.8 years) have significantly higher blood Klotho (952 ± 282 pg/ml). In addition, plasma Kl is correlated negatively with age and serum Ca, and positively with serum Pi.[43] To date, there is no publication on plasma Klotho levels in CKD patients.

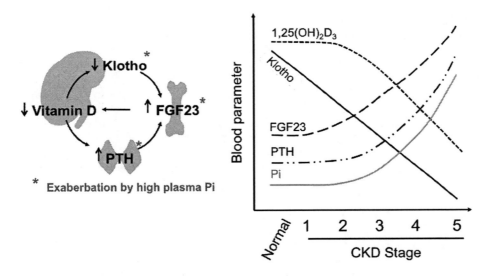

Figure 3. Proposed model and time profile of changes in Klotho and hormones relevant to mineral metabolism in CKD. Left panel depicts a model where the decline in Klotho is an early event which then leads to other secondary changes. The secondary changes constitute a downhill spiral that exacerbates each other. The decline of Klotho protein in the kidney, urine and blood is an early event, and continues to decrease with CKD progression. Low Klotho consequently induces FGF23 resistance causing compensatory increase in blood FGF23 levels to maintain Pi homeostasis in CKD. The compensatory increase in FGF23 then suppresses vitamin D synthesis. In addition, low vitamin D and high blood Pi increases PTH. The high PTH may contribute to high FGF23 at late stage of CKD. The proposed time profile is shown on the right.

POTENTIAL STRATEGIES TO INCREASE KLOTHO PROTEIN

The animal data to date is overwhelmingly strong to indicate that Klotho is not merely an early biomarker for CKD, but also a pathogenetic intermediate for accelerating CKD progression and for development of complications. Restoration of endogenous Klotho or administration of exogenous Klotho might provide novel treatment strategies for CKD patients. There are two ways to increase soluble Klotho- administering exogenous Klotho or enhancing endogenous Klotho (Fig. 4).

Administration of Exogenous Klotho

Ideally, administration of exogenous Klotho to CKD subjects is one simple and effective means to treat an endocrine deficiency similar to replacement of erythropoietin and vitamin D. Klotho can potentially reverse or retard the progression of CKD. Even in advanced stages of CKD, Klotho supplementation can alleviate extrarenal complications of CKD.

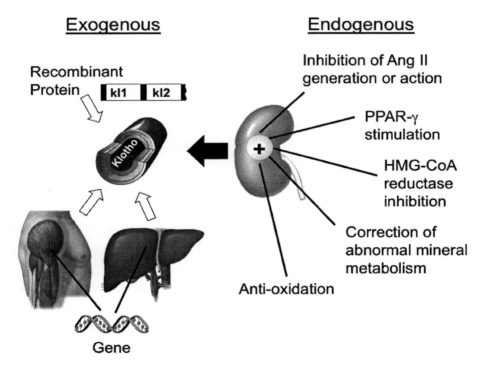

Figure 4. Potential strategies to increase Klotho protein. Prevention and retardation of progression of CKD could theoretically block the decline of endogenous Klotho protein in the kidney. Administration of exogenous Klotho protein or Klotho gene could directly provide high levels of Klotho in the blood. This is potentially useful for advanced CKD or ESRD. For early or moderate CKD patients where endogenous Klotho-producing cells are not destroyed completely or simply suppressed, stimulation of endogenous renal Klotho is of potential therapeutic benefit.

Delivery of Klotho Gene to CKD Subjects

Gene therapy is broadly defined as the insertion, alteration, or removal of genes within an individual's to change the phenotype. One form of gene therapy involves the insertion of functional genes into an unspecified genomic location to replace a mutated gene or enhance or silence the target gene. Other forms involve directly correcting the mutation or modifying normal gene via a viral carrier. Although the technology is still rather experimental, it has been used with some success.

Theoretically, gene therapy has distinct potential to treat CKD and complications at the most fundamental level. The present viral vector systems seem to have limitations for clinical use because of uncertainties regarding their toxicity and immunogenicity. But animal experiments have shown encouraging results. Adenovirus-mediated gene transfer has succeeded in gene expression in the kidney. Adeno-associated virus has a potential to be utilized as a vector targeting both kidney and skeletal muscle. Non-viral vectors such as the haemagglutinating virus of Japan (HVJ)-liposome method and cationic liposome are possibilities, but their efficiency needs to be improved. Electric pulse is emerging as a new and less harmful strategy of gene transfer to various tissues, including the kidney.[201,202]

Klotho gene delivery via adeno-associated virus (AAV) carrying mouse klotho full-length cDNA (AAV.mKl) efficiently attenuates the progression of spontaneous hypertension and renal damage in spontaneous hypertensive rats (SHR). A single dose of AAV. mKl prevents the progression of spontaneous hypertension for at least 12 weeks and reverses reduced Klotho expression in SHR rats. AAV.mKl attenuates renal tubular atrophy and dilation, tubular deposition of protein aceous material, glomerular collapse, and collagen deposition seen in SHRs, indicating that *Klotho* gene delivery limits renal damage induced by hypertension.[203]

Administration of Soluble Klotho Protein

Administration of exogenous Klotho protein is more direct, safe, and easier modality to correct endocrine Klotho deficiency in CKD compared with delivery Klotho gene. Following the history of use of Calcitriol and active derivative,[204-207] and erythropoiesis-stimulating agents[208-210] for CKD patients, Klotho protein may be a viable option in the horizon.

Compared with viral delivery system, fewer studies were reported for Klotho protein delivery. Recombinant Klotho protein of the full length of extracellular domain is able to inhibit IGF signal transduction to prolong animal life span,[20] suppress Wnt signal pathway to decrease cell senescence and retain more stem cells,[11] modulate renal ion channel or transporter[10,14-16] or control FGF23 signal transduction.[211] Klotho administration has been proven successful in the setting of acute kidney injury in animals which is a state of acute Klotho deficiency.[50] Klotho protein was injected intraperitoneally into rats after ischemia-reperfusion injury. Rats given Klotho had better renal function, less kidney damage, and lower neutrophil gelatinase-associated lipocalin.[50] More interestingly, this Klotho preparation inhibits renal fibrosis in UUO model by suppressing expression of fibrosis markers (α-Smooth muscle actin, Vimentin, Collagen-1, Metalloproteases) and TGF-β1 target genes (Snail, Twist) in a dose dependent manner. Klotho does not reduce TGF-β1 production but rather inhibits TGF-β1 signaling suggesting that Klotho prevents renal fibrosis primarily through inhibiting TGF-β1 signaling.[131] The authors'

SECRETED KLOTHO AND CHRONIC KIDNEY DISEASE 147

laboratories found that administration of Klotho protein does not only ameliorate kidney functions and histology, but also alleviates vascular calcification in CKD rats (personal observation). Thus, exogenous Klotho protein supplementation is a potentially feasible way of replacement therapy in Klotho deficient states (Fig. 4).

Up-Regulation of Endogenous Klotho Protein

In instances where endogenous Klotho-producing cells are not destroyed but simply suppressed, strategies to increase its production will be of therapeutic benefit, especially to early CKD patients (Fig. 4).

Blockage of Angiotensin II

Ang II is known to be an important mediator for initiation and progression of kidney diseases. Ang II contributes to hypertension, intraglomerular hyperfiltration, oxidant stress and lipid peroxidation and tubulointerstitial inflammation and fibrosis in acute or chronic kidney diseases.[212-222] Angiotensin converting enzyme inhibitor and angiotensin receptor antagonist are extensively used in the treatment of chronic kidney diseases. Recent studies have demonstrated that Ang II contributes to the pathogenesis of kidney diseases by reducing Klotho expression in the kidney.[21] Induction of endogenous Klotho protein expression may have potential as a therapeutic agent in treating Ang II-related kidney disease.[21,223] Similar therapeutic effect was observed in Ang II-perfused rats when treated with angiotensin receptor antagonist.

Angiotensin II Type I receptor antagonist, Losartan, does not only block the down-regulation of Klotho in the kidney of rats perfused with Ang II, but also in the kidney of rats treated with Cyclosporin A with improvement of kidney function, histology and less oxidative stress,[42] suggesting that (1) CsA triggers oxidative stress and downregulates Klotho expression; both amplifying each other to cause nephropathy; (2) angiotensin Type I receptor antagonist interrupts the vicious cycle and attenuates kidney damage induced by CsA.

Free radical scavenger is able to suppress Ang II-induced downregulation of Klotho in the kidneys of rats, to decrease plasma oxidative stress marker, and to block the decline in creatinine clearance, indicating oxidative stress is involved in downregulation of renal Klotho induced by Ang II.[39]

HMG-CoA Reductase Inhibitor

3-hydroxy-3-methylglutaryl CoA (HMG-CoA) reductase inhibitors are effective in lowering plasma concentration of LDL-cholesterol and are widely used in patients with hypercholesterolemia. Recently, this group of drugs called statins, have been shown to decrease the incidence of myocardial infarction and other ischemic vascular events independent of their lipid lowering properties.[119] An ischemic hindlimb model in mice achieved by ligation of femoral and saphenous arteries showed higher percentage of leg loss in *Kl+/-* mice compared to *WT* mice suggesting that lower angiogenesis and impairing blood flow recovery and diminishing capillary growth in *Kl+/-* mice.[119] A HMG CoA reductase inhibitor, cerivastatin, enhances angiogenesis and restores the impaired neovascularization in *Kl+/-* mice after ischemic lindlimb illustrated by similar blood perfusion between *Kl+/-* mice treated with cerivastatin and *WT* mice.[118] The mechanism by which statins accelerate angiogenesis and vasculogenesis in Klotho-deficient mice is at least partly related to maintenance of Klotho,[181,224] and secondary to improvement of endothelial function through endothelium-derived NO production.[117]

An in vitro study reveals that pretreatment with atorvastatin ameliorates the Ang II-induced reduction of *Klotho* mRNA.[181] However, whether this in vitro effect of HMG CoA reductase inhibitors translates to an in vivo effect on the kidneys of whole animals or humans needs to be further defined.

PPAR-γ Agonists

Peroxisome proliferator-activated receptors (PPAR) are a group of nuclear receptors that function as transcription regulators of metabolic pathways. PPAR-gamma (PPAR-γ), also known as the glitazone receptor, is the molecular target of thiazolidinediones (TZDs), and has been used for treatment of Type II diabetes mellitus (DM) in human.[225]

In addition to therapeutic benefits of improved insulin sensitivity, thiazolidinediones upregulate Klotho expression both in vivo and in vitro. The induction of both Klotho mRNA and protein expression in HEK293 cells by PPAR-γ agonist is blocked by PPAR-γ antagonists or siRNA-mediated gene silencing of PPAR-γ. This induction is mediated by binding of PPAR-γ to 5′-flanking region of *Klotho* gene. Moreover, thiazolidinediones or adenovirus-mediated overexpression of PPAR-γ increases Klotho expression in mouse kidneys while renal klotho expression was attenuated in mice treated with PPAR-gamma antagonists. These results demonstrate that Klotho is a downstream target gene along PPAR-γ signal transduction.[226] This effect may be one of mechanisms of PPAR-γ action on aging, DM and bone disease.[227]

Oral administration of troglitazone for 10 weeks significantly up-regulates renal *Klotho* mRNA expression, enhances endothelium-dependent aortic relaxation, and reduces systolic blood pressure, plasma glucose and triglyceride levels in OLETF rats, suggesting that improvement of the vascular endothelium function and dyslipidemia by troglitazone might be accompanied by up-regulation of renal Klotho.[80] Pioglitazone not only improves metabolic abnormalities of diabetes and consequent diabetic nephropathy, but also protects against nondiabetic chronic kidney disease in experimental models of aging kidney. It reduces proteinuria, improves GFR, decreases sclerosis, and alleviates cell senescence in the kidneys of aged rats. A similar effect is observed by increased the expression of PPAR-γ in the kidney.[108] Proposed underlying mechanisms include increased expression of Klotho, decreased systemic and renal oxidative stress, and decreased mitochondrial injury.[108]

PPAR-α is another member of the PPAR family and is highly expressed in the kidney. PPAR-α activation is able to protect the kidney from acute injury either by cisplatin or ischemic reperfusion.[228,229] But whether PPAR-α modulates renal Klotho as PPAR-γ does has not been studied yet.

Anti-Oxidant and Free Radical Scavenger

Oxidative stress in the kidney plays an important role in the development and progression of kidney disease. Oxidative stress directly suppresses Klotho expression in a kidney epithelial cell line in vitro[22] and in the kidney in vivo.[22,39,230] Oxidative stress is also implicated in Ang II[39] or CsA-induced Klotho downregulation.[42] The moderate efficacy of antioxidants used for treatment of acute and chronic kidney disease has been shown most in animal studies and not in CKD patients.[39,168,169,231]

SECRETED KLOTHO AND CHRONIC KIDNEY DISEASE

Animals with genetic or secondary Klotho deficiency have low anti-oxidants or/and over production of free radicals and lipid peroxidation in the kidneys.[23,39,40,170,172] In contrast, Klotho-overexpressing mice and animals who received Klotho gene, or treated for up-regulation of endogenous Klotho have exactly the opposite changes.[18,42,108,172,232]

Thus it appears that Klotho deficiency increases oxidative stress or makes cells more prone to oxidative stress induced injury, and oxidative stress further down-regulates Klotho expression. Anti-oxidants are potentially useful in interrupting the deterioration spiral.

CONCLUSION AND PERSPECTIVES

In animal models, CKD is a sustained state of pan Klotho deficiency in the kidney, plasma, and urine. This fact remains to be established in humans. Klotho plays a pathogenetic role in kidney disease progression, and development of disturbed mineral metabolism such as secondary hyperthyroidism and vascular calcification. As such, it is more than a biomarker. Early administration of exogenous Klotho protein, delivery of Klotho gene, or enhancement of endogenous Klotho could correct Klotho deficiency and improve kidney function in CKD (Fig. 4). The potential utility of Klotho in clinical practice is at least two-fold. First, Klotho could serve as an early and sensitive biomarker of presence of kidney diseases. But its specificity and its prognostic value and differential diagnostic value remain to be studied in humans. Second, Klotho supplementation may provide novel therapy for AKI patients to retard or block its progression to CKD and for CKD by slowing progression as well as preventing and reversing complications.

The therapeutic efficacy of Klotho in kidney disease has been unequivocally demonstrated in several animal models. One needs to validate the efficacy of Klotho in larger variety of kidney diseases and most importantly how Klotho exerts its effects which in variably will be pleiotropic. The mechanism of decline of renal Klotho in kidney diseases is not completely clarified. The upstream regulators of Klotho need to be identified. However, some therapeutic modalities including ACE inhibitor, HMG CoA reductase inhibitor, and anti-oxidants could sustain or increase endogenous Klotho expression to normal levels. The immediate challenge is to test whether human CKD resembles the rodent counterpart and if so, how to more efficiently increase Klotho levels in patients with kidney disease by stimulating endogenous Klotho or giving recombinant Klotho.

REFERENCES

1. Koh N, Fujimori T, Nishiguchi S et al. Severely reduced production of klotho in human chronic renal failure kidney. Biochem Biophys Res Commun 2001; 280(4):1015-1020.
2. Hu MC, Shi M, Zhang J et al. Klotho deficiency causes vascular calcification in chronic kidney disease. J Am Soc Nephrol 2011; 22(1):124-136.
3. Aizawa H, Saito Y, Nakamura T et al. Downregulation of the Klotho gene in the kidney under sustained circulatory stress in rats. Biochem Biophys Res Commun 1998; 249(3):865-871.
4. Matsumura Y, Aizawa H, Shiraki-Iida T et al. Identification of the human klotho gene and its two transcripts encoding membrane and secreted klotho protein. Biochem Biophys Res Commun 1998; 242(3):626-630.
5. Shiraki-Iida T, Aizawa H, Matsumura Y et al. Structure of the mouse klotho gene and its two transcripts encoding membrane and secreted protein. FEBS Lett 1998; 424(1-2):6-10.
6. Kuro-o M, Matsumura Y, Aizawa H et al. Mutation of the mouse klotho gene leads to a syndrome resembling ageing. Nature 1997; 390(6655):45-51.

7. Tohyama O, Imura A, Iwano A et al. Klotho is a novel beta-glucuronidase capable of hydrolyzing steroid beta-glucuronides. J Biol Chem 2004; 279(11):9777-9784.
8. Chen CD, Podvin S, Gillespie E et al. Insulin stimulates the cleavage and release of the extracellular domain of Klotho by ADAM10 and ADAM17. Proc Natl Acad Sci USA 2007; 104(50):19796-19801.
9. Bloch L, Sineshchekova O, Reichenbach D et al. Klotho is a substrate for alpha-, beta- and gamma-secretase. FEBS Lett 2009; 583(19):3221-3224.
10. Huang CL. Regulation of ion channels by secreted Klotho: mechanisms and implications. Kidney Int 2010; 77(10):855-860.
11. Liu H, Fergusson MM, Castilho RM et al. Augmented Wnt signaling in a mammalian model of accelerated aging. Science 2007; 317(5839):803-806.
12. Goetz R, Nakada Y, Hu MC et al. Isolated C-terminal tail of FGF23 alleviates hypophosphatemia by inhibiting FGF23-FGFR-Klotho complex formation. Proc Natl Acad Sci USA 2010; 107(1):407-412.
13. Carpenter TO, Insogna KL, Zhang JH et al. Circulating Levels of Soluble Klotho and FGF23 in X-Linked Hypophosphatemia: Circadian Variance, Effects of Treatment and Relationship to Parathyroid Status. J Clin Endocrinol Metab 2010; 95(11):E352-E357.
14. Hu MC, Shi M, Zhang J et al. Klotho: a novel phosphaturic substance acting as an autocrine enzyme in the renal proximal tubule. FASEB J 2010; 24(9):3438-3450.
15. Cha SK, Hu MC, Kurosu H et al. Regulation of renal outer medullary potassium channel and renal K(+) excretion by Klotho. Mol Pharmacol 2009; 76(1):38-46.
16. Cha SK, Ortega B, Kurosu H et al. Removal of sialic acid involving Klotho causes cell-surface retention of TRPV5 channel via binding to galectin-1. Proc Natl Acad Sci USA 2008; 105(28):9805-9810.
17. Kato Y, Arakawa E, Kinoshita S et al. Establishment of the anti-Klotho monoclonal antibodies and detection of Klotho protein in kidneys. Biochem Biophys Res Commun 2000; 267(2):597-602.
18. Haruna Y, Kashihara N, Satoh M et al. Amelioration of progressive renal injury by genetic manipulation of Klotho gene. Proc Natl Acad Sci USA 2007; 104(7):2331-2336.
19. Saito Y, Nakamura T, Ohyama Y et al. In vivo klotho gene delivery protects against endothelial dysfunction in multiple risk factor syndrome. Biochem Biophys Res Commun 2000; 276(2):767-772.
20. Kurosu H, Yamamoto M, Clark JD et al. Suppression of aging in mice by the hormone Klotho. Science 2005; 309(5742):1829-1833.
21. Mitani H, Ishizaka N, Aizawa T et al. In vivo klotho gene transfer ameliorates angiotensin II-induced renal damage. Hypertension 2002; 39(4):838-843.
22. Mitobe M, Yoshida T, Sugiura H et al. Oxidative stress decreases klotho expression in a mouse kidney cell line. Nephron Exp Nephrol 2005; 101(2):e67-e74.
23. Sugiura H, Yoshida T, Tsuchiya K et al. Klotho reduces apoptosis in experimental ischaemic acute renal failure. Nephrol Dial Transplant 2005; 20(12):2636-2645.
24. Chang Q, Hoefs S, van der Kemp AW et al. The beta-glucuronidase klotho hydrolyzes and activates the TRPV5 channel. Science 2005; 310(5747):490-493.
25. Yoshida T, Fujimori T, Nabeshima Y. Mediation of unusually high concentrations of 1,25-dihydroxy vitamin D in homozygous klotho mutant mice by increased expression of renal 1alpha-hydroxylase gene. Endocrinology 2002; 143(2):683-689.
26. Tsujikawa H, Kurotaki Y, Fujimori T et al. Klotho, a gene related to a syndrome resembling human premature aging, functions in a negative regulatory circuit of vitamin D endocrine system. Mol Endocrinol 2003; 17(12):2393-2403.
27. Silver J, Naveh-Many T. FGF23 and the parathyroid glands. Pediatr Nephrol 2010; 25(11):2241-2245.
28. Ben-Dov IZ, Galitzer H, Lavi-Moshayoff V et al. The parathyroid is a target organ for FGF23 in rats. J Clin Invest 2007; 117(12):4003-4008.
29. Invidia L, Salvioli S, Altilia S et al. The frequency of Klotho KL-VS polymorphism in a large Italian population, from young subjects to centenarians, suggests the presence of specific time windows for its effect. Biogerontology 2010; 11(1):67-73.
30. Zhang F, Zhai G, Kato BS et al. Association between KLOTHO gene and hand osteoarthritis in a female Caucasian population. Osteoarthritis Cartilage 2007.
31. Rhee EJ, Oh KW, Lee WY et al. The differential effects of age on the association of KLOTHO gene polymorphisms with coronary artery disease. Metabolism 2006; 55(10):1344-1351.
32. Rhee EJ, Oh KW, Yun EJ et al. Relationship between polymorphisms G395A in promoter and C1818T in exon 4 of the KLOTHO gene with glucose metabolism and cardiovascular risk factors in Korean women. J Endocrinol Invest 2006; 29(7):613-618.
33. Nolan VG, Adewoye A, Baldwin C et al. Sickle cell leg ulcers: associations with haemolysis and SNPs in Klotho, TEK and genes of the TGF-beta/BMP pathway. Br J Haematol 2006; 133(5):570-578.
34. Arking DE, Becker DM, Yanek LR et al. KLOTHO allele status and the risk of early-onset occult coronary artery disease. Am J Hum Genet 2003; 72(5):1154-1161.

SECRETED KLOTHO AND CHRONIC KIDNEY DISEASE

35. Arking DE, Krebsova A, Macek M Sr et al. Association of human aging with a functional variant of klotho. Proc Natl Acad Sci USA 2002; 99(2):856-861.
36. Kawano K, Ogata N, Chiano M et al. Klotho gene polymorphisms associated with bone density of aged postmenopausal women. J Bone Miner Res 2002; 17(10):1744-1751.
37. Yu J, Deng M, Zhao J et al. Decreased expression of klotho gene in uremic atherosclerosis in apolipoprotein E-deficient mice. Biochem Biophys Res Commun 2010; 391(1):261-266.
38. Friedman DJ, Afkarian M, Tamez H et al. Klotho variants and chronic hemodialysis mortality. J Bone Miner Res 2009; 24(11):1847-1855.
39. Saito K, Ishizaka N, Mitani H et al. Iron chelation and a free radical scavenger suppress angiotensin II-induced downregulation of klotho, an anti-aging gene, in rat. FEBS Lett 2003; 551(1-3):58-62.
40. Cheng MF, Chen LJ, Cheng JT. Decrease of Klotho in the kidney of streptozotocin-induced diabetic rats. J Biomed Biotechnol 2010; 2010:513853.
41. Han DH, Piao SG, Song JH et al. Effect of sirolimus on calcineurin inhibitor-induced nephrotoxicity using renal expression of KLOTHO, an antiaging gene. Transplantation 2010; 90(2):135-141.
42. Yoon HE, Ghee JY, Piao S et al. Angiotensin II blockade upregulates the expression of Klotho the anti-ageing gene, in an experimental model of chronic cyclosporine nephropathy. Nephrol Dial Transplant 2011; 26(3):800-813.
43. Yamazaki Y, Imura A, Urakawa I et al. Establishment of sandwich ELISA for soluble alpha-Klotho measurement: Age-dependent change of soluble alpha-Klotho levels in healthy subjects. Biochem Biophys Res Commun 2010; 398(3):513-518.
44. Zuo Z, Lei H, Wang X et al. Aging-related kidney damage is associated with a decrease in klotho expression and an increase in superoxide production. Age (Dordr) 2010; Epub ahead of press.
45. Ichikawa S, Imel EA, Kreiter ML et al. A homozygous missense mutation in human KLOTHO causes severe tumoral calcinosis. J Clin Invest 2007; 117(9):2684-2691.
46. Manya H, Akasaka-Manya K, Endo T. Klotho protein deficiency and aging. Geriatr Gerontol Int 2010; 10 Suppl 1:S80-S87.
47. Brownstein CA, Adler F, Nelson-Williams C et al. A translocation causing increased alpha-klotho level results in hypophosphatemic rickets and hyperparathyroidism. Proc Natl Acad Sci USA 2008; 105(9):3455-3460.
48. Takeshita K, Fujimori T, Kurotaki Y et al. Sinoatrial node dysfunction and early unexpected death of mice with a defect of klotho gene expression. Circulation 2004; 109(14):1776-1782.
49. Imura A, Iwano A, Tohyama O et al. Secreted Klotho protein in sera and CSF: implication for posttranslational cleavage in release of Klotho protein from cell membrane. FEBS Lett 2004; 565(1-3):143-147.
50. Hu MC, Shi M, Zhang J et al. Klotho deficiency is an early biomarker of renal ischemia-reperfusion injury and its replacement is protective. Kidney Int 2010; 78 (12):1240-1251.
51. Schramme A, Abdel-Bakky MS, Gutwein P et al. Characterization of CXCL16 and ADAM10 in the normal and transplanted kidney. Kidney Int 2008; 74(3):328-338.
52. Kuro-o M. Klotho as a regulator of oxidative stress and senescence. Biol Chem 2008; 389(3):233-241.
53. Huang BS, Leenen FH. Dietary Na, age and baroreflex control of heart rate and renal sympathetic nerve activity in rats. Am J Physiol 1992; 262(5 Pt 2):H1441-H1448.
54. Yamazaki M, Ozono K, Okada T et al. Both FGF23 and extracellular phosphate activate Raf/MEK/ERK pathway via FGF receptors in HEK293 cells. J Cell Biochem 2010; 111(5):1210-1221.
55. Mazzaferro S, Pasquali M, Pirro G et al. The bone and the kidney. Arch Biochem Biophys 2010; 503(1):95-102.
56. Keisala T, Minasyan A, Lou YR et al. Premature aging in vitamin D receptor mutant mice. J Steroid Biochem Mol Biol 2009; 115(3-5):91-97.
57. Chattopadhyay N, Baum M, Bai M et al. Ontogeny of the extracellular calcium-sensing receptor in rat kidney. Am J Physiol 1996; 271(3 Pt 2):F736-F743.
58. Biber J, Hernando N, Forster I et al. Regulation of phosphate transport in proximal tubules. Pflugers Arch 2009; 458(1):39-52.
59. Moe OW. PiT-2 coming out of the pits. Am J Physiol Renal Physiol 2009; 296(4):F689-F690.
60. Panda DK, Al Kawas S, Seldin MF et al. 25-hydroxyvitamin D 1alpha-hydroxylase: structure of the mouse gene, chromosomal assignment, and developmental expression. J Bone Miner Res 2001; 16(1):46-56.
61. Lai WP, Chau TS, Cheung PY et al. Adaptive responses of 25-hydroxyvitamin D3 1-alpha hydroxylase expression to dietary phosphate restriction in young and adult rats. Biochim Biophys Acta 2003; 1639(1):34-42.
62. Shimada T, Urakawa I, Isakova T et al. Circulating fibroblast growth factor 23 in patients with end-stage renal disease treated by peritoneal dialysis is intact and biologically active. J Clin Endocrinol Metab 2010; 95(2):578-585.
63. Fukagawa M, Kazama JJ. With or without the kidney: the role of FGF23 in CKD. Nephrol Dial Transplant 2005; 20(7):1295-1298.
64. Weber TJ, Liu S, Indridason OS et al. Serum FGF23 levels in normal and disordered phosphorus homeostasis. J Bone Miner Res 2003; 18(7):1227-1234.

65. Nakatani T, Sarraj B, Ohnishi M et al. In vivo genetic evidence for klotho-dependent, fibroblast growth factor 23 (Fgf23)-mediated regulation of systemic phosphate homeostasis. FASEB J 2009; 23(2):433-441.
66. Jono S, Shioi A, Ikari Y et al. Vascular calcification in chronic kidney disease. J Bone Miner Metab 2006; 24(2):176-181.
67. London GM. Cardiovascular calcifications in uremic patients: clinical impact on cardiovascular function. J Am Soc Nephrol 2003; 14(9 Suppl 4):S305-S309.
68. Weinstein JR, Anderson S. The aging kidney: physiological changes. Adv Chronic Kidney Dis 2010; 17(4):302-307.
69. Martin JE, Sheaff MT. Renal ageing. J Pathol 2007; 211(2):198-205.
70. Lerma EV. Anatomic and physiologic changes of the aging kidney. Clin Geriatr Med 2009; 25(3):325-329.
71. Chen G, Bridenbaugh EA, Akintola AD et al. Increased susceptibility of aging kidney to ischemic injury: identification of candidate genes changed during aging, but corrected by caloric restriction. Am J Physiol Renal Physiol 2007; 293(4):F1272-F1281.
72. Musso CG, Liakopoulos V, Ioannidis I et al. Acute renal failure in the elderly: particular characteristics. Int Urol Nephrol 2006; 38(3-4):787-793.
73. Campbell KH, O'Hare AM. Kidney disease in the elderly: update on recent literature. Curr Opin Nephrol Hypertens 2008; 17(3):298-303.
74. Hu MC, Bankir L, Michelet S et al. Massive reduction of urea transporters in remnant kidney and brain of uremic rats. Kidney Int 2000; 58(3):1202-1210.
75. Tamura K, Suzuki Y, Matsushita M et al. Prevention of aortic calcification by etidronate in the renal failure rat model. Eur J Pharmacol 2007; 558(1-3):159-166.
76. Peralta CA, Kurella M, Lo JC et al. The metabolic syndrome and chronic kidney disease. Curr Opin Nephrol Hypertens 2006; 15(4):361-365.
77. Nagase M, Yoshida S, Shibata S et al. Enhanced aldosterone signaling in the early nephropathy of rats with metabolic syndrome: possible contribution of fat-derived factors. J Am Soc Nephrol 2006; 17(12):3438-3446.
78. Locatelli F, Pozzoni P, Del Vecchio L. Renal manifestations in the metabolic syndrome. J Am Soc Nephrol 2006; 17(4 Suppl 2):S81-S85.
79. Kawano K, Hirashima T, Mori S et al. Spontaneous long-term hyperglycemic rat with diabetic complications. Otsuka Long-Evans Tokushima Fatty (OLETF) strain. Diabetes 1992; 41(11):1422-1428.
80. Yamagishi T, Saito Y, Nakamura T et al. Troglitazone improves endothelial function and augments renal klotho mRNA expression in Otsuka Long-Evans Tokushima Fatty (OLETF) rats with multiple atherogenic risk factors. Hypertens Res 2001; 24(6):705-709.
81. Ohyama Y, Kurabayashi M, Masuda H et al. Molecular cloning of rat klotho cDNA: markedly decreased expression of klotho by acute inflammatory stress. Biochem Biophys Res Commun 1998; 251(3):920-925.
82. Thurston RD, Larmonier CB, Majewski PM et al. Tumor necrosis factor and interferon-gamma down-regulate Klotho in mice with colitis. Gastroenterology 2010; 138(4):1384-1394.
83. Webster AC, Woodroffe RC, Taylor RS et al. Tacrolimus versus ciclosporin as primary immunosuppression for kidney transplant recipients: meta-analysis and meta-regression of randomised trial data. BMJ 2005; 331(7520):810.
84. Jiang H, Sakuma S, Fujii Y et al. Tacrolimus versus cyclosporin A: a comparative study on rat renal allograft survival. Transpl Int 1999; 12(2):92-99.
85. Gulati S, Prasad N, Sharma RK et al. Tacrolimus: a new therapy for steroid-resistant nephrotic syndrome in children. Nephrol Dial Transplant 2008; 23(3):910-913.
86. Tang S, Tang AW, Tam MK et al. Use of tacrolimus in steroid- and cyclophosphamide-resistant minimal change nephrotic syndrome. Am J Kidney Dis 2003; 42(5):E13-E15.
87. Ziolkowski J, Paczek L, Senatorski G et al. Renal function after liver transplantation: calcineurin inhibitor nephrotoxicity. Transplant Proc 2003; 35(6):2307-2309.
88. Naesens M, Kuypers DR, Sarwal M. Calcineurin inhibitor nephrotoxicity. Clin J Am Soc Nephrol 2009; 4(2):481-508.
89. Formica RN Jr, Lorber KM, Friedman AL et al. Sirolimus-based immunosuppression with reduce dose cyclosporine or tacrolimus after renal transplantation. Transplant Proc 2003; 35(3 Suppl):95S-98S.
90. Kestenbaum B, Sampson JN, Rudser KD et al. Serum phosphate levels and mortality risk among people with chronic kidney disease. J Am Soc Nephrol 2005; 16(2):520-528.
91. Cannata-Andia JB, Rodriguez-Garcia M. Hyperphosphataemia as a cardiovascular risk factor — how to manage the problem. Nephrol Dial Transplant 2002; 17 Suppl 11:16-19.
92. Isakova T, Gutierrez OM, Chang Y et al. Phosphorus binders and survival on hemodialysis. J Am Soc Nephrol 2009; 20(2):388-396.
93. Kuro-o M. A potential link between phosphate and aging—lessons from Klotho-deficient mice. Mech Ageing Dev 2010; 131(4):270-275.
94. Morishita K, Shirai A, Kubota M et al. The progression of aging in klotho mutant mice can be modified by dietary phosphorus and zinc. J Nutr 2001; 131(12):3182-3188.

SECRETED KLOTHO AND CHRONIC KIDNEY DISEASE

153

95. Lips P. Vitamin D deficiency and secondary hyperparathyroidism in the elderly: consequences for bone loss and fractures and therapeutic implications. Endocr Rev 2001; 22(4):477-501.
96. Gutierrez OM. Fibroblast growth factor 23 and disordered vitamin D metabolism in chronic kidney disease: updating the "trade-off" hypothesis. Clin J Am Soc Nephrol 2010; 5(9):1710-1716.
97. Liu S, Tang W, Zhou J et al. Fibroblast growth factor 23 is a counter-regulatory phosphaturic hormone for vitamin D. J Am Soc Nephrol 2006; 17(5):1305-1315.
98. Palm F, Nangaku M, Fasching A et al. Uremia induces abnormal oxygen consumption in tubules and aggravates chronic hypoxia of the kidney via oxidative stress. Am J Physiol Renal Physiol 2010; 299(2):F380-F386.
99. Polichnowski AJ, Jin C, Yang C et al. Role of renal perfusion pressure versus angiotensin II on renal oxidative stress in angiotensin II-induced hypertensive rats. Hypertension 2010; 55(6):1425-1430.
100. Yu M, Kim YJ, Kang DH. Indoxyl Sulfate-Induced Endothelial Dysfunction in Patients with Chronic Kidney Disease via an Induction of Oxidative Stress. Clin J Am Soc Nephrol 2011: 6(1):30-39.
101. Barreto FC, Barreto DV, Liabeuf S et al. Serum indoxyl sulfate is associated with vascular disease and mortality in chronic kidney disease patients. Clin J Am Soc Nephrol 2009; 4(10):1551-1558.
102. Niwa T, Adijiang A, Higuchi Y et al. Indoxyl Sulfate Reduces Klotho Expression and Promotes Senescence in the Kidney of Hypertensive Rats (Abstract). J Am Soc Nephrol 2010; 21:746A.
103. Jennings P, Koppelstaetter C, Aydin S et al. Cyclosporine A induces senescence in renal tubular epithelial cells. Am J Physiol Renal Physiol 2007; 293(3):F831-F838.
104. Kailong L, Du X, Yani H et al. P53-Rb signaling pathway is involved in tubular cell senescence in renal ischemia/reperfusion injury. Biocell 2007; 31(2):213-223.
105. Dmitrieva NI, Burg MB. High NaCl promotes cellular senescence. Cell Cycle 2007; 6(24):3108-3113.
106. Yang H, Fogo AB. Cell senescence in the aging kidney. J Am Soc Nephrol 2010; 21(9):1436-1439.
107. Nakano-Kurimoto R, Ikeda K, Uraoka M et al. Replicative senescence of vascular smooth muscle cells enhances the calcification through initiating the osteoblastic transition. Am J Physiol Heart Circ Physiol 2009; 297(5):H1673-H1684.
108. Yang HC, Deleuze S, Zuo Y et al. The PPARgamma agonist pioglitazone ameliorates aging-related progressive renal injury. J Am Soc Nephrol 2009; 20(11):2380-2388.
109. Shimizu H, Bolati D, Adijiang A et al. Senescence and dysfunction of proximal tubular cells are associated with activated p53 expression by indoxyl sulfate. Am J Physiol Cell Physiol 2010; 299(5):C1110-C1117.
110. De Oliveira RM. Klotho RNAi induces premature senescence of human cells via a p53/p21 dependent pathway. FEBS Lett 2006; 580(24):5753-5758.
111. Ikushima M, Rakugi H, Ishikawa K et al. Anti-apoptotic and anti-senescence effects of Klotho on vascular endothelial cells. Biochem Biophys Res Commun 2006; 339(3):827-832.
112. Taniyama Y, Morishita R. Does therapeutic angiogenesis overcome CKD? Hypertens Res 2010; 33(2):114-115.
113. Mu W, Long DA, Ouyang X et al. Angiostatin overexpression is associated with an improvement in chronic kidney injury by an anti-inflammatory mechanism. Am J Physiol Renal Physiol 2009; 296(1):F145-F152.
114. Westerweel PE, Hoefer IE, Blankestijn PJ et al. End-stage renal disease causes an imbalance between endothelial and smooth muscle progenitor cells. Am J Physiol Renal Physiol 2007; 292(4):F1132-F1140.
115. Chade AR, Zhu X, Mushin OP et al. Simvastatin promotes angiogenesis and prevents microvascular remodeling in chronic renal ischemia. FASEB J 2006; 20(10):1706-1708.
116. Reinders ME, Rabelink TJ, Briscoe DM. Angiogenesis and endothelial cell repair in renal disease and allograft rejection. J Am Soc Nephrol 2006; 17(4):932-942.
117. Nakamura T, Saito Y, Ohyama Y et al. Production of nitric oxide, but not prostacyclin, is reduced in klotho mice. Jpn J Pharmacol 2002; 89(2):149-156.
118. Shimada T, Takeshita Y, Murohara T et al. Angiogenesis and vasculogenesis are impaired in the precocious-aging klotho mouse. Circulation 2004; 110(9):1148-1155.
119. Fukino K, Suzuki T, Saito Y et al. Regulation of angiogenesis by the aging suppressor gene klotho. Biochem Biophys Res Commun 2002; 293(1):332-337.
120. Maekawa Y, Ishikawa K, Yasuda O et al. Klotho suppresses TNF-alpha-induced expression of adhesion molecules in the endothelium and attenuates NF-kappaB activation. Endocrine 2009; 35(3):341-346.
121. Zeisberg M, Duffield JS. Resolved: EMT produces fibroblasts in the kidney. J Am Soc Nephrol 2010; 21(8):1247-1253.
122. Liu Y. New insights into epithelial-mesenchymal transition in kidney fibrosis. J Am Soc Nephrol 2010; 21(2):212-222.
123. Ardura JA, Rayego-Mateos S, Ramila D et al. Parathyroid hormone-related protein promotes epithelial-mesenchymal transition. J Am Soc Nephrol 2010; 21(2):237-248.
124. Kalluri R, Neilson EG. Epithelial-mesenchymal transition and its implications for fibrosis. J Clin Invest 2003; 112(12):1776-1784.
125. Zeisberg M, Hanai J, Sugimoto H et al. BMP-7 counteracts TGF-beta1-induced epithelial-to-mesenchymal transition and reverses chronic renal injury. Nat Med 2003; 9(7):964-968.

126. Iwano M, Plieth D, Danoff TM et al. Evidence that fibroblasts derive from epithelium during tissue fibrosis. J Clin Invest 2002; 110(3):341-350.
127. Iwano M. EMT and TGF-beta in renal fibrosis. Front Biosci (Schol Ed) 2010; 2:229-238.
128. Hills CE, Squires PE. TGF-beta1-induced epithelial-to-mesenchymal transition and therapeutic intervention in diabetic nephropathy. Am J Nephrol 2010; 31(1):68-74.
129. Sato M, Muragaki Y, Saika S et al. Targeted disruption of TGF-beta1/Smad3 signaling protects against renal tubulointerstitial fibrosis induced by unilateral ureteral obstruction. J Clin Invest 2003; 112(10):1486-1494.
130. Sugiura H, Yoshida T, Kohei J et al. TGF-β Was Upregulated in Renal Fibrosis Model of Klotho Defect Mouse and Affected Renal Klotho Expression Level (Abstract). J Am Soc Nephrol 2010; 21:376A.
131. Doi S, Yorioka N, Kuroo M. Secreted Klotho Protein Counteracts Renal Fibrosis through Inhibiting TGF-b1 Signaling (Abstract). J Am Soc Nephrol 2010; 21:270A.
132. Yamamoto K, Loskutoff DJ. The kidneys of mice with autoimmune disease acquire a hypofibrinolytic/procoagulant state that correlates with the development of glomerulonephritis and tissue microthrombosis. Am J Pathol 1997; 151(3):725-734.
133. Yamamoto K, Loskutoff DJ, Saito H. Renal expression of fibrinolytic genes and tissue factor in a murine model of renal disease as a function of age. Semin Thromb Hemost 1998; 24(3):261-268.
134. Eddy AA, Fogo AB. Plasminogen activator inhibitor-1 in chronic kidney disease: evidence and mechanisms of action. J Am Soc Nephrol 2006; 17(11):2999-3012.
135. Takeshita K, Yamamoto K, Ito M et al. Increased expression of plasminogen activator inhibitor-1 with fibrin deposition in a murine model of aging, "Klotho" mouse. Semin Thromb Hemost 2002; 28(6):545-554.
136. Moe SM, Chen NX. Mechanisms of vascular calcification in chronic kidney disease. J Am Soc Nephrol 2008; 19(2):213-216.
137. Cannata-Andia JB, Rodriguez-Garcia M, Carrillo-Lopez N et al. Vascular calcifications: pathogenesis, management, and impact on clinical outcomes. J Am Soc Nephrol 2006; 17(12 Suppl 3):S267-S273.
138. Mizobuchi M, Towler D, Slatopolsky E. Vascular calcification: the killer of patients with chronic kidney disease. J Am Soc Nephrol 2009; 20(7):1453-1464.
139. Li X, Yang HY, Giachelli CM. Role of the sodium-dependent phosphate cotransporter, Pit-1, in vascular smooth muscle cell calcification. Circ Res 2006; 98(7):905-912.
140. Lim K, Lu T-S, Zehnder D et al. Development of Klotho-FGFR1/3 Dependent Resistance to FGF-23 in Human Aortic Smooth Muscle Cells Exposed to Calcifying Stress. J Am Soc Nephrol 2010; 21:140A.
141. London GM, Marchais SJ, Guerin AP et al. Contributive factors to cardiovascular hypertrophy in renal failure. Am J Hypertens 1989; 2(11 Pt 2):261S-263S.
142. Guerin AP, Pannier B, Marchais SJ et al. Cardiovascular disease in the dialysis population: prognostic significance of arterial disorders. Curr Opin Nephrol Hypertens 2006; 15(2):105-110.
143. Cerasola G, Nardi E, Palermo A et al. Epidemiology and pathophysiology of left ventricular abnormalities in chronic kidney disease: a review. J Nephrol 2011; 24(1):1-10.
144. Rinat C, Becker-Cohen R, Nir A et al. A comprehensive study of cardiovascular risk factors, cardiac function and vascular disease in children with chronic renal failure. Nephrol Dial Transplant 2010; 25(3):785-793.
145. Edwards NC, Steeds RP, Stewart PM et al. Effect of spironolactone on left ventricular mass and aortic stiffness in early-stage chronic kidney disease: a randomized controlled trial. J Am Coll Cardiol 2009; 54(6):505-512.
146. Burrell LM, Johnston CI. Angiotensin II receptor antagonists. Potential in elderly patients with cardiovascular disease. Drugs Aging 1997; 10(6):421-434.
147. Achinger SG, Ayus JC. The role of vitamin D in left ventricular hypertrophy and cardiac function. Kidney Int 2005; 95 Suppl:S37-S42.
148. Mirza MA, Larsson A, Melhus H et al. Serum intact FGF23 associate with left ventricular mass, hypertrophy and geometry in an elderly population. Atherosclerosis 2009; 207(2):546-551.
149. Galitzer H, Ben-Dov IZ, Silver J et al. Parathyroid cell resistance to fibroblast growth factor 23 in secondary hyperparathyroidism of chronic kidney disease. Kidney Int 2010; 77(3):211-218.
150. Khan S, Vitamin D deficiency and secondary hyperparathyroidism among patients with chronic kidney disease. Am J Med Sci 2007; 333(4):201-207.
151. Krajisnik T, Olauson H, Mirza MA et al. Parathyroid Klotho and FGF-receptor 1 expression decline with renal function in hyperparathyroid patients with chronic kidney disease and kidney transplant recipients. Kidney Int 2010; 78(10):1024-1032.
152. Canalejo R, Canalejo A, Martinez-Moreno JM et al. FGF23 fails to inhibit uremic parathyroid glands. J Am Soc Nephrol 2010; 21(7):1125-1135.
153. Hofman-Bang J, Martuseviciene G, Santini MA et al. Increased parathyroid expression of klotho in uremic rats. Kidney Int 2010; 78(11):1119-1127.
154. Gattineni J, Baum M. Regulation of phosphate transport by fibroblast growth factor 23 (FGF23): implications for disorders of phosphate metabolism. Pediatr Nephrol 2010; 25(4):591-601.

SECRETED KLOTHO AND CHRONIC KIDNEY DISEASE

155. Carter JL, O'Riordan SE, Eaglestone GL et al. Bone mineral metabolism and its relationship to kidney disease in a residential care home population: a cross-sectional study. Nephrol Dial Transplant 2008; 23(11):3554-3565.
156. Oliveira RB, Cancela AL, Graciolli FG et al. Early control of PTH and FGF23 in normophosphatemic CKD patients: a new target in CKD-MBD therapy? Clin J Am Soc Nephrol 2010; 5(2):286-291.
157. Fukagawa M, Kazama JJ. FGF23: its role in renal bone disease. Pediatr Nephrol 2006; 21(12):1802-1806.
158. Tomida K, Hamano T, Mikami S et al. Serum 25-hydroxyvitamin D as an independent determinant of 1-84 PTH and bone mineral density in nondiabetic predialysis CKD patients. Bone 2009; 44(4):678-683.
159. Jassal SK, von Muhlen D, Barrett-Connor E. Measures of renal function, BMD, bone loss, and osteoporotic fracture in older adults: the Rancho Bernardo study. J Bone Miner Res 2007; 22(2):203-210.
160. Kawaguchi H, Manabe N, Miyaura C et al. Independent impairment of osteoblast and osteoclast differentiation in klotho mouse exhibiting low-turnover osteopenia. J Clin Invest 1999; 104(3):229-237.
161. Kawaguchi H, Manabe N, Chikuda H et al. Cellular and molecular mechanism of low-turnover osteopenia in the klotho-deficient mouse. Cell Mol Life Sci 2000; 57(5):731-737.
162. Shiraki-Iida T, Iida A, Nabeshima Y et al. Improvement of multiple pathophysiological phenotypes of klotho (kl/kl) mice by adenovirus-mediated expression of the klotho gene. J Gene Med 2000; 2(4):233-242.
163. Segawa H, Yamanaka S, Ohno Y et al. Correlation between hyperphosphatemia and type II Na-Pi cotransporter activity in klotho mice. Am J Physiol Renal Physiol 2007; 292(2):F769-F779.
164. Sorribas V, Markovich D, Hayes G et al. Cloning of a Na/Pi cotransporter from opossum kidney cells. J Biol Chem 1994; 269(9):6615-6621.
165. Ramos AM, Albalate M, Vazquez S et al. Hyperphosphatemia and hyperparathyroidism in incident chronic kidney disease patients. Kidney Int 2008; 111(Suppl):S88-S93.
166. Koizumi T, Murakami K, Nakayama H et al. Role of dietary phosphorus in the progression of renal failure. Biochem Biophys Res Commun 2002; 295(4):917-921.
167. Tonelli M, Wang W, Hemmelgarn B et al. Phosphate removal with several thrice-weekly dialysis methods in overweight hemodialysis patients. Am J Kidney Dis 2009; 54(6):1108-1115.
168. Marshall CB, Pippin JW, Krofft RD et al. Puromycin aminonucleoside induces oxidant-dependent DNA damage in podocytes in vitro and in vivo. Kidney Int 2006; 70(11):1962-1973.
169. Diamond JR. Reactive oxygen species and progressive glomerular disease. J Lab Clin Med 1994; 124(4):468-469.
170. Shih PH, Yen GC. Differential expressions of antioxidant status in aging rats: the role of transcriptional factor Nrf2 and MAPK signaling pathway. Biogerontology 2007; 8(2):71-80.
171. Kachiwala SJ, Harris SE, Wright AF et al. Genetic influences on oxidative stress and their association with normal cognitive ageing. Neurosci Lett 2005; 386(2):116-120.
172. Yamamoto M, Clark JD, Pastor JV et al. Regulation of oxidative stress by the anti-aging hormone klotho. J Biol Chem 2005; 280(45):38029-38034.
173. Feng R, He W, Ochi H. A new murine oxidative stress model associated with senescence. Mech Ageing Dev 2001; 122(6):547-559.
174. Melk A, Schmidt BM, Braun H et al. Effects of donor age and cell senescence on kidney allograft survival. Am J Transplant 2009; 9(1):114-123.
175. Yang F, Huang XR, Chung AC et al. Essential role for Smad3 in angiotensin II-induced tubular epithelial-mesenchymal transition. J Pathol 2010; 221(4):390-401.
176. Liang XB, Ma LJ, Naito T et al. Angiotensin type 1 receptor blocker restores podocyte potential to promote glomerular endothelial cell growth. J Am Soc Nephrol 2006; 17(7):1886-1895.
177. Park SY, Song CY, Kim BC et al. Angiotensin II mediates LDL-induced superoxide generation in mesangial cells. Am J Physiol Renal Physiol 2003; 285(5):F909-F915.
178. Stevenson KM, Edgley AJ, Bergstrom G et al. Angiotensin II infused intrarenally causes preglomerular vascular changes and hypertension. Hypertension 2000; 36(5):839-844.
179. Peters H, Border WA, Noble NA. Angiotensin II blockade and low-protein diet produce additive therapeutic effects in experimental glomerulonephritis. Kidney Int 2000; 57(4):1493-1501.
180. Zhou Q, Lin S, Tang R et al. Role of Fosinopril and Valsartan on Klotho Gene Expression Induced by Angiotensin II in Rat Renal Tubular Epithelial Cells. Kidney Blood Press Res 2010; 33(3):186-192.
181. Narumiya H, Sasaki S, Kuwahara N et al. HMG-CoA reductase inhibitors up-regulate anti-aging klotho mRNA via RhoA inactivation in IMCD3 cells. Cardiovasc Res 2004; 64(2):331-336.
182. Rosner MH. Urinary biomarkers for the detection of renal injury. Adv Clin Chem 2009; 49:73-97.
183. Perco P, Wilflingseder J, Bernthaler A et al. Biomarker candidates for cardiovascular disease and bone metabolism disorders in chronic kidney disease: a systems biology perspective. J Cell Mol Med 2008; 12(4):1177-1187.
184. Kovesdy CP, Kalantar-Zadeh K. Review article: Biomarkers of clinical outcomes in advanced chronic kidney disease. Nephrology (Carlton) 2009; 14(4):408-415.
185. Devarajan P. The use of targeted biomarkers for chronic kidney disease. Adv Chronic Kidney Dis 2010; 17(6):469-479.

186. Guh JY. Proteinuria versus albuminuria in chronic kidney disease. Nephrology (Carlton) 2010; 15(Suppl 2):53-56.
187. Bokenkamp A, Domanetzki M, Zinck R et al. Reference values for cystatin C serum concentrations in children. Pediatr Nephrol 1998; 12(2):125-129.
188. Galteau MM, Guyon M, Gueguen R et al. Determination of serum cystatin C: biological variation and reference values. Clin Chem Lab Med 2001; 39(9):850-857.
189. Shlipak MG, Wassel Fyr CL, Chertow GM et al. Cystatin C and mortality risk in the elderly: the health, aging, and body composition study. J Am Soc Nephrol 2006; 17(1):254-261.
190. Djousse L, Kurth T, Gaziano JM. Cystatin C and risk of heart failure in the Physicians' Health Study (PHS). Am Heart J 2008; 155(1):82-86.
191. Drechsler C, Krane V, Winkler K et al. Changes in adiponectin and the risk of sudden death, stroke, myocardial infarction, and mortality in hemodialysis patients. Kidney Int 2009; 76(5):567-575.
192. Saraheimo M, Forsblom C, Thorn L et al. Serum adiponectin and progression of diabetic nephropathy in patients with type 1 diabetes. Diabetes Care 2008; 31(6):1165-1169.
193. Baggio B, Gambaro G, Briani G et al. Urinary excretion of glycosaminoglycans and brush border and lysosomal enzymes as markers of glomerular and tubular involvement in kidney diseases. Contrib Nephrol 1984; 42:107-110.
194. Ryu S, Chang Y, Kim DI et al. gamma-Glutamyltransferase as a predictor of chronic kidney disease in nonhypertensive and nondiabetic Korean men. Clin Chem 2007; 53(1):71-77.
195. Dhaun N, Lilitkarntakul P, Macintyre IM et al. Urinary endothelin-1 in chronic kidney disease and as a marker of disease activity in lupus nephritis. Am J Physiol Renal Physiol 2009; 296(6):F1477-F1483.
196. Cottone S, Mule G, Guarneri M et al. Endothelin-1 and F2-isoprostane relate to and predict renal dysfunction in hypertensive patients. Nephrol Dial Transplant 2009; 24(2):497-503.
197. Chang HY, Tung CW, Lee PH et al. Hyperuricemia as an independent risk factor of chronic kidney disease in middle-aged and elderly population. Am J Med Sci 2010; 339(6):509-515.
198. Feig DI. Uric acid: a novel mediator and marker of risk in chronic kidney disease? Curr Opin Nephrol Hypertens 2009; 18(6):526-530.
199. Madero M, Sarnak MJ, Wang X et al. Uric acid and long-term outcomes in CKD. Am J Kidney Dis 2009; 53(5):796-803.
200. Yilmaz MI, Sonmez A, Saglam M et al. FGF-23 and vascular dysfunction in patients with stage 3 and 4 chronic kidney disease. Kidney Int 2010; 78(7):679-685.
201. Appledorn DM, Seregin S, Amalfitano A. Adenovirus vectors for renal-targeted gene delivery. Contrib Nephrol 2008; 159:47-62.
202. Imai E. Gene therapy approach in renal disease in the 21st century. Nephrol Dial Transplant 2001; 16 Suppl 5:26-34.
203. Wang Y, Sun Z. Klotho gene delivery prevents the progression of spontaneous hypertension and renal damage. Hypertension 2009; 54(4):810-817.
204. Brown AJ, Slatopolsky E. Drug insight: vitamin D analogs in the treatment of secondary hyperparathyroidism in patients with chronic kidney disease. Nat Clin Pract Endocrinol Metab 2007; 3(2):134-144.
205. Gal-Moscovici A, Sprague SM. Role of vitamin D deficiency in chronic kidney disease. J Bone Miner Res 2007; 22(Suppl 2):V91-V94.
206. Gal-Moscovici A, Sprague SM. Use of vitamin D in chronic kidney disease patients. Kidney Int 2010; 78(2):146-151.
207. Andress DL. Vitamin D treatment in chronic kidney disease. Semin Dial 2005; 18(4):315-321.
208. Chen HH, Tarng DC, Lee KF et al. Epoetin alfa and darbepoetin alfa: effects on ventricular hypertrophy in patients with chronic kidney disease. J Nephrol 2008; 21(4):543-549.
209. Coles GA, Cavill I. Erythropoiesis in the anaemia of chronic renal failure: the response to CAPD. Nephrol Dial Transplant 1986; 1(3):170-174.
210. Dharmarajan TS, Widjaja D. Erythropoiesis-stimulating agents in anemia: use and misuse. J Am Med Dir Assoc 2009; 10(9):607-616.
211. Kurosu H, Ogawa Y, Miyoshi M et al. Regulation of fibroblast growth factor-23 signaling by klotho. J Biol Chem 2006; 281(10):6120-6123.
212. Hall RL, Wilke WL, Fettman MJ. Captopril slows the progression of chronic renal disease in partially nephrectomized rats. Toxicol Appl Pharmacol 1985; 80(3):517-526.
213. Brezis M, Greenfeld Z, Shina A et al. Angiotensin II augments medullary hypoxia and predisposes to acute renal failure. Eur J Clin Invest 1990; 20(2):199-207.
214. Hall JE. The renin-angiotensin system: renal actions and blood pressure regulation. Compr Ther 1991; 17(5):8-17.
215. Nabokov A, Amann K, Gassmann P et al. The renoprotective effect of angiotensin-converting enzyme inhibitors in experimental chronic renal failure is not dependent on enhanced kinin activity. Nephrol Dial Transplant 1998; 13(1):173-176.
216. Tojo A, Onozato ML, Kobayashi N et al. Angiotensin II and oxidative stress in Dahl Salt-sensitive rat with heart failure. Hypertension 2002; 40(6):834-839.
217. Muller DN, Mullally A, Dechend R et al. Endothelin-converting enzyme inhibition ameliorates angiotensin II-induced cardiac damage. Hypertension 2002; 40(6):840-846.

218. Agarwal R. Proinflammatory effects of oxidative stress in chronic kidney disease: role of additional angiotensin II blockade. Am J Physiol Renal Physiol 2003; 284(4):F863-F869.
219. Remuzzi G, Perico N, Macia M et al. The role of renin-angiotensin-aldosterone system in the progression of chronic kidney disease. Kidney Int 2005; 99(Suppl):S57-S65.
220. Wenzel RR. Renal protection in hypertensive patients: selection of antihypertensive therapy. Drugs 2005; 65 Suppl 2:29-39.
221. Mizuno M, Sada T, Kato M et al. The effect of angiotensin II receptor blockade on an end-stage renal failure model of type 2 diabetes. J Cardiovasc Pharmacol 2006; 48(4):135-142.
222. Lebel M, Rodrigue ME, Agharazii M et al. Antihypertensive and renal protective effects of renin-angiotens in system blockade in uremic rats treated with erythropoietin. Am J Hypertens 2006; 19(12):1286-1292.
223. Negri AL. The klotho gene: a gene predominantly expressed in the kidney is a fundamental regulator of aging and calcium/phosphorus metabolism. J Nephrol 2005; 18(6):654-658.
224. Kuwahara N, Sasaki S, Kobara M et al. HMG-CoA reductase inhibition improves anti-aging klotho protein expression and arteriosclerosis in rats with chronic inhibition of nitric oxide synthesis. Int J Cardiol 2008; 123(2):84-90.
225. Wang X, Liu X, Zhan Y et al. Pharmacogenomic, physiological, and biochemical investigations on safety and efficacy biomarkers associated with the peroxisome proliferator-activated receptor-gamma activator rosiglitazone in rodents: a translational medicine investigation. J Pharmacol Exp Ther 2010; 334(3):820-829.
226. Zhang H, Li Y, Fan Y et al. Klotho is a target gene of PPAR-gamma. Kidney Int 2008; 74(6):732-739.
227. Zhang R, Zheng F. PPAR-gamma and aging: one link through klotho? Kidney Int 2008; 74(6):702-704.
228. Lopez-Hernandez FJ, Lopez-Novoa JM. Potential utility of PPARalpha activation in the prevention of ischemic and drug-induced acute renal damage. Kidney Int 2009; 76(10):1022-1024.
229. Li S, Nagothu KK, Desai V et al. Transgenic expression of proximal tubule peroxisome proliferator-activated receptor-alpha in mice confers protection during acute kidney injury. Kidney Int 2009; 76(10):1049-1062.
230. Rakugi H, Matsukawa N, Ishikawa K et al. Anti-oxidative effect of Klotho on endothelial cells through cAMP activation. Endocrine 2007; 31(1):82-87.
231. Shah SV, Baliga R, Rajapurkar M et al. Oxidants in chronic kidney disease. J Am Soc Nephrol 2007; 18(1):16-28.
232. Sugiura H, Yoshida T, Mitobe M et al. Klotho reduces apoptosis in experimental ischaemic acute kidney injury via HSP-70. Nephrol Dial Transplant 2010; 25(1):60-68.

CHAPTER 10

FGF23 AS A NOVEL THERAPEUTIC TARGET

Takashi Shimada*,[1] and Seiji Fukumoto[2]

[1]*Innovative Drug Research Laboratories, Kyowa Hakko Kirin Co., Ltd, Tokyo, [2]Japan Division of Nephrology and Endocrinology, Department of Medicine, University of Tokyo Hospital, Tokyo, Japan.*
Corresponding Author: Takashi Shimada—Email: takashi.shimada@kyowa-kirin.co.jp

Abstract: Fibroblast growth factor (FGF) 23 is a hormone that acts to decrease phosphate, 1,25-dihydroxyvitamin D and parathyroid hormone levels in circulation. Particularly, appropriate actions of FGF23 are essential for maintaining physiological phosphate and vitamin D metabolism. Therefore, either gain or loss of function of FGF23 can impair these homeostatic regulations, causing several metabolic bone diseases. The measurement of circulating levels of FGF23 in patients with various types of hypophosphatemic rickets and/or osteomalacia has revealed that several of them are FGF23-dependent diseases, highlighting a novel therapeutic concept that the inhibition of the excess activity of FGF23 could be beneficial for patients with these diseases. Indeed, preliminary studies with a mouse disease model have validated this concept. On the other hand, replacement therapy with recombinant FGF23 may be applied to the disease caused by loss of function of FGF23. Although these concepts still need to be proven with more detailed analyses, the latest knowledge on the FGF23-related diseases and the development of methods to appropriately regulate FGF23 actions may synergistically create novel therapeutic maneuvers.

INTRODUCTION

Fibroblast growth factor (FGF) 23 was originally identified as a novel member of the FGF family with high homology to FGF15/19.[1] Since FGF23 has also been shown to be a humoral factor involved in the development of both autosomal dominant hypophosphatemic rickets/osteomalacia (ADHR)[2] and tumor-induced rickets/osteomalacia (TIO),[3,4] much attention has been paid on this last member of the FGF family in terms of its relevance to these diseases.

Endocrine FGFs and Klothos, edited by Makoto Kuro-o.
©2012 Landes Bioscience and Springer Science+Business Media.

FGF23 AS A NOVEL THERAPEUTIC TARGET

Animal studies, such as those conducted in transgenic mice, have demonstrated that excess activity of FGF23 can cause hypophosphatemic rickets/osteomalacia.[4-9] Indeed, clinical studies have revealed that several hypophosphatemic rickets/osteomalacia diseases, including the aforementioned two, are associated with inappropriately elevated FGF23 levels in circulation.[10-14] Conversely, there are hypophosphatemic diseases that exhibit low levels of FGF23 in circulation, such as Fanconi's syndrome and vitamin D deficiency.[15] Because levels of circulating FGF23 are known to decrease in response to a low phosphate diet,[16,17] it is likely that there is a physiological mechanism to decrease FGF23 in the situation of chronic hypophosphatemia and/or phosphate depletion. These observations have illuminated a novel concept that hypophosphatemic rickets/osteomalacia can be classified into FGF23-dependent and -independent diseases, emphasizing the importance of measuring blood levels of FGF23. Furthermore, it is conceivable that an intervention therapy that blocks FGF23 bioactivity with neutralizing antibodies or other antagonistic reagents can be applied to FGF23-dependent hypophosphatemic rickets/osteomalacia.

In contrast, the studies on *Fgf23* KO mice have revealed that FGF23 is essential for mammals to maintain normal phosphate and vitamin D metabolism.[18,19] Indeed, the indispensable role of FGF23 in mineral metabolism has been strengthened by a heterogeneous group of inherited disorders called familial hyperphosphatemic tumoral calcinosis characterized by hyperphosphatemia and ectopic calcification.[20-22] In this condition, the impaired function of FGF23 has been suggested to be the underlying cause,[23-27] and replacement therapy with recombinant FGF23 (or some agonistic reagents for FGF23 signaling) is likely to be effective.

Here we overview possible therapeutic benefits of targeting this molecule in FGF23-related diseases by summarizing the data from animal studies which demonstrate the biological and pharmacological effects of either exogenously injected recombinant FGF23 or neutralizing antibodies to FGF23.

BIOACTIVITIES OF RECOMBINANT PROTEIN OF FGF23

Rapid Actions

To elucidate the function of FGF23, several studies with animals have been conducted, in which pharmacological doses of recombinant protein of FGF23 were given to normal mice or rats as a rapid bolus injection. To date, three different actions have been reported as primary actions of FGF23 (Table 1), activities that take place within several hours.

The earliest change by FGF23 is observed in the parathyroid glands.[28,29] FGF23 can suppress the mRNA expression of PTH within an hour after the administration. FGF23 also decreases circulating PTH level within several minutes. Because reduction of PTH secretion was seen not only in mice, which received recombinant FGF23 systemically, but also in isolated parathyroid glands, this effect by recombinant FGF23 appears to be a result of direct action on parathyroid glands.

Other events that occur rapidly after the delivery of recombinant FGF23 are simultaneous changes in the degrees of renal mRNA expression of vitamin D-metabolizing enzymes, Cyp27b1 and Cyp24.[4,30,31] Cyp27b1 hydroxylates 25-hydroxyvitamin D [25(OH)D] and converts it to the active form of vitamin D, 1,25-dihydroxyvitamin D [1,25(OH)$_2$D].[32] Therefore, Cyp27b1 works to increase circulatory 1,25(OH)$_2$D level. On the other hand, Cyp24 hydroxylates 1,25(OH)$_2$D to form 1,24,25-trihydroxyvitamin D, which

Table 1. Summary of changes induced by FGF23

	Rapid Action	Long-Term Action
Blood phosphate level	low	very low
Urinary phosphate excretion	moderately increased	significantly increased
Npt2a, Npt2c protein level	decreased	significantly decreased
Blood 1,25-(OH)$_2$D level	very low	low normal
Cyp27b1 expression	significantly decreased	moderately decreased
Cyp24 expression	significantly increased	significantly increased
Blood PTH level	low	high
PTH mRNA in parathyroid	decreased	ND
Bone mineralization	not changed	severely impaired
Growth plate structure	not changed	moderately disorganized

ND: not determined

is readily recruited to a sequential degradation pathway.[32] The injection of recombinant FGF23 simultaneously induces a decrease in Cyp27b1 expression and an increase in Cyp24 expression, suppressing the activating step and promoting the degradation step of 1,25(OH)$_2$D, respectively. These changes were detected in the kidney within one hour after the injection of recombinant FGF23 and immediately followed by a significant decrease in circulating levels of 1,25-(OH)$_2$D.[31]

The third rapid action of recombinant FGF23 is a phosphaturic effect in the kidney. The kidney is the most important site for regulation of serum phosphate levels in the chronic state.[33] The blood phosphate is easily filtered through glomeruli and 80-90% of filtered phosphate is reabsorbed back to the circulation by proximal tubular epithelial cells. This reabsorption process is mostly carried out by two renal sodium-dependent phosphate cotransporters, NPT2a and NPT2c, which are localized in the apical brush border membrane of proximal tubular cells.[34,35] Because the targeted ablations of genes encoding these transporters (*Slc34a1, Slc34a3*) in mice[36,37] and mutations in NPT2c gene (*SLC34A3*) in humans[38-44] result in severe hypophosphatemia, the renal phosphate reabsorption conducted by these carriers is the most crucial process in determining serum phosphate levels.

FGF23 decreases protein levels of NPT2a and NPT2c.[4,30,31,45-47] Indeed, the mice, which received recombinant FGF23, exhibited downregulated NPT2a and NPT2c protein levels in kidney, which was accompanied with an increase in urinary phosphate excretion and a decrease in serum phosphate thereafter. This action appears to require a longer time (hours) than the effects on vitamin D metabolism and PTH secretion (minutes).[31] At the moment, the molecular mechanism by which FGF23 regulates NPT2a and NPT2c in the kidney remains unclear.

As it is well known that PTH can also downregulate protein levels of NPT2a and NPT2c, the contribution of PTH to the action of recombinant FGF23 was evaluated in parathyroidectomized mice/rats.[30,31] Consequently, recombinant FGF23 could downregulate both NPT2a and NPT2c in the animals without PTH. In addition, while FGF23 requires a few hours to regulate NPT2a levels, the PTH action on this transporter is quite rapid.[48] Therefore, FGF23 and PTH are assumed to play different physiological roles in phosphate metabolism, though either can function as a phosphaturic hormone that reduces the expression of NPT2a and NPT2c.

FGF23 AS A NOVEL THERAPEUTIC TARGET

161

The contribution of vitamin D to the action of FGF23 on the phosphate transporter was also evaluated. As 1,25(OH)$_2$D-responsive element has recently been reported in the promoter region of *Npt2a* gene[49]and the phosphaturic action of FGF23 is observed after a significant reduction in the circulating levels of 1,25(OH)$_2$D,[31] it is possible that the lowered 1,25(OH)$_2$D contributes to the downregulation of NPT2a and NPT2c. However, recombinant FGF23 could decrease both NPT2a protein levels and blood phosphate levels even in the vitamin D receptor knockout mice,[50,51]confirming that FGF23 can regulate the expression of NPT2a in a 1,25(OH)$_2$D-independent manner.

Long-Term Actions

The rapid decreases in phosphate, 1,25(OH)$_2$D and PTH levels were induced by a bolus injection of recombinant FGF23, which can be considered as primary actions of FGF23. In contrast, when animals are continuously exposed to excess amount of FGF23, as seen in transgenic mice[7-9] and nude mice bearing Chinese hamster ovary (CHO) cells stably expressing recombinant FGF23,[4-6] different and additional responses can be observed (Table 1). These studies have provided many insights into the understanding how excess activity of FGF23 causes hypophosphatemic disease or secondary hyperparathyroidism.

The most prominent difference between rapid and long-term actions of FGF23 appeared in bones. Either transgenic mice or nude mice with FGF23-expressing CHO cells demonstrated poorly mineralized bones, increased unmineralized osteoids, abnormal growth plate development and growth retardation. This experimental observation is consistent with the condition of hypophosphatemic rickets/osteomalacia.

Another remarkable difference was an increased circulating level of PTH, which was accompanied by hyperplasia of the parathyroid glands in mice bearing FGF23-expressing CHO cells.[6] This observation is contradictory to the finding seen in mice receiving a rapid bolus injection of recombinant FGF23,[28,29] which has not yet been fully explained. However, it is postulated that the continuously lowered 1,25(OH)$_2$D levels induced by FGF23 should work to decrease serum calcium levels, which can be a stronger stimulus for PTH secretion. In addition, low 1,25(OH)$_2$D itself can also play a role in enhancing PTH secretion. These situations likely give a significant change into the power balance between FGF23, PTH and 1,25(OH)$_2$D in the physiological condition, resulting in the enhanced PTH secretion, even while the FGF23 signal is still actively suppressing the secretion of PTH. Though further studies should be done to address this hypothesis, a similar pathophysiological pathway has been proposed in the development of secondary hyperparathyroidism in chronic kidney disease.[52-54]

HYPOPHOSPHATEMIC RICKETS/OSTEOMALACIA DUE TO EXCESS ACTION OF FGF23

Immunometric assays measuring concentrations of circulating FGF23 have enabled researchers to explore the pathophysiological role of FGF23 in various types of diseases with abnormal mineral and bone metabolism. In normal subjects, the concentration of FGF23 is maintained below a certain level, while patients with TIO show higher circulating levels of FGF23.[13,14] TIO is a paraneoplastic disease with sporadic hypophosphatemic rickets/osteomalacia, characterized by hypophosphatemia and low circulating levels of 1,25(OH)$_2$D.[55,56] As surgical removal of the tumors generally results in a complete resolution of symptoms, it had been postulated that some causative factor for this disease is secreted from the responsible tumors. Indeed, *FGF23* was identified as a dominantly expressed gene in the tumors resected from patients with TIO,[4] and elevated circulating levels of FGF23 were shown to decrease rapidly postresection, followed by recovery of blood phosphate levels.[14] This confirmed that FGF23 is a responsible factor for this disease.

In addition to TIO, surveys on circulating FGF23 levels in different types of hypophosphatemic rickets/osteomalacia have unveiled that there are several kinds of hypophosphatemic rickets/osteomalacia associated with high levels of FGF23 in circulation (Table 2). Given that the animals continuously exposed to excess amounts of recombinant FGF23 reproduced hypophosphatemic rickets/osteomalacia, it is conceivable that high levels of circulating FGF23 seen in various types of hypophosphatemic rickets/osteomalacia play a causative role, as is the case in TIO. These results raised two novel concepts: (1) that hypophosphatemic rickets/osteomalacia can be classified into FGF23-dependent and -independent types and (2) that the FGF23-dependent type can be improved by blocking the inappropriately excessive action of FGF23.

INHIBITION OF BIOACTIVITY OF CIRCULATING FGF23

The Structure of FGF23 and Its Receptor

Though there may be many strategies for blocking FGF23 activity, here we highlight three examples with neutralizing antibodies. Structurally, FGF23 consists of two different parts (Fig. 1): N-terminal portion, which shows high homology to other members of the FGF family, and C-terminal domain, which is unique to FGF23 but relatively similar to FGF19 and FGF21.[57,58] While there are only a few studies addressing the structure-function relationship of FGF23, these have provided the important insights that N- and C-terminal domains seem to be responsible for binding to the canonical FGF receptors and a

Table 2. Diseases caused by aberrant actions of FGF23

Disease	Serum Phosphate Level	*Responsible Gene* (Causative Factor)
Excess FGF23 action		
Tumor-induced rickets/osteomalacia (TIO)	P↓	FGF23
Autosomal dominant hypophosphatemic rickets/osteomalacia (ADHR)	P↓	*FGF23*
X-linked hypophosphatemic rickets/osteomalacia (XLH)	P↓	*PHEX*
Autosomal recessive hypophosphatemic rickets (ARHR)	P↓	*DMP1*
McCune-Albright syndrome/ fibrous dysplasia	P↓*	*GNAS*
Impaired FGF23 action		
Familial hyperphosphatemic tumoral calcinosis	P↑	*GALNT3, FGF23, KLOTHO*

* Only applicable to the case with high FGF23 level.

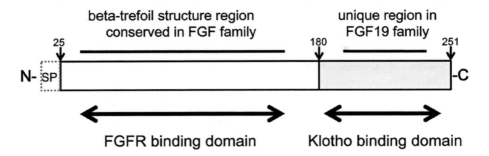

Figure 1. Schematic representation of FGF23 protein FGF23 consists of 251 amino acids. The N-terminal 24 amino acid-sequence functions as signal peptide (SP). FGF23 is highly susceptible to a protease cleavage that occurs between Arg-179 and Ser-180. The N-terminal domain forms a beta-trefoil structure and is relatively similar to other members of the FGF family, while the C-terminal domain is homologous only to the FGF19 family.

coreceptor, Klotho, respectively.[57-60] Klotho is a Type-I membrane protein and was originally identified as an aging-related molecule.[61] In the last few years, however, it has been proven that Klotho binds to FGF23 and enables it to activate the MAP kinase cascade via FGFR.[59,60] Importantly, FGF23 has shown only minimal affinity to any type of canonical FGF receptors in the absence of Klotho;[62] however, in the presence of Klotho, FGF23 is able to bind to FGFRs, thereby triggering the activation of downstream MAPK cascade; phosphorylations of FGFRs and ERK1/2 and upregulation of *Egr-1* mRNA expression, even in nonrenal cells such as CHO in the presence of Klotho.[60] Furthermore, when the expression of *Egr-1* mRNA was examined in various tissues after the injection of recombinant FGF23, increased expression was observed in quite a limited number of tissues (e.g., kidney, parathyroid and pituitary) where Klotho had been shown to be expressed.[60] Thus, Klotho plays a definitive role in the tissue-specific actions of FGF23. In other words, thanks to Klotho, FGF23, a systemic factor produced by bone,[63,64] can target the limited tissues without inappropriately activating canonical FGFRs that are ubiquitously expressed in various tissues.

Neutralizing Antibody to FGF23; Inhibition of FGF23-Klotho Interaction

Given the indispensable role of Klotho in FGF23 signaling, inhibiting the interaction between FGF23 and Klotho appears a conceivable strategy to block the FGF23 actions. Indeed, a mouse monoclonal antibody recognizing the C-terminal portion of FGF23 was shown by both immunoprecipitation and BIACORE analyses to hamper FGF23 binding to Klotho.[65] This resulted in the inhibition of the upregulation of *Egr-1* expression assessed by the reporter system using luciferase driven by the *Egr-1* promoter. Furthermore, administration of this antibody into normal mice resulted in increased serum phosphate and 1,25(OH)$_2$D levels, which were accompanied by increased renal expressions of NPT2a and Cyp27b1. These actions are mirror images of those seen in mice receiving recombinant FGF23, indicating that the antibody inhibited the endogenous FGF23 action by hampering the FGF23-Klotho interaction.

Neutralizing Antibody to FGF23; Inhibition of FGF23-FGFR Interaction

It may also be plausible that blocking the interaction between FGF23 and FGFRs is an effective way to inhibit the FGF23 action. At the moment, it remains controversial which FGF receptor(s) is/are involved in the FGF23/Klotho complex. So far FGFR1c, 3c and 4 have been reported to be coprecipitated with FGF23/Klotho complex.[59,60] However, there are some discrepancies between the in vitro and in vivo results obtained from different groups.[66,67] Given that high structural similarity exists between FGF23 and other members of the FGF family in the N-terminal region and that in vitro findings showed that FGFRs were phosphorylated in response to FGF23, it is quite likely that FGF23 requires canonical FGFRs in addition to Klotho for its action. Indeed, a mouse monoclonal antibody, whose epitope was in the conserved putative receptor-interacting region of the N-terminal portion of FGF23, not only inhibited the in vitro reporter activity using the *Egr-1* promoter, but also increased serum phosphate and $1,25(OH)_2D$ levels in normal mice.[65] It is of note that the changes induced by this antibody were the same as those seen after the administration of the antibody recognizing the C-terminus of FGF23.[65] Specifically, the similarities in efficacy and time-course suggest that inhibiting either FGF23-FGFR or FGF23-Klotho interaction can similarly affect FGF23 action.

Neutralizing Antibody to FGF23; Inhibition of Klotho Function

There is one report showing that a monoclonal antibody recognizing Klotho can function as a neutralizing antibody.[60] The mice treated with this antibody exhibited hyperphosphatemia and elevated $1,25(OH)_2D$ levels, as well as increased renal expressions of NPT2a and Cyp27b1.[60] These changes are consistent with the observations in mice treated with neutralizing antibodies to FGF23. It is of note that the authors also found that the mice treated with the Klotho antibody demonstrated a significant increase in circulating levels of FGF23. A similar elevation of blood FGF23 level was reported in Klotho-deficient mice.[60,68] Thus, it is obvious that the injection of antiKlotho antibody disturbed the function of Klotho, inducing an impaired action of FGF23. Based on the function of Klotho as a coreceptor for FGF23/FGFR, the Klotho-antibody mediated impairment of FGF23 action is likely the result of FGF23 insensitivity. This illustrates well that blocking Klotho function by an antibody (or some other tool) can affect FGF23 activity. In this regard, this antibody acted as a receptor antagonist for FGF23, reinforcing the importance of Klotho in the FGF23 signaling/activity.

The concept that blockades for Klotho function can impair FGF23 activity has been proven by a clinical finding that a homozygous missense mutation in *KLOTHO* gene leads to loss of function of FGF23.[69] This mutation was identified in a patient with severe tumoral calcinosis characterized by hyperphosphatemia, high circulating level of FGF23 and ectopic calcification. The authors' interpretation was that the mutation (H193R) which is localized within a putative glycosidase domain of Klotho, destabilized the protein. Actually, the expression level of recombinant mutant Klotho was markedly reduced compared to that of wild-type protein in vitro.

THE PHARMACOLOGICAL EFFECT OF NEUTRALIZING ANTIBODIES TO FGF23 IN HYPOPHOSPHATEMIC MICE

X-Linked Hypophosphatemic Rickets/Osteomalacia and FGF23 (XLH)

X-linked hypophosphatemic rickets/osteomalacia (XLH) is the most common form of inherited hypophosphatemic rickets characterized by hypophosphatemia associated with renal phosphate wasting.[70] The responsible gene for XLH is *PHEX* (phosphate-regulating gene with homologies to endopeptidases on the X-chromosome), which encodes a 749-amino acid Type I membrane protein and is believed to possess endopeptidase activity.[71] However, its physiological function remains unclear. Hypophosphatemic rickets/osteomalacia in patients with XLH, as well as its murine homologue, *Hyp* mouse,[72] had been postulated to be caused by a humoral factor derived from extrarenal tissue, termed as phosphatonin.[73-77] In both XLH patients and *Hyp* mice, the serum concentration of FGF23 is higher than that in normal subjects and wild-type mice, respectively.[11,14,64] In addition, mRNA expression of *Fgf23* has been shown to be upregulated in osteocytes in *Hyp* mice.[64] Furthermore, it has been demonstrated that the hypophosphatemic character of *Hyp* mice was reversed by crossing *Hyp* with *Fgf23* KO mice.[19] Taken together, excess amount of FGF23 in XLH/*Hyp* is considered to play a substantial, causative role as "phosphatonin".

Concept Validation Study for Therapeutic Maneuver of FGF23 Antibody in Hyp Mice

Administration of neutralizing antibodies against FGF23 or Klotho to wild-type mice has been shown to result in a successful blockade of FGF23 action in vivo. Based on this evidence, the hypothesis was proposed that such an FGF23-blockade could be a therapeutic maneuver in FGF23-dependent hypophosphatemic rickets/osteomalacia. To assess this, a concept validation study was conducted in *Hyp* mice, in which juvenile mice were treated with anti-FGF23 antibodies once weekly for one month.[78] The obtained results were striking (Table 3). Hypophosphatemia, which is one of the most characteristic phenotypes in this mutant strain, was corrected in a dose-dependent manner accomplished by recovery of renal NPT2a protein expression. Inappropriately low $1,25(OH)_2D$ levels and abnormally elevated alkaline phosphate levels were also improved. It is of note that blood PTH levels, which tend to be high in both *Hyp* mice and XLH patients, were suppressed after the one-month treatment. As expected from the improvement of renal phosphate wasting, most of rachitic features of *Hyp* bones were corrected. Increased osteoid was replaced by mineralized trabecular and cortical bone and organized columnar structures of chondrocytes were observed in metaphyseal growth plates. Thus, the therapy with anti-FGF23 antibody was very effective and did not show any life-threatening side effects in mice. This study has not only shown the potential of therapeutic use of an FGF23-blockade in patients with XLH, but also provided solid evidence that FGF23 is a long-sought phosphatonin playing a causative role in XLH/Hyp.

It may be also conceivable to hypothesize that therapy with anti-FGF23 antibodies could be effective in other FGF23-dependent hypophosphatemic rickets/osteomalacia

Table 3. Summary of pharmacological experiment with anti-FGF23 antibodies in *Hyp* mice

	Untreated *Hyp*	Treated *Hyp*
Blood phosphate level	very low	normal-high
Urinary phosphate excretion	high	normal
Blood 1,25(OH)$_2$D level	normal	high
Blood PTH level	moderately high	low
Blood ALP	very high	normal
NPT2a protein level	hardly detected	low normal
Cyp27b1 expression	normal	high
Cyp24 expression	normal	low normal
Growth	slightly retarded	almost normal
Bone		
Growth plate development	expanded, distorted	normal
Bone mineralization	impaired	normal
Unmineralized osteoid	increased	normal
Bone length	short	almost normal
BFR	hardly calculated	low normal

listed in Table. 2. Indeed, a preliminary study has demonstrated that hypophosphatemic rickets/osteomalacia of *Dmp1* knockout mouse, which is considered to be a murine model of ARHR, was corrected by repeated injections of anti-FGF23 antibodies.[79] Because *Dmp1* KO mice, as well as ARHR patients, are known to have high FGF23 levels in circulation, these conditions were assumed to be caused by excess actions of FGF23.[80,81] The fact that the neutralizing antibodies to FGF23 corrected this condition in the *Dmp1* KO mice reinforces the concept that the usefulness of this novel therapeutic strategy can be extended to other FGF23-dependent hypophosphatemic disorders.

THE POSSIBLE HORMONE REPLACEMENT THERAPY OF FGF23 IN TUMORAL CALCINOSIS

Familial hyperphosphatemic tumoral calcinosis is characterized by ectopic calcification in soft tissues and accompanied by hyperphosphatemia and high 1,25(OH)$_2$D levels in circulation.[21] To date, three genes (*GALNT3*, *FGF23* and *KLOTHO*), have been reported to be responsible for this disease and various mutations in these genes have been identified in affected family members.[22,24,69] Inactivating mutations in *GALNT3* or *FGF23* cause low level of full-length biologically active FGF23.[82] *GALNT3* gene product, UDP-N-acetyl-alpha-D-galactosamine: polypeptide N-acetylgalactosaminyltransferase 3, is required for the initiation of *O*-glycosylation and FGF23 has been shown to be *O*-glycosylated at Thr-178 by this enzyme.[83] Mutations in *GALNT3* gene are postulated to induce the susceptibility of FGF23 protein to processing between Arg-179 and Ser-180. This is likely due to the *O*-glycan at Thr-178 becoming attached by *GALNT3* gene product, which could sterically prevent the processing. Indeed, *Galnt3* KO mouse showed a lowered level of intact FGF23 in circulation, while the C-terminal fragment of FGF23 was highly accumulated in bone.[84] Mutations in *FGF23* gene have also been

FGF23 AS A NOVEL THERAPEUTIC TARGET

interpreted to impair the secretion of full-length FGF23. Therefore, replacement therapy with recombinant FGF23 appears to be ideal for these patients with low active FGF23. On the other hand, the other case of familial tumor calcinosis caused by a mutation in *KLOTHO* gene does not seem to be suitable for the replacement therapy because this patient is presumably resistant to FGF23 action as described above.

CONCLUSION

The blocking reagents for FGF23 activity can be novel therapeutic tools for hypophosphatemic rickets/osteomalacia caused by inappropriately enhanced FGF23 activity. This maneuver is simple and quite convincing as judged from the rationale of pathophysiology. The blockades can be anti-FGF23 antibodies recognizing either the receptor- or Klotho-binding domain, antiKlotho antibodies or small compounds interfering with FGF23/Klotho/FGFR interaction. In contrast, recombinant FGF23 or its agonistic reagents may be useful as replacement therapy for tumoral calcinosis with low full-length FGF23.

REFERENCES

1. Yamashita T, Yoshioka M, Itoh N. Identification of a novel fibroblast growth factor, FGF-23, preferentially expressed in the ventrolateral thalamic nucleus of the brain. Biochem Biophys Res Commun 2000; 277:494-498.
2. The ADHR Consortium. Autosomal dominant hypophosphataemic rickets is associated with mutations in FGF23. Nat Genet 2000; 26:345-348.
3. White KE, Jonsson KB, Carn G et al. The autosomal dominant hypophosphatemic rickets (ADHR) gene is a secreted polypeptide overexpressed by tumors that cause phosphate wasting. J Clin Endocrinol Metab 2001; 86:497-500.
4. Shimada T, Mizutani S, Muto T et al. Cloning and characterization of FGF23 as a causative factor of tumor-induced osteomalacia. Proc Natl Acad Sci USA 2001; 98:6500-6505.
5. Shimada T, Muto T, Urakawa I et al. Mutant FGF-23 responsible for autosomal dominant hypophosphatemic rickets is resistant to proteolytic cleavage and causes hypophosphatemia in vivo. Endocrinology 2002; 143:3179-3182.
6. Bai XY, Miao D, Goltzman D et al. The autosomal dominant hypophosphatemic rickets R176Q mutation in fibroblast growth factor 23 resists proteolytic cleavage and enhances in vivo biological potency. J Biol Chem 2003; 278:9843-9849.
7. Bai X, Miao D, Li J et al. Transgenic mice overexpressing human fibroblast growth factor 23 (R176Q) delineate a putative role for parathyroid hormone in renal phosphate wasting disorders. Endocrinology 2004; 145:5269-5279.
8. Larsson T, Marsell R, Schipani E et al. Transgenic mice expressing fibroblast growth factor 23 under the control of the alpha1(I) collagen promoter exhibit growth retardation, osteomalacia and disturbed phosphate homeostasis. Endocrinology 2004; 145:3087-3094.
9. Shimada T, Urakawa I, Yamazaki Y et al. FGF-23 transgenic mice demonstrate hypophosphatemic rickets with reduced expression of sodium phosphate cotransporter type IIa. Biochem Biophys Res Commun 2004; 314:409-414.
10. Yamamoto T, Imanishi Y, Kinoshita E et al. The role of fibroblast growth factor 23 for hypophosphatemia and abnormal regulation of vitamin D metabolism in patients with McCune-Albright syndrome. J Bone Miner Metab 2005; 23:231-237.
11. Weber TJ, Liu S, Indridason OS et al. Serum FGF23 levels in normal and disordered phosphorus homeostasis. J Bone Miner Res 2003; 18:1227-1234.
12. Riminucci M, Collins MT, Fedarko NS et al. FGF-23 in fibrous dysplasia of bone and its relationship to renal phosphate wasting. J Clin Invest 2003; 112:683-692.
13. Jonsson KB, Zahradnik R, Larsson T et al. Fibroblast growth factor 23 in oncogenic osteomalacia and X-linked hypophosphatemia. N Engl J Med 2003; 348:1656-1663.

14. Yamazaki Y, Okazaki R, Shibata M et al. Increased circulatory level of biologically active full-length FGF-23 in patients with hypophosphatemic rickets/osteomalacia. J Clin Endocrinol Metab 2002; 87:4957-4960.
15. Endo I, Fukumoto S, Ozono K et al. Clinical usefulness of measurement of fibroblast growth factor 23 (FGF23) in hypophosphatemic patients: proposal of diagnostic criteria using FGF23 measurement. Bone 2008; 42:1235-1239.
16. Burnett SM, Gunawardene SC, Bringhurst FR et al. Regulation of C-terminal and intact FGF-23 by dietary phosphate in men and women. J Bone Miner Res 2006; 21:1187-1196.
17. Ferrari SL, Bonjour JP, Rizzoli R. Fibroblast growth factor-23 relationship to dietary phosphate and renal phosphate handling in healthy young men. J Clin Endocrinol Metab 2005; 90:1519-1524.
18. Shimada T, Kakitani M, Yamazaki Y et al. Targeted ablation of Fgf23 demonstrates an essential physiological role of FGF23 in phosphate and vitamin D metabolism. J Clin Invest 2004; 113:561-568.
19. Sitara D, Razzaque MS, Hesse M et al. Homozygous ablation of fibroblast growth factor-23 results in hyperphosphatemia and impaired skeletogenesis and reverses hypophosphatemia in Phex-deficient mice. Matrix Biol 2004; 23:421-432.
20. Lyles KW, Burkes EJ, Ellis GJ et al. Genetic transmission of tumoral calcinosis: autosomal dominant with variable clinical expressivity. J Clin Endocrinol Metab 1985; 60:1093-1096.
21. Slavin RE, Wen J, Kumar D et al. Familial tumoral calcinosis. A clinical, histopathologic and ultrastructural study with an analysis of its calcifying process and pathogenesis. Am J Surg Pathol 1993; 17:788-802.
22. Topaz O, Shurman DL, Bergman R et al. Mutations in GALNT3, encoding a protein involved in O-linked glycosylation, cause familial tumoral calcinosis. Nat Genet 2004; 36:579-581.
23. Araya K, Fukumoto S, Backenroth R et al. A novel mutation in fibroblast growth factor 23 gene as a cause of tumoral calcinosis. J Clin Endocrinol Metab 2005; 90:5523-5527.
24. Benet-Pages A, Orlik P, Strom TM et al. An FGF23 missense mutation causes familial tumoral calcinosis with hyperphosphatemia. Hum Mol Genet 2005; 14:385-390.
25. Chefetz I, Heller R, Galli-Tsinopoulou A et al. A novel homozygous missense mutation in FGF23 causes Familial Tumoral Calcinosis associated with disseminated visceral calcification. Hum Genet 2005; 118:261-266.
26. Larsson T, Davis SI, Garringer HJ et al. Fibroblast growth factor-23 mutants causing familial tumoral calcinosis are differentially processed. Endocrinology 2005; 146:3883-3891.
27. Larsson T, Yu X, Davis SI et al. A novel recessive mutation in fibroblast growth factor-23 causes familial tumoral calcinosis. J Clin Endocrinol Metab 2005; 90:2424-2427.
28. Ben-Dov IZ, Galitzer H, Lavi-Moshayoff V et al. The parathyroid is a target organ for FGF23 in rats. J Clin Invest 2007; 117:4003-4008.
29. Krajisnik T, Bjorklund P, Marsell R et al. Fibroblast growth factor-23 regulates parathyroid hormone and 1alpha-hydroxylase expression in cultured bovine parathyroid cells. J Endocrinol 2007; 195:125-131.
30. Saito H, Kusano K, Kinosaki M et al. Human fibroblast growth factor-23 mutants suppress Na+-dependent phosphate cotransport activity and 1alpha, 25-dihydroxyvitamin D3 production. J Biol Chem 2003; 278:2206-2211.
31. Shimada T, Hasegawa H, Yamazaki Y et al. FGF-23 is a potent regulator of vitamin D metabolism and phosphate homeostasis. J Bone Miner Res 2004; 19:429-435.
32. Bikle D, Adams J, Christakos S. Vitamin D: Production, metabolism, mechanism of action and clinical requirements. In: Seeman E, ed. Primer on the Metabolic Bone Diseases and Disorders of Mineral Metabolism, 7th ed. Washington D.C.: The American Society for Bone and Research, 2008; 141-149.
33. Favus M, Bushinsky DA, Lemann J. Regulation of calcium, magnesium and phosphate metabolism. In: Christakos S, Holick MF, eds. Primer on the Metabolic Bone Diseases and Disorders of Mineral Metabolism, sixth ed. Washington, D.C.: the American Society for Bone and Mineral Research, 2006; 76-83.
34. Miyamoto K, Ito M, Tatsumi S et al. New aspect of renal phosphate reabsorption: the type IIc sodium-dependent phosphate transporter. Am J Nephrol 2007; 27:503-515.
35. Tenenhouse HS. Phosphate transport: molecular basis, regulation and pathophysiology. J Steroid Biochem Mol Biol 2007; 103:572-577.
36. Beck L, Karaplis AC, Amizuka N et al. Targeted inactivation of Npt2 in mice leads to severe renal phosphate wasting, hypercalciuria and skeletal abnormalities. Proc Natl Acad Sci USA 1998; 95:5372-5377.
37. Segawa H, Onitsuka A, Furutani J et al. Npt2a and Npt2c in mice play distinct and synergistic roles in inorganic phosphate metabolism and skeletal development. Am J Physiol Renal Physiol 2009; 297:F671-678.
38. Bergwitz C, Roslin NM, Tieder M et al. SLC34A3 mutations in patients with hereditary hypophosphatemic rickets with hypercalciuria predict a key role for the sodium-phosphate cotransporter NaPi-IIc in maintaining phosphate homeostasis. Am J Hum Genet 2006; 78:179-192.
39. Ichikawa S, Sorenson AH, Imel EA et al. Intronic deletions in the SLC34A3 gene cause hereditary hypophosphatemic rickets with hypercalciuria. J Clin Endocrinol Metab 2006; 91:4022-4027.
40. Lorenz-Depiereux B, Benet-Pages A, Eckstein G et al. Hereditary hypophosphatemic rickets with hypercalciuria is caused by mutations in the sodium-phosphate cotransporter gene SLC34A3. Am J Hum Genet 2006; 78:193-201.

FGF23 AS A NOVEL THERAPEUTIC TARGET

41. Yamamoto T, Michigami T, Aranami F et al. Hereditary hypophosphatemic rickets with hypercalciuria: a study for the phosphate transporter gene type IIc and osteoblastic function. J Bone Miner Metab 2007; 25:407-413.
42. Bastepe M, Juppner H. Inherited hypophosphatemic disorders in children and the evolving mechanisms of phosphate regulation. Rev Endocr Metab Disord 2008; 9:171-180.
43. Jaureguiberry G, Carpenter TO, Forman S et al. A novel missense mutation in SLC34A3 that causes hereditary hypophosphatemic rickets with hypercalciuria in humans identifies threonine 137 as an important determinant of sodium-phosphate cotransport in NaPi-IIc. Am J Physiol Renal Physiol 2008; 295:F371-379.
44. Levi M. Novel NaPi-2c mutations that cause mistargeting of NaPi-2c protein and uncoupling of Na-Pi cotransport cause HHRH. Am J Physiol Renal Physiol 2008; 295:F369-370.
45. Yamashita T, Konishi M, Miyake A et al. Fibroblast growth factor (FGF)-23 inhibits renal phosphate reabsorption by activation of the mitogen-activated protein kinase pathway. J Biol Chem 2002; 277:28265-28270.
46. Segawa H, Kawakami E, Kaneko I et al. Effect of hydrolysis-resistant FGF23-R179Q on dietary phosphate regulation of the renal type-II Na/Pi transporter. Pflugers Arch 2003; 446:585-592.
47. Baum M, Schiavi S, Dwarakanath V et al. Effect of fibroblast growth factor-23 on phosphate transport in proximal tubules. Kidney Int 2005; 68:1148-1153.
48. Pfister MF, Ruf I, Stange G et al. Parathyroid hormone leads to the lysosomal degradation of the renal type II Na/Pi cotransporter. Proc Natl Acad Sci USA 1998; 95:1909-1914.
49. Yamamoto H, Tani Y, Kobayashi K et al. Alternative promoters and renal cell-specific regulation of the mouse type IIa sodium-dependent phosphate cotransporter gene. Biochim Biophys Acta 2005; 1732:43-52.
50. Inoue Y, Segawa H, Kaneko I et al. Role of the vitamin D receptor in FGF23 action on phosphate metabolism. Biochem J 2005; 390:325-331.
51. Shimada T, Yamazaki Y, Takahashi M et al. Vitamin D receptor-independent FGF23 actions in regulating phosphate and vitamin D metabolism. Am J Physiol Renal Physiol 2005; 289:F1088-1095.
52. Fukagawa M, Nii-Kono T, Kazama JJ. Role of fibroblast growth factor 23 in health and in chronic kidney disease. Curr Opin Nephrol Hypertens 2005; 14:325-329.
53. Gutierrez O, Isakova T, Rhee E et al. Fibroblast growth factor-23 mitigates hyperphosphatemia but accentuates calcitriol deficiency in chronic kidney disease. J Am Soc Nephrol 2005; 16:2205-2215.
54. Kazama JJ, Gejyo F, Shigematsu T et al. Role of circulating fibroblast growth factor 23 in the development of secondary hyperparathyroidism. Ther Apher Dial 2005; 9:328-330.
55. Drezner MK, Feinglos MN. Osteomalacia due to 1 alpha,25-dihydroxycholecalciferol deficiency. Association with a giant cell tumor of bone. J Clin Invest 1977; 60:1046-1053.
56. Econs MJ, Drezner MK. Tumor-induced osteomalacia—unveiling a new hormone. N Engl J Med 1994; 330:1679-1681.
57. Goetz R, Beenken A, Ibrahimi OA et al. Molecular insights into the klotho-dependent, endocrine mode of action of fibroblast growth factor 19 subfamily members. Mol Cell Biol 2007; 27:3417-3428.
58. Beenken A, Mohammadi M. The FGF family: biology, pathophysiology and therapy. Nat Rev Drug Discov 2009; 8:235-253.
59. Kurosu H, Ogawa Y, Miyoshi M et al. Regulation of fibroblast growth factor-23 signaling by klotho. J Biol Chem 2006; 281:6120-6123.
60. Urakawa I, Yamazaki Y, Shimada T et al. Klotho converts canonical FGF receptor into a specific receptor for FGF23. Nature 2006; 444:770-774.
61. Kuro-o M, Matsumura Y, Aizawa H et al. Mutation of the mouse klotho gene leads to a syndrome resembling ageing. Nature 1997; 390:45-51.
62. Yu X, Ibrahimi OA, Goetz R et al. Analysis of the biochemical mechanisms for the endocrine actions of fibroblast growth factor-23. Endocrinology 2005; 146:4647-4656.
63. Mirams M, Robinson BG, Mason RS et al. Bone as a source of FGF23: regulation by phosphate? Bone 2004; 35:1192-1199.
64. Liu S, Zhou J, Tang W et al. Pathogenic role of Fgf23 in Hyp mice. Am J Physiol Endocrinol Metab 2006; 291:E38-49.
65. Yamazaki Y, Tamada T, Kasai N et al. Anti-FGF23 neutralizing antibodies show the physiological role and structural features of FGF23. J Bone Miner Res 2008; 23:1509-1518.
66. Liu S, Vierthaler L, Tang W et al. FGFR3 and FGFR4 do not mediate renal effects of FGF23. J Am Soc Nephrol 2008; 19:2342-2350.
67. Gattineni J, Bates C, Twombley K et al. FGF23 decreases renal NaPi-2a and NaPi-2c expression and induces hypophosphatemia in vivo predominantly via FGF receptor 1. Am J Physiol Renal Physiol 2009; 297:F282-291.
68. Segawa H, Yamanaka S, Ohno Y et al. Correlation between hyperphosphatemia and type II Na-Pi cotransporter activity in klotho mice. Am J Physiol Renal Physiol 2007; 292:F769-779.
69. Ichikawa S, Imel EA, Kreiter ML et al. A homozygous missense mutation in human KLOTHO causes severe tumoral calcinosis. J Musculoskelet Neuronal Interact 2007; 7:318-319.
70. Drezner MK. PHEX gene and hypophosphatemia. Kidney Int 2000; 57:9-18.

71. The HYP Consortium. A gene (PEX) with homologies to endopeptidases is mutated in patients with X-linked hypophosphatemic rickets. Nat Genet 1995; 11:130-136.
72. Beck L, Soumounou Y, Martel J et al. Pex/PEX tissue distribution and evidence for a deletion in the 3' region of the Pex gene in X-linked hypophosphatemic mice. J Clin Invest 1997; 99:1200-1209.
73. Meyer RA Jr, Meyer MH, Gray RW. Parabiosis suggests a humoral factor is involved in X-linked hypophosphatemia in mice. J Bone Miner Res 1989; 4:493-500.
74. Meyer RA Jr, Tenenhouse HS, Meyer MH et al. The renal phosphate transport defect in normal mice parabiosed to X-linked hypophosphatemic mice persists after parathyroidectomy. J Bone Miner Res 1989; 4:523-532.
75. Nesbitt T, Coffman TM, Griffiths R et al. Crosstransplantation of kidneys in normal and Hyp mice. Evidence that the Hyp mouse phenotype is unrelated to an intrinsic renal defect. J Clin Invest 1992; 89:1453-1459.
76. Grieff M. New insights into X-linked hypophosphatemia. Curr Opin Nephrol Hypertens 1997; 6:15-19.
77. Nelson AE, Mason RS, Robinson BG. The PEX gene: not a simple answer for X-linked hypophosphataemic rickets and oncogenic osteomalacia. Mol Cell Endocrinol 1997; 132:1-5.
78. Aono Y, Yamazaki Y, Yasutake J et al. Therapeutic Effects of Anti-FGF23 Antibodies in Hypophosphatemic Rickets/Osteomalacia. J Bone Miner Res 2009; 24:1879-1888.
79. Kang Y, Zhang R, Lu Y et al. Crossing Talk between Pi homeostasis and Bone. J Bone Miner Res 2009; 24.
80. Feng JQ, Ward LM, Liu S et al. Loss of DMP1 causes rickets and osteomalacia and identifies a role for osteocytes in mineral metabolism. Nat Genet 2006; 38:1310-1315.
81. Lorenz-Depiereux B, Bastepe M, Benet-Pages A et al. DMP1 mutations in autosomal recessive hypophosphatemia implicate a bone matrix protein in the regulation of phosphate homeostasis. Nat Genet 2006; 38:1248-1250.
82. Fukumoto S. Physiological regulation and disorders of phosphate metabolism—pivotal role of fibroblast growth factor 23. Intern Med 2008; 47:337-343.
83. Frishberg Y, Ito N, Rinat C et al. Hyperostosis-hyperphosphatemia syndrome: a congenital disorder of O-glycosylation associated with augmented processing of fibroblast growth factor 23. J Bone Miner Res 2007; 22:235-242.
84. Ichikawa S, Sorenson AH, Austin AM et al. Ablation of the Galnt3 gene leads to low-circulating intact fibroblast growth factor 23 (Fgf23) concentrations and hyperphosphatemia despite increased Fgf23 expression. Endocrinology 2009; 150:2543-2550.

CHAPTER 11

PHYSIOLOGY OF FGF15/19

Stacey A. Jones

Drug Discovery, Research and Development, GlaxoSmithKline, Research Triangle Park, North Carolina, USA.
Email: stacey.a.jones@gsk.com

Abstract: This chapter will review the various biological actions of the mouse fibroblast growth factor 15 (Fgf15) and human fibroblast growth factor 19 (FGF19). Unlike other members of the fibroblast growth factor (FGF) family, the Fgf15 and FGF19 orthologs do not share a high degree of sequence identity. Fgf15 and FGF19 are members of an atypical subfamily of FGFs that function as hormones. Due to subtle changes in tertiary structure, these FGFs have low heparin binding affinity enabling them to diffuse away from their site of secretion and signal to distant cells. FGF signaling through the FGF receptors is also different for this sub-family, requiring klotho protein cofactors rather than heparin sulfate proteoglycan. Mouse Fgf15 and human FGF19 play key roles in enterohepatic signaling, regulation of liver bile acid biosynthesis, gallbladder motility and metabolic homeostasis.

INTRODUCTION

The fibroblast growth factor (FGF) family consists of 22 structurally related proteins that are involved in various biological processes including embryonic development, wound healing, angiogenesis and metabolic signaling.[1-4] Many of the FGFs are secreted proteins involved in paracrine cell signaling. Because these FGFs display high affinity for heparan sulfate proteoglycans in the extracellular matrix of tissues, their signaling tends to remain spatially near their site of release. The typical FGF signals through the four cell surface tyrosine receptor kinase fibroblast growth factor receptors (FGFR): FGFR1, FGFR2, FGFR3 and FGFR4.

Within the family of FGFs exist seven subfamilies consisting of FGFs that display increased structural and functional similarity. Some of these subfamilies exhibit important differences from the typical FGF. One such subfamily consists of FGF11, FGF12, FGF13 and FGF14, also known as FGF homology factors, which are intracellular signaling

---

Endocrine FGFs and Klothos, edited by Makoto Kuro-o.
©2012 Landes Bioscience and Springer Science+Business Media.

molecules that do not signal through the FGFRs.[5,6] Another subfamily, known as the FGF19 subfamily, is comprised of the human FGF19, FGF21, FGF23 and mouse Fgf15 proteins. This subfamily is the most divergent of the FGF subfamilies with only 2 to 3% greater identity between the family members compared to members of other subfamilies.[1,7] Importantly, members of the FGF19 subfamily have reduced affinity for heparan sulfate proteoglycan and are thus able to serve as endocrine signaling molecules.[7,8] Recent work has uncovered important hormone actions of these proteins. The role of FGF21 in regulation of energy homeostasis and of FGF23 in phosphate and calcium homeostasis will be covered elsewhere in this volume. The endocrine actions of FGF19 and Fgf15 will be the focus of this chapter.

DISCOVERY OF MOUSE Fgf15 AND HUMAN FGF19

The gene encoding the mouse Fgf15 protein was identified in 1997 in a search for target genes of the chimeric homeodomain oncoprotein E2A-Pbx1.[9] Subtractive cDNA cloning identified a novel FGF expressed in mouse NIH3T3 fibroblasts following infection with a recombinant E2A-Pbx1-expressing retrovirus. The full-length clone encoded a 218 amino acid protein that was efficiently secreted from cells.[9] Although the new clone shared only 15-30% amino acid identity to other FGF family members, its genomic organization was similar to other members of the family, with three exons and two introns, increasing the confidence that this was indeed a divergent new member of the FGF family.[9] Expression analysis of the new Fgf15 in mouse embryo suggested a role in embryonic development. Fgf15 first appears in neuroectoderm between E7.5 and E8 and is predominantly expressed in the developing nervous system throughout development.[9] The role of Fgf15 in brain development has been recently reviewed and will not be discussed further in this chapter.[10]

The human FGF19 gene was first identified in 1999 following a search for cDNAs similar to mouse Fgf15 among GenBank expressed sequence tags (EST).[11] The EST from a Ntera-2 human neuroepithelial cell line encoded 95 amino acids with around 43% identity to the mouse Fgf15.[11] Rapid Amplification of cDNA Ends methodology was used to isolate the full-length cDNA from a human fetal brain cDNA library. The resulting full-length cDNA encoded a 216 amino acid protein containing an amino-terminal signal sequence, two highly conserved cysteine residues and a clear evolutionary relationship to other members of the FGF family.[11] The new FGF was most closely related to the mouse Fgf15 with 50% amino acid identity.[11,12] Because most mouse FGFs share greater than 90% sequence identity with their human orthologs and because distinct members of the FGF family typically share 50-70% amino acid identity, the new FGF was presumed to be a new member of the family and given the name FGF19.[11,12] Expression analysis of FGF19 revealed expression in brain, skin, cartilage, retina, gallbladder, small intestine, kidney, placenta and umbilical cord.[11,12] Binding studies revealed a unique aspect of this new FGF. Unlike other members of the FGF family, FGF19 bound only one of the fibroblast growth factor receptors, FGFR4.[12] Interestingly, FGF19 also demonstrated little mitogenic activity compared to other FGFs.[12]

SIGNALING MECHANISMS

FGFs bind and signal through the cell surface FGFRs. Typically this binding interaction requires stabilization by heparan sulfate glycosaminoglycan (HSGAG) cofactors that are present in the extracellular matrix.[1,4,8] The interaction with HSGAG also limits the diffusion of FGFs from their site of release and creates a low-affinity FGF reservoir that enhances paracrine signaling.[4,8] Early binding studies aimed at discriminating FGF-FGFR interactions documented the low affinity binding of FGF19 sub-family members compared to the nanomolar affinity of typical FGF-FGFR interactions.[13] In fact, FGF19, Fgf15 and the other members of the FGF19 subfamily have reduced affinity for HSGAG, which increases the radius of diffusion and enables endocrine signaling.[7,8] Yet without the needed stabilization of the FGF-FGFR interaction by HSGAGs, how does FGF19/Fgf15 binding and signaling proceed?

Recently a new family of cofactors has been identified that interacts with the FGF19 subfamily to promote FGFR signaling.[14-22] Klotho and βKlotho are single-pass transmembrane proteins with restricted tissue distribution.[19] The Klotho proteins were identified as cofactors of the FGF19 subfamily due to the striking similarity of phenotypes seen in knockout mice.[19] In the case of FGF19/Fgf15, βKlotho$^{-/-}$, FGFR4$^{-/-}$ and Fgf15$^{-/-}$ mice show overlapping phenotypes suggesting these proteins may function in a cooperative manner.

Studies to determine the role of βKlotho in FGF19 signaling utilizing cell-based systems demonstrated robust binding of FGF19 to FGFR4 in the presence, but not in the absence of βKlotho.[20-22] Binding to FGFR1c and weak binding to FGFR2c and FGFR3c was also seen under some conditions.[20] Furthermore, downstream phosphorylation of ERK1/2 and FRS2 was seen following FGF19 exposure only in the presence of βKlotho.[20] Such detailed studies have not been reported with Fgf15. Many assumptions about FGF19 biology that were based on early reports that FGF19 bound and signaled only through FGFR4 are now being reassessed in light of the discovery of the new binding cofactors.

Fgf19 AND Fgf15 ARE ORTHOLOGS

Following the discovery of the human FGF19 and mouse Fgf15 genes, efforts to find the mouse FGF19 and human Fgf15 orthologs were unsuccessful. Examination of the evolutionary relationships and genomic organization of the FGF family members revealed that FGF19 and Fgf15 reside on syntenic regions of the zebrafish, mouse and human genomes.[1,23] These genomic regions also contain the FGF3 and FGF4 genes. Indeed, evidence suggests the entire FGF19 subfamily, consisting of FGF19, FGF21, FGF23 and mouse Fgf15, arose from an ancestral FGF4 gene through local gene duplication around the time of the emergence of vertebrates.[2,24] Subsequent divergence of these genes gave rise to the four members of the FGF19 sub-family.[2,24]

Additional evidence beyond the genomic organization of FGF19 and Fgf15 has emerged supporting the orthologous relationship of these two genes. Molecular modeling and crystal structures indicate FGF19 and Fgf15 share important structural features that are divergent from the prototypical FGF and that these structural features are more tightly conserved between FGF19 and Fgf15 than the other members of the FGF19 sub-family. The crystal structure of FGF19, resolved to 1.3 Å, revealed two distinguishing features that set it apart from the typical FGF.[25] The first feature was a disordered region replacing the loop that typically forms the heparin-binding region of most FGFs.[7,25] Modeling of other FGF sequences onto the FGF19 structure revealed that only the FGF19 subfamily members, including Fgf15, showed disorder in this important region and consequently may also be missing the heparin-binding loop.[25] Indeed, reduced heparin binding of the FGF19 subfamily has been demonstrated and likely enables the endocrine signaling seen with these FGFs.[7,8] The second distinguishing feature of the FGF19 structure was the presence of two disulfide bonds that stabilize two extended βhairpin loops in the βtrefoil motif.[7] The only other FGF to contain both of the disulfide bonds seen in the FGF19 crystal structure is Fgf15.[7] Thus both protein structure and genomic structure suggest Fgf15 is the mouse ortholog of the human FGF19. Research into the biological role of these proteins further affirms that FGF19 and Fgf15 are orthologs.

REGULATION OF BILE ACID BIOSYNTHESIS

Further evidence of the orthologous relationship of Fgf15 and FGF19 is found in the transcriptional regulation and biological activities of these proteins. Two early clues gave insight into the role of Fgf15 and FGF19 in bile acid biology. The first clue was the discovery that both FGF19 and Fgf15 are direct target genes of Farnesoid X Receptor (FXR, NR1H4), a member of the nuclear hormone receptor class of ligand-activated transcription factors.[26,27] FXR is the master regulator of bile acid homeostasis and regulates genes involved in bile acid biosynthesis and transport.[28,29] FXR is most highly expressed in liver and intestine and functions as a heterodimer with the nuclear receptor Retinoid X Receptor (RXR), a common feature of nuclear receptor subfamily 1 members.[30,31] FXR is activated by endogenous bile acid ligands including the primary bile acid chenodeoxycholic acid (CDCA).[32-34] GW4064 is a commonly used synthetic agonist of FXR.[35]

Early work to uncover the biological role of FXR utilized CDCA and GW4064 to identify FXR target genes. FGF19 was the most highly induced gene following CDCA or GW4064-treatment of primary human hepatocytes in culture.[26] Additional work confirmed FGF19 as a direct target gene of FXR by identifying an FXR responsive element within the second intron of FGF19.[26] Interestingly, the same element, a perfect inverted-repeat separated by 1 nucleotide, or IR1, was also found in the second intron of Fgf15.[27] When these regions of intron 2 were cloned into plasmids containing a downstream reporter gene and then transfected into cells, responsiveness to FXR agonists was conferred.[26,27] Mutation of the IR1 sequence eliminated the FXR response. Binding experiments utilizing Electrophoretic Mobility Shift Assay (EMSA) demonstrated binding of the FXR/RXR heterodimer to the IR1 sequence of both FGF19 and Fgf15, confirming them as bona fide FXR target genes.[26,27,36,37]

A second important clue into the potential contribution of FGF19 to bile acid biology came from the phenotype of FGFR4 null mice. FGF19 had been reported to bind only to FGFR4. FGFR4 is the only FGFR expressed in mature hepatocytes.[12,38] FGFR4 null mice have an enlarged bile pool and elevated expression of CYP7A1, the gene encoding cholesterol 7-hydroxylase, the enzyme that catalyzes the first and rate-limiting step in bile acid biosynthesis.[39] Later, mice expressing a constitutively active form of FGFR4 were shown to have decreased CYP7A1 expression and reduced bile pool size.[40] This information, coupled with the knowledge that FGF19 expression was induced by bile acid-mediated activation of the FXR bile acid receptor lead to the intriguing possibility that FGF19 might play a role in feedback regulation of bile acid biosynthesis.

Regulation of CYP7A1 expression is controlled by a complex network of transcription factors and nuclear receptors (see recent reviews by Gilardi et al and Hylemon et al).[41,42] Bile acids are key mediators of transcriptional regulation of CYP7A1, working through multiple signaling pathways to ensure tight control of bile acid biosynthesis. Feedback repression of bile acid biosynthesis following activation of FXR was already known to occur through a regulatory cascade of nuclear receptors including Short Heterodimer Partner (SHP) and Liver Receptor Homolog-1 (LRH1).[43,44] SHP, another direct target gene of FXR, is an atypical nuclear receptor that lacks a DNA binding domain and acts as a transcriptional repressor.[45,46] LRH1 is an essential regulator of CYP7A1 expression and binds to the Bile Acid Responsive Element (BARE) in the CYP7A1 promoter.[47,48] The nuclear receptor HNF4 also drives CYP7A1 expression via the BARE.[49-52] FXR activation in hepatocytes by bile acid or GW4064 gave rise to increased SHP levels that bound and repressed LRH1 or HNF4 transcriptional activity on the CYP7A1 promoter.[43] It was important to investigate whether the FXR-FGF19 signaling pathway represented a potential new mechanism of FXR-mediated CYP7A1 regulation.

Treatment of primary human hepatocytes with FGF19 resulted in a striking repression of CYP7A1.[26] The repression of CYP7A1 by FGF19 was dependent on activation of c-Jun N-terminal kinase (JNK) but was shown to be distinct from the pathway of direct bile acid activation of JNK.[26,40] Additional studies have also shown involvement of the phosphatidylinositol (PI) 3-kinase signaling pathway.[53] Initially it was proposed that FGF19 acted in a paracrine manner in liver, signaling local feedback repression of bile acid biosynthesis, yet FGF19 did not appear to be expressed in normal liver, only in cultured hepatoctyes and cholestatic liver.[11,12,26,54] Expression of Fgf15 was not reported in liver either, suggesting the proposed paracrine mechanism was not active under normal physiologic conditions.[9,27]

The next advance in understanding FGF19/Fgf15 biology came following treatment of mice with the FXR agonists cholic acid and GW4064. Fgf15 was not induced in liver following treatment, however it was induced in intestine, especially in ileum where FXR expression is high.[27] FGFR4, however, was not expressed in ileum.[27] Could intestinal FGF19/Fgf15 act to repress bile acid biosynthesis through FGFR4 in the liver? Indeed, an "intestinal factor" regulating CYP7A1 transcription had been postulated in 1995 to reconcile the need for bile acids to pass through the intestine in order to affect transcriptional activity of CYP7A1 in the liver.[27,41,55] Inagaki et al. elegantly demonstrated that Fgf15 most likely is this enigmatic "intestinal factor."[27] Bile duct ligation (BDL) restricts bile acid flow to the intestine and results in increased CYP7A1 activity in liver. [27,56,57] BDL also eliminated Fgf15 expression in ileum.[27] Treatment of BDL mice with the FXR agonist GW4064 restored Fgf15 expression in ileum and reduced CYP7A1 mRNA 10-fold.[27] Exogenously applied Fgf15 also repressed CYP7A1 mRNA in wild-type mice, demonstrating Fgf15 could regulate CYP7A1 expression in an endocrine manner.[27,58] Inagaki et al. went on to demonstrate that both liver FGFR4 and SHP were necessary for Fgf15 mediated endocrine regulation of CYP7A1 by demonstrating the loss of Fgf15 repression of CYP7A1 in FGFR4[-/-] or SHP[-/-] mice. In addition, Fgf15[-/-] mice were shown to have elevated CYP7A1 expression, elevated cholesterol 7-hydroxylase protein and activity and increased fecal bile acid excretion.[27] Later, βKlotho mice were described that also had elevated bile acid synthesis and secretion and impaired repression of CYP7A1 expression by bile acids.[59] Thus Fgf15 is a bile acid- and FXR-regulated entero-hepatic signal that mediates feedback repression of hepatic bile acid biosynthesis in a βKlotho-, FGFR4- and SHP-dependent manner (Fig. 1).

Importantly, FGF19 appears to play the same role in human physiology. Serum FGF19 levels increase postprandially, at a time when bile acids are reabsorbed in the distal ileum for recirculation through the entero-hepatic circulation.[60] Shortly after serum FGF19 levels begin to rise, cholesterol 7-hydroxylase activity in the liver, as measured by the ratio of 7-hydroxy-4-cholesten-3-one to total serum cholesterol, declines steeply.[60] In the fasted state, serum bile acids and FGF19 levels remain low while bile acid biosynthesis is elevated.[60] Upon feeding of CDCA, FGF19 levels rise and administration of bile acid-binding resins decrease serum FGF19.[60] Therefore, Fgf15 and FGF19 are important endocrine signaling molecules that are tightly regulated by bile acid concentrations and, in turn, regulate bile acid biosynthesis.

REGULATION OF GALLBLADDER FILLING

Bile acids synthesized in the liver are stored in the gallbladder until needed for digestion. Upon delivery of protein or fat-rich chyme from the stomach to the proximal duodenum, the endocrine hormone cholecystokinin (CCK) is released from cells in the mucosal epithelium of the duodenum. CCK signals to the gallbladder and Sphincter of Oddi to induce gallbladder contraction, sphincter relaxation and delivery of bile to the intestine where it aids in the emulsification and digestion of fats.

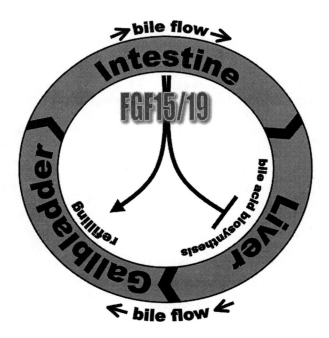

Figure 1. FGF15/19 Regulation of Enterohepatic Circulation. FGF15/19 is released from enterocytes of the intestine into the portal circulation during bile acid reabsorption in the ileum. FGF15/19 represses hepatic bile acid biosynthesis and promotes gallbladder refilling.

Hormonal regulation of gallbladder filling is mediated by Fgf15 and FGF19.[61] Gallbladders of Fgf15[-/-] mice do not fill with bile, yet have normal gallbladder epithelium.[61] Administration of recombinant FGF19 or Fgf15 to Fgf15[-/-] mice restored gallbladder filling, increasing gallbladder volume 10-fold within 15 minutes of treatment.[61] Gallbladder volume even increased by 2-fold in wild-type mice.[61] Interestingly, FGF19 administration increased gallbladder volume in FGFR4[-/-] mice that also have the depleted gallbladder phenotype, suggesting a different FGFR may be involved in this effect.[61,39] βKlotho[/] mice also have a depleted gallbladder, underscoring the importance of the βKlotho cofactor for Fgf15/FGF19 in regulating gallbladder filling.[59] Understanding of the downstream signaling pathways and the FGF receptors responsible for mediating these effects is still unclear, although relaxation of gallbladder smooth muscle by increasing cAMP levels likely plays a role.[61] Thus FGF19 and Fgf15 signal to both liver and gallbladder in the postprandial state when bile acid recycling through the ileum stimulates release of these hormones from the intestine, resulting in repression of hepatic bile acid biosynthesis and stimulation of gallbladder refilling (Fig. 1).[62]

CHRONIC DIARRHEA AND DEFECTIVE ILEAL FGF19 RELEASE

An intricate network of signaling pathways regulate bile acid biosynthesis and enteroheptic bile flow. The importance of maintaining bile acid concentrations within physiological ranges is underscored by these complex cascades. Sufficient availability of bile acids is key for nutrient absorption, but due to the detergent properties of these molecules, high levels of bile acids are toxic. Bile acid reabsorption in the distal ileum is a key step in enterhepatic circulation and bile acid recycling. Bile acid malabsorption (BAM) occurs in various pathophysiological conditions and leads to chronic watery diarrhea due to high colonic bile acid concentrations. Distinct clinical BAM syndromes have been identified. In Type 1 BAM surgical resection or inflammation, as in Crohn's disease, is the cause of the malabsorption and diarrhea. Type 2 BAM is of idiopathic origin with increased liver bile acid biosynthesis but no known ileal pathology.[63,64] Type 3 BAM describes various conditions including malabsorption secondary to the disturbed timing of bile flow following cholecystectomy.[65]

Mice lacking the intestinal apical sodium-dependent bile acid transporter, ASBT, provide a genetic model of BAM disease.[66] These mice do not express intestinal Fgf15 since bile acids are unable to enter the enterocytes and activate FXR.[67] As a consequence, the mice have elevated bile acid biosynthesis and elevated fecal bile acid excretion. Upon treatment with either Fgf15 or the FXR agonist GW4064, ASBT[-/-] mice had reduced bile acid biosynthesis and fecal bile acid excretion.[67] This suggests restoration of the FGF15 pathway can correct the fecal bile acid seen in BAM (Fig. 2). Further exploration into the etiology of Type 2 BAM, also called primary BAM, in light of the new mechanisms of bile acid regulation by Fgf15 and FGF19 led Walters et al to investigate FGF19 in patients with bile acid malabsorption.[68,69] Patients with primary idiopathic BAM had significantly reduced FGF19 levels compared to controls. Reduced FGF19 was also seen in Type 3 BAM patients following cholecystectomy and in Type 1 BAM patients following ileal surgery. Bile acid biosynthesis correlated inversely with FGF19 levels in these patients.[68] Because diagnostic tests for BAM are not widely available, Walters et al have suggested the investigation of plasma FGF19 as a diagnostic test for BAM.[68]

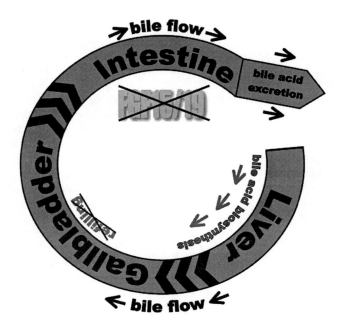

Figure 2. Bile Acid Malabsorption and Lost FGF15/19 Signaling. Interruption of bile acid reabsorption in the intestine causes a loss of FGF15/19 release and signaling. Consequently, hepatic bile acid biosynthesis increases and the gallbladder doesn't fill with bile. Increased intestinal flow of bile acids to the colon leads to bile acid diarrhea.

Low Fgf15 levels have also recently been shown to occur in antibiotic-treated mice.[70] Mice given ampicillin or the non-absorbed antibiotic cocktail bacitracin neomycin streptomycin have strongly diminished expression of Fgf15 mRNA resulting in increased bile acid biosynthesis and increased intestinal bile acid levels.[70] These findings raise the intriguing possibility that antibiotics cause intestinal side effects in part by disrupting the normal hormonal feedback regulation of bile acid homeostasis. The mechanism whereby antibiotics reduce Fgf15 expression needs further elucidation, but may involve the loss of secondary bile acids produced by enterobacteria that act as FXR agonists in the intestine.[70] Other species, including humans, make primary bile acids that can activate FXR[32-34] so it will be interesting to see if antibiotics affect FGF19 and bile acid homeostasis in additional species.

METABOLIC EFFECTS

The metabolic effects of FGF19 were appreciated soon after its discovery due to studies with transgenic mice in which FGF19 was expressed under the control of the muscle-specific myosin light chain promoter. Within the first month of life, FGF19 transgenic mice show a lean phenotype with decreased body weight due to decreased

PHYSIOLOGY OF FGF15/19

fat mass in spite of increased food intake.[71] Interestingly, the mice did not have impaired nutrient absorption as might be expected if FGF19 repression of bile acid biosynthesis significantly contracted the bile pool. Instead, the mice had an increased metabolic rate due to increased energy expenditure.[71] They also had lower serum glucose and insulin levels, improved glucose tolerance and improved insulin sensitivity compared to wild-type littermates.[71] Similar effects were seen after administration of recombinant FGF19 for 7 days to mice maintained on a high fat diet for 4-6 weeks.[72] Two potential mechanisms to explain the actions of FGF19 were proposed. First, FGF19 increased interscapular brown adipose tissue (BAT) and uncoupling protein-2 (UCP-2), which could contribute to the lean phenotype.[71,73,74] FGF19 also decreased levels of liver acetyl CoA carboxylase 2 (ACC2) which could increase mitochondrial oxidation of fatty acids.[71,72] ACC2 null mice also have a lean phenotype.[75] Like the ACC2 null mice, FGF19 transgenic mice do not become obese on a high fat diet. [71,75]

Recent work on the metabolic effects of Fgf15/FGF19 has demonstrated activation of the PI 3-kinase pathway and subsequent decreases in forkhead transcription factor 1 (FOXO1)-dependent gene expression in liver.[53] Activation of the PI3-kinase pathway leads to phosphorylation of the serine/threonine protein kinase Akt. Activated Akt phosphorylates FoxO1, causing its translocation to the cytoplasm thus reducing expression of FOXO1 target genes. Liver FOXO1 target genes include important components of energy metabolism including phosphoenolpyruvate carboxykinase (PEPCK), glucose-6-phosphatase, peroxisome proliferator-activated receptor gamma coactivator 1-alpha (PGC-1) and recently, CYP7A1.[53,76,77] It has been suggested that Fgf15/FGF19 may be able to activate the PI3-kinase/Akt/FOXO1 pathway to limit hepatic glucose production in diabetic conditions where insulin activation of the pathway is impaired.[53]

Consistent with FGF19 acting through FGFR4, FGFR4 null mice are insulin resistant, hypertriglyceridemic and have increased fast mass.[78] Restoration of FGFR4 selectively to hepatocytes of FGFR4 null mice decreased plasma lipid levels but failed to restore normal insulin sensitivity,[78] raising the possibility that FGF19 acts on FGFR4 not only in liver but in other tissues as well. In this regard, Fu et al. showed that intracerebroventricular injection of FGF19 in mice increased metabolic rate to a comparable degree as systemic administration, indicating that FGF19 can act through the central nervous system.[72] Additional studies in which FGFR4 is selectively eliminated in different tissues will be required to determine precisely where FGF19 acts to exert its metabolic effects.

CONCLUSION AND FUTURE DIRECTIONS

Much has been learned in the decade following the discovery of FGF19. Reconciling the orthologous relationship of Fgf15 and FGF19 has enabled study of the biological roles of these signaling molecules in animal models. The use of knockout mice has been pivotal in understanding the multi-organ signaling systems of these hormones. Following the identification of the Klotho binding partner for these FGFs, it will be useful to clarify the receptor utilization for both Fgf15 and FGF19, especially in tissues such as gallbladder.

Additional clarification on the mitogenic potential of Fgf15 and FGF19 in the presence of the Klotho binding partners will also be important areas of future research. FGF19 transgenic mice develop hepatocellular carcinoma,[79-81] but due to the significant divergence between these orthologs it is unclear whether this model is predictive of either Fgf15 or FGF19 biology in their natural context. It bears repeating that Fgf15 and FGF19 are only marginally more similar to each other than to the other members of the sub-family, FGF21 and FGF23. Furthermore, FGF19 has also been identified as a tumor suppressor gene, underscoring the need for further research in this area.[82]

The therapeutic potential of FGF19 signaling pathways has been recognized and will certainly be a focus of future study. Intervention in diabetes and obesity,[71,72,53] gallstone disease,[61,62] liver disease[26] and bile acid diarrhea[68] in BAM has been proposed and will certainly continue to be investigated.

REFERENCES

1. Ornitz DM, Itoh N. Fibroblast growth factors. Genome Biology 2001; 2(3): reviews3005.1-3005.12.
2. Itoh N, Ornitz DM. Evolution of the Fgf and Fgfr gene families. Trends in Genetics 2004; 20(11):563-569.
3. Popovici C, Roubin R, Coulier F et al. An evolutionary history of the FGF superfamily. BioEssays 2005; 27:849-857.
4. Beenken A, Mohammadi M. The FGF family: biology, pathophysiology and therapy. Nature Reviews Drug Discovery 2009; 8:235-253.
5. Olsen SK, Garbi M, Zampieri N et al. Fibroblast growth factor (FGF) homologous factors share structural but not functional homology with FGFs. JBC 2003; 278(36):34226-34236.
6. Goldfarb M. Fibroblast growth factor homologous factors: evolution, structure and function. Cytokine Growth Factor Rev 2005; 16(2):215-220.
7. Goetz R, Beenken A, Ibrahimi OA et al. Molecular insights in the Klotho-dependent, endocrine mode of action of fibroblast growth factor 19 subfamily members. Mol Cell Biol 2007; 27:3417-3428.
8. Asada M, Shinomiya M, Suzuki M et al. Glycosaminoglycan affinity for the complete fibroblast growth factor family. BBA 2009; 1790:40-48.
9. McWhirter JR, Goulding M, Weiner JA et al. A novel fibroblast growth factor gene expressed in the developing nervous system is a downstream target of the chimeric homeodomain oncoprotein E2A-Pbx1. Development 1997; 124:3221-3232.
10. Iwata T, Hevner RF. Fibroblast growth factor signaling in development of the cerebral cortex. Develop. Growth Differ 2009; 51:299-323.
11. Nishimura T, Utsunomiya Y, Hoshikawa M et al. Structure and expression of a novel human FGF, FGF-19, expressed in the fetal brain. BBA 1999; 1444:148-151.
12. Xie M, Holcomb I, Deuel B et al. FGF-19, A novel fibroblast growth factor with unique specificity for FGFR4. Cytokine 1999; 11(10):729-735.
13. Zhang X, Ibrahimi OA, Olsen SK et al. Receptor specificity of the fibroblast growth factor family. The complete mammalian FGF family. JBC 2006; 281:15694-15700.
14. Kurosu H, Ogawa Y, Miyoshi M et al. Regulation of Fibroblast growth factor-23 signaling by Klotho. JBC 2006; 281(10):6120-6123.
15. Urakawa I, Yamazaki Y, Shimada T et al. Klotho converts canonical FGF receptor into a specific receptor for FGF23. Nature 2006; 444:770-774.
16. Kurosu H, Kuro-o M. The klotho gene family and the endocrine fibroblast growth factors. Curr Opin Nephrol Hypertens 2008; 17:368-372.
17. Kuro-o M. Endocrine FGFs and klothos: emerging concepts. Trends Endocrin Metab 2008; 19(7):239-245.
18. Wu X, Lemon B, Si X et al. C-terminal tail of FGF19 determines its specificity toward klotho coreceptors. JBC 2008; 283(48):33304-33309.
19. Kurosu H, Kuro-o M. The klotho gene family as a regulator of endocrine fibroblast growth factors. Mol Cell Endo 2009; 299:72-78.
20. Kurosu H, Choi M, Ogawa Y et al. Tissue-specific expression of Klotho and fibroblast growth factor (FGF) receptor isoforms determines metabolic activity of FGF19 and FGF21. JBC 2007; 282(37):26687-26695.
21. Lin BC, Wang M, Blackmore C et al. Liver-specific activities of FGF19 require klotho beta. JBC 2007; 282(37):27277-27284.
22. Wu X, Ge H, Gupte J et al. Co-receptor requirements for fibroblast growth factor-19 signaling. JBC 2007; 282(40):29069-29072.
23. Katoh Y, Katoh M. Evolutionary conservation of CCND1-ORAOV1-FGF19-FGF4 locus from zebrafish to human. Int J Mol Med 2003; 12:45-50.
24. Itoh N, Ornitz DM. Functional evolutionary history of the mouse Fgf gene family. Developmental Dynamics 2008; 237:18-27.

PHYSIOLOGY OF FGF15/19

25. Harmer NJ, Pellegrini L, Chirgadze D et al. The crystal structure of fibroblast growth factor (FGF) 19 reveals novel features of the FGF family and offers a structural basis for its unusual receptor affinity. Biochemistry 2004; 43:629-640.

26. Holt JA, Luo G, Billin AN. Definition of a novel growth factor-dependent signal cascade for the suppression of bile acid biosynthesis. Genes and Dev 2003; 17:1581-1591.

27. Inagaki T, Choi M, Moschetta A et al. Fibroblast growth factor 15 functions as an enterohepatic signal to regulate bile acid homeostasis. Cell Metabolism 2005; 2:217-225.

28. Chiang JYL. Bile acid regulation of gene expression: roles of nuclear hormone receptors. Endocrine Reviews 2002; 23(4): 443-463.

29. Kuipers F, Claudel T, Sturm E et al. The Farnesoid X Receptor (FXR) as modulator of bile acid metabolism. Reviews in Endocrine and Metabolic Disorders 2004; 5:319-326.

30. Forman BM, Goode E, Chen J et al. Identification of a nuclear receptor that is activated by farnesol metabolites. Cell 1995; 81:687-693.

31. Seol W, Choi HS, Moore DD. Isolation of proteins that interact specifically with the retinoid X receptor: Two novel orphan receptors. Mol Endocrinol 1995; 9:72-85.

32. Makishima M, Okamoto AY, Repa JJ et al. Identification of a nuclear receptor for bile acids. Science 1999; 284:1362-1365.

33. Parks DJ, Blanchard SG, Bledsoe RK et al. Bile acids: natural ligands for an orphan nuclear receptor. Science 1999; 284:1365-1368.

34. Wang H, Chen J, Hollister K et al. Endogenous bile acids are ligands for the nuclear receptor FXR/BAR. Mol Cell 1999; 3:543-553.

35. Maloney PR, Parks DJ, Haffner CD et al. Identification of a chemical tool for the orphan nuclear receptor FXR. J Med Chem 2000; 43:2971-2974.

36. Li J, Pircher PC, Schulman IG et al. Regulation of complement C3 expression by the bile acid receptor FXR. JBC 2005; 280:7427-7434.

37. Song KH, Li T, Owsley E et al. Bile acids activate fibroblast growth factor 19 signaling in human hepatocytes to inhibit cholesterol 7-hydroxylase gene expression. Hepatology 2009; 49:297-305.

38. Kan M, Wu X, Wang F et al. Specificity for fibroblast growth factors determined by heparan sulfate in a binary complex with the receptor kinase. JBC 1999; 274(22):15947-15952.

39. Yu C, Wang F, Kan M et al. Elevated cholesterol metabolism and bile acid synthesis in mice lacking membrane tyrosine kinase receptor FGFR4. JBC 2000; 275(20):15482-15489.

40. Yu C, Wang F, Jin C et al. Independent repression of bile acid synthesis and activation of c-Jun N-terminal kinase (JNK) by activated hepatocyte fibroblast growth factor receptor 4 (FGFR4) and bile acids. JBC 2005; 280:17707-17714.

41. Gilardi F, Mitro N, Godio C et al. The pharmacological exploitation of cholesterol 7-hydroxylase, the key enzyme in bile acid synthesis: from binding resins to chromatin remodeling to reduce plasma cholesterol. Pharmacol and Therapeutics 2007; 116449-472.

42. Hylemon PB, Zhou H, Pandak WM et al. Bile acids as regulatory molecules. J Lipid Res 2009; 50:1509-1520.

43. Goodwin B, Jones SA, Price RR et al. A regulatory cascade of the nuclear receptors FXR, SHP-1 and LRH-1 repress bile acid biosynthesis. Mol Cell 2000; 6:517-526.

44. Lu TT, Makishima M, Repa JJ et al. Molecular basis for feedback regulation of bile acid synthesis by nuclear receptors. Mol Cell 2000; 6:507-515.

45. Seol W, Choi HS, Moore DD. An orphan nuclear hormone receptor that lacks a DNA binding domain and heterodimerizes with other receptors. Science 1996; 272:1336-1339.

46. Seol W, Chung M, Moore DD. Novel receptor interaction and repression domains in the orphan receptor SHP. Mol Cell Biology 1997; 17(12):7126-7131.

47. Becker-Andre M, Andre E, DeLamarter JF. Identification of nuclear receptor mRNAs by RT-PCR amplification of conserved zinc-finger motif sequences. Biochem Biophys Res Commun 1993; 194:1371-1379.

48. Nita M, Du S, Brown C et al. CPF: an orphan nuclear receptor that regulates liver-specific expression of the human cholesterol 7alpha-hydroxylase gene. PNAS 1999; 96:6660-6665.

49. Cooper AD, Chen J, Botelho-Yetkinler MJ et al. Characterization of hepatic-specific regulatory elements in the promoter region of the human cholesterol 7-hydroxylase gene. JBC 1997; 272(6):3444-3452.

50. Lee Y-K, Dell H, Dowhan DH et al. The orphan nuclear receptor SHP inhibits hepatocyte nuclear factor 4 and retinoid X receptor transactivation: Two mechanisms for repression. Mol Cell Bio 2000; 20(1):187-195.

51. DeFabiani E, Nitro N, Anzulovich AC et al. The negative effects of bile acids and tumor necrosis factor- on the transcription of cholesterol 7-hydroxylase gene (CYP7A1) converge to hepatic nuclear factor-4: a novel mechanism of feedback regulation of bile acid synthesis mediated by nuclear receptors. JBC 2001; 276:30708-30716.

52. Stroup D, Chiang JYL. HNF4 and COUP-TFII interact to modulate transcription of the cholesterol 7-hydroxyase gene (CYP7A1). J Lipid Res 2000; 41:1-11.

53. Shin DJ, Osborne TF. FGF15/FGFR4 Integrates growth factor signaling with hepatic bile acid metabolism and insulin action. JBC 2009; 284(17):1110-11120.

54. Schaap FG, van der Gaag NA, Gouma KJ et al. High expression of the bile salt-homeostatic hormone fibroblast growth factor 19 in the liver of patients with extrahepatic cholestasis. Hepatology 2009; 49:1228-1235.
55. Pandak WM, Heuman DM, Redford K et al. Hormonal regulation of cholesterol 7-hydroxylase specific activity, mRNA levels and transcriptional activity in vivo in the rat. J Lipid Res 1997; 38:2483-2491.
56. Dueland S, Reichen J, Everson GT et al. Regulation of cholesterol and bile acid homeostasis in bile-obstructed rats. Biochem J 1991; 280:373-377.
57. Gustafsson J. Effect of biliary obstruction on 26-hydroxylation of C27-steroids in bile acid synthesis. J Lipid Res 1978; 19:237-243.
58. Kim I, Ahn SH, Inagaki T et al. Differential regulation of bile acid homeostasis by the farnesoid X receptor in liver and intestine. J Lipid Res 2007; 48:2664-2672.
59. Ito S, Fujimori T, Furuya A et al. Impaired negative feedback suppression of bile acid synthesis in mice lacking Klotho. J Clinical Invest 2005; 115(8):2202-2208.
60. Lundåsen T, Gälman C, Angelin B et al. Circulating intestinal fibroblast growth factor 19 has a pronounced diurnal variation and modulates hepatic bile acid synthesis in man. J Intern Med 2006; 260:530-536.
61. Choi M, Moschetta A, Bookout AL et al. Identification of a hormonal basis for gallbladder filling. Nature Med 2006; 12(11):1253-1255.
62. Portincasa P, Di Ciaula A, Wang HH et al. Coordinate regulation of gallbladder motor function in the gut-liver axis. Hepatology 2008; 47:2112-2126.
63. Thaysen EH, Pedersen L Diarrhoea associated with idiopathic bile acid malabsorption. Fact or fantasy? Dan Med Bull 1973; 20:174-177.
64. Thaysen EH, Pedersen L. Idiopathic bile acid catharsis. Gut 1976; 17:965-970.
65. Sauter GH, Moussavian AC, Meyer G et al. Bowel habits and bile acid malabsorption in the months after cholecystectomy. Am J Gastroenterol 2002; 97:1732-1735.
66. Dawson PA, Haywood J, Craddock AL et al. Targeted deletion of the ileal bile acid transporter eliminates enterohepatic cycling of bile acids in mice. JBC 2003; 278:33920-33927.
67. Jung D, Inagaki T, Gerard RD et al. FXR agonists and FGF15 reduce fecal bile acid excretion in a mouse model of bile acid malabsorption. J Lipid Res 2007; 48(12):2693-2700.
68. Walters JRF, Tasleem AM, Omer OS et al. A new mechanism for bile acid diarrhea: effective feedback inhibition of bile acid biosynthesis. Clin Gastroenterol Hepatol 2009; 7(11):1189-94 2009.
69. Hofmann AF, Mangelsdorf DJ, Kliewer SA. Chronic diarrhea due to excessive bile acid synthesis and not defective ileal transport: a new syndrome of defective fibroblast growth factor 19 release. Clin Gastroenterol Hepatol 2009; 7(11): 1151–1154.
70. Miyata M, Yakamatsu Y, Kuribayashi H et al. Administration of ampicillin elevates hepatic primary bile acid synthesis through suppression of ileal FGF15 expression. JPET, doi:10.1124/jpet/109.160093
71. Tomlinson E, Fu L, John L et al. Transgenic mice expressing human fibroblast growth factor-19 display increased metabolic rate and decreased adiposity. Endocrinology 2002; 143(5):1741-1747.
72. Fu L, John LM, Adams SH et al. Fibroblast growth factor 19 increases metabolic rate and reverses dietary and leptin-deficient diabetes. Endocrinology 2004; 145(6):2594-2603.
73. Fleury C, Neverova M, Collins S et al. Uncoupling protein-1: a novel gene linked to obesity and hyperinsulinemia. Nat Genet 1997; 15:269-272.
74. Pinkney JH, Boss O, Bray GA et al. Physiological relationships of uncoupling protein-2 gene expression in human adipose tissue in vivo. J Clin Endocrinol Metab 2000; 85:2312-2317.
75. Abu-Elheiga L, Matzuk MM, Abo-Hashema KA et al. Continuous fatty acid oxidation and reduced fat storage in mice lacking acetyl-CoA carboxylase 2. Science 2001; 291:2613-2616.
76. Li T, Kong X, Owsley E et al. Insulin regulation of cholesterol 7-hydroxylase expression in human hepatocytes: roles of forkhead box O1 and sterol regulatory element-binding protein 1c. JBC 2006; 281(39):28745-28754.
77. Li T, Ma H, Chiang JY. TGF1, TNF and insulin signaling crosstalk in regulation of the rat cholesterol 7-hydroxylase gene expression. J Lipid Res 2008; 49(9):1981-1989.
78. Huang X, Yang C, Luo Y et al. FGFR4 prevents hyperlipidemia and insulin resistance but underlies high-fat diet-induced fatty liver. Diabetes 2007; 56:2501-2510.
79. Nicholes K, Guillet S, Tomlinson E et al. A mouse model of hepatocellular carcinoma. Ectopic expression of fibroblast growth factor 19 in skeletal muscle of transgenic mice. Am J Pathology 2002; 160(6):2295-2307.
80. Pai R, Dunlap D, Qing J et al. Inhibition of fibroblast growth factor 19 reduces tumor growth by modulating -catenin signaling. Cancer Res 2008; 68(13):5086-5095.
81. Desnoyers LR, Pai R, Ferrando RE et al. Targeting FGF19 inhibits tumor growth in colon cancer xenograft and FGF19 transgenic hepatocellular carcinoma models. Oncogene 2008; 27:85-97.
82. Lopez-Serra L, Ballestar E, Ropero S et al. Unmasking of epigenetically silenced candidate tumor suppressor genes by removal of methyl-CpG-binding domain proteins. Oncogene 2008; 27:3556-3566.

CHAPTER 12

FGF19 AND CANCER

Benjamin C. Lin and Luc R. Desnoyers*

*Oncology Biomarker Development/Pharmacodynamic Biomarkers, Genentech, Inc.,
South San Francisco, California, USA.
Corresponding Author: Luc R. Desnoyers—Email: desnoyer@gene.com

Abstract: Fibroblast growth factors (FGFs) and their cognate receptors, FGF receptors (FGFRs), play critical roles in a variety of normal developmental and physiological processes. Numerous reports support a role for deregulation of FGF-FGFR signaling, whether it is at the ligand and/or receptor level, in tumor development and progression. The FGF19-FGFR4 signaling axis has been implicated in the pathogenesis of several cancers, including hepatocellular carcinomas in mice and potentially in humans. This chapter focuses on recent progress in the understanding of the molecular mechanisms of FGF19 action and its potential involvement in cancer.

INTRODUCTION

Fibroblast growth factors (FGFs) comprise a family of 22 structurally related polypeptides with diverse biological activities.[1] Most of these signaling molecules bind to and activate members of the FGF receptor (FGFR) family. The FGFR family is composed of four receptor tyrosine kinases, designated FGFR1 - FGFR4, and one receptor which lacks a cytoplasmic tyrosine kinase domain, designated FGFR5.[2,3] The interaction of FGFs with FGFR1-4 results in receptor dimerization and autophosphorylation, recruitment of membrane-associated and cytosolic accessory proteins, and initiation of multiple signaling cascades.[4] The FGF-FGFR signaling network plays important roles in development and tissue repair by regulating cellular functions/processes such as growth, differentiation, migration, morphogenesis, and angiogenesis. Not surprisingly, dysregulation of this signaling system appears to be important for tumor development and progression.

Endocrine FGFs and Klothos, edited by Makoto Kuro-o.
©2012 Landes Bioscience and Springer Science+Business Media.

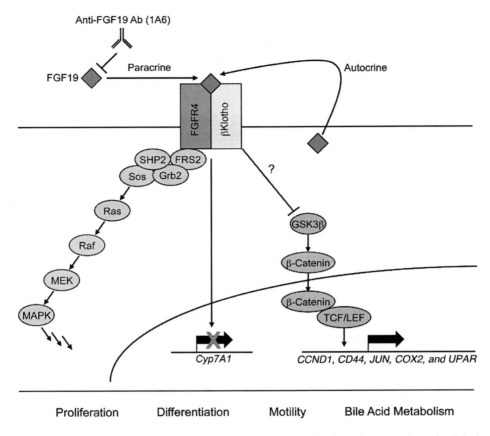

Figure 1. FGF19-FGFR4 Signaling Pathway. The growth factor FGF19 regulates a variety of cellular/physiologic functions in normal and possibly neoplastic states (e.g., bile acid metabolism, proliferation, differentiation, and cellular motility). Signaling by this ligand is mediated by its cognate receptor FGFR4 and a recently identified coreceptor βKlotho. FGF19-mediated activation of mitogen-activated protein kinase (MAPK) and β-catenin pathways may be involved in the development and progression of cancers, such as hepatocellular carcinoma. Blocking the interaction of FGF19 with FGFR4 using a high affinity, high specificity anti-FGF19 antibody (1A6) inhibits FGF19 signaling and tumor growth in vivo.

Most FGFs function primarily in a paracrine and/or autocrine fashion. However, the FGF19 subfamily members (FGF19/Fgf15, FGF21, and FGF23) can function as endocrine factors or hormones. FGF19 is an important regulator of metabolism under normal physiological conditions. In disease states, FGF19 may be critical for the development and progression of a number of cancers. Of particular note, FGF19 signaling has been proposed to be important in promoting hepatocellular carcinomas (HCCs) in mice and potentially in humans.[5-7] As illustrated in Figure 1, the effects of FGF19 on downstream signaling (e.g., mitogen-activated protein kinase and β-catenin pathways) are mediated by its cognate receptor FGFR4 and a recently identified coreceptor βKlotho (KLB).[8,9] Thus, FGF19 may serve as an important therapeutic target in the treatment of relevant cancers.

FGF19/FGF15

FGF19 was cloned by homology to its mouse ortholog *Fgf15*.[10,11] *FGF19* messenger RNA (mRNA) is found in brain, skin, cartilage, retina, gall bladder, small intestine, and kidney.[10,11] The expression of FGF19/Fgf15 is induced in the ileum in response to bile acids that are released into the intestinal lumen after feeding. FGF19/Fgf15 then circulates to the liver to suppress the expression of CYP7A1, the rate-limiting enzyme for bile acid synthesis.[12] FGF19/Fgf15 also limits bile acid release into the intestine by triggering gall bladder filling.[13] In this manner, FGF19/Fgf15 serves as a key regulator in a postprandial negative feedback loop modulating bile acid synthesis and release.

FGF19 BINDING SPECIFICITY: FGFR4 AND THE βKLOTHO CORECEPTOR

Among the FGF family of ligands, FGF19 has a unique receptor binding specificity. Initially, FGF19 was shown to bind exclusively to FGFR4.[11] Using co-immunoprecipitation, Xie et al assessed the binding of FGF19 to immunoadhesin chimeric constructs of a subset of the human FGFRs (i.e., FGFR1, FGFR2-3 IIIb and IIIc isoforms, and FGFR4). These authors observed a heparin-dependent interaction of FGF19 that was restricted to FGFR4.[11] Utilizing a panel of FGFR immunoadhesin chimeric constructs that included all of the known splice variants of FGFR1-3 (i.e., IIIb and IIIc isoforms) and FGFR4, we also demonstrated that FGF19 bound exclusively to FGFR4 using both co-immunoprecipitation and solid phase receptor binding assays.[9] Although FGF19 exhibits little or no binding to glycosaminoglycan (the endocrine FGFs as a whole have a low affinity for heparin),[14] its interaction with FGFR4 is exquisitely dependent on the presence of heparin.[11] This interaction could be explained by the unique heparin binding ability that FGFR4 has among the FGFR family of receptors.[15,16] Zhang et al demonstrated a different receptor specificity for FGF19 in proliferation assays using the BaF3 murine pro-B-cell line. Although FGF19 had the greatest activity with the BaF3 cells stably transfected with mouse Fgfr4, this ligand also promoted the proliferation of the mouse Fgfr1 IIIc, Fgfr2 IIIc, and Fgfr3 IIIc expressing cells.[17] The activity of FGF19 in this context was also dependent on the presence of heparin.[17] FGF19 is not expressed in mice and it shares only 51% to 53% amino acid sequence identity with its closest mouse ortholog, Fgf15.[10,11] In contrast, all of the remaining human FGFs, with the exception of FGF19, share 75% to 100% amino acid sequence identity with their corresponding mouse orthologs. Given the reduced sequence identity between the human and mouse orthologs, it is possible that FGF19 binds with a different specificity to the human FGFRs as compared to the mouse Fgfrs.

While FGF19 may be able to bind and function through a number of other FGFRs, its activity is primarily transmitted through the binding and activation of FGFR4. FGFR4 is the most widely distributed member of the FGFR family. Under normal circumstances, FGFR4 is expressed in liver, lung, gall bladder, small intestine, pancreas, colorectal, lymphoid, ovary, and breast tissues.[9] It has recently been demonstrated that endocrine FGFs, such as FGF19, also require the Klotho family of proteins as coreceptors to promote the binding of these ligands to their corresponding FGFRs.[8,9,18,19]

The Klotho family of proteins, comprised of Klotho (KL) and KLB, are Type I transmembrane glycoproteins containing extracellular regions that contain two beta-glycosidase-like repeats. FGFR4 binds to both KL and KLB.[8,9,19] We and others have recently identified KLB as a coreceptor for FGFR4 that is required for the high affinity binding and activity of FGF19.[8,9] Using biochemical and cell based assays we demonstrated that human KLB promotes the exclusive interaction of FGF19 with FGFR4.[9] As previously mentioned, FGF19 has low affinity for heparin. KLB appears to stabilize the interaction of this ligand with its receptor, perhaps acting as a surrogate for heparin. Intriguingly, Kurosu et al used cells stably transfected with the various Fgfr isoforms to show that mouse Klb promotes the binding of FGF19 not only to mouse Fgfr4, but also to mouse Fgfr1 IIIc, Fgfr2 IIIc, and Fgfr3 IIIc.[8] This role of KLB in FGF19 binding and activity may in part explain the ability of FGF19 to act through other FGFRs. Previous binding studies demonstrating exclusive binding of FGF19 to FGFR4 were done in the absence of KLB.[11] It has been reported that FGF19 can bind and activate FGFR4 in the absence of KLB.[20] However, KLB significantly increased FGF19 potency on FGFR4 activation.[20] The authors speculated that due to the low normal plasma concentration of FGF19/Fgf15, KLB is critical in conferring FGFR4 activation under physiological conditions; the experiments reported by Wu et al were performed using supraphysiological levels of recombinant FGF19. Although these studies showed differential receptor binding specificities they reach identical conclusions. KLB reduces or alleviates the need of FGF19 for heparin to bind to the FGFRs, KLB promotes the interaction of FGF19 with the FGFRs without altering its binding specificity, and FGF19 binds preferentially to FGFR4.

Given that FGF19 activity requires the co-expression of FGFR4 and KLB, the distribution and the relative levels of these proteins will dictate the target organ site of FGF19 endocrine action. Unlike the broad distribution of *FGFR4* expression, the distribution of *KLB* is more restricted. *KLB* is expressed in adipose, liver, pancreas, and breast tissues.[9] *FGFR4* and *KLB* are both highly co-expressed in the liver and, not surprisingly, mediate the specific activities of FGF19 in this tissue.[9,21] Much lower levels of *FGFR4* and *KLB* co-expression are also found in pancreas and breast tissues.[9] However, FGF19 activity has not been described in these tissues thus far. *KLB* is highly expressed in adipose tissues, but the absence of *FGFR4* precludes the activity of FGF19 in this tissue.[9] These findings taken together indicate that the liver is expected to be the main target organ for the endocrine actions of FGF19.

FGF19 AND CANCER

FGF19 Activity

FGF-FGFR signaling networks play important roles in cell proliferation and the activities stimulated by some of these ligand-receptor combinations have been linked to the development and progression of cancer.[22,23] FGF19 may promote hepatocellular carcinoma by utilizing a number of potential molecular mechanisms (summarized in Table 1). The ectopic expression of *FGF19* in transgenic mice led to tumor formation in the liver.[6] These *FGF19* transgenic mice developed liver tumors by 10 to 12 months of age.[6] Tumors arose from pericentral hepatocytes after increased proliferation and dysplasia.[6] This increase in hepatocellular proliferation is believed to be a prerequisite for neoplastic transformation.[24] Consistent with these findings, the injection of wild-type mice with recombinant FGF19 also promoted hepatocyte proliferation.[6] Elevated expression of α-fetoprotein (*AFP*) mRNA, an oncofetal protein used as a marker for the detection of liver cancer,[25] accompanied the increased hepatocellular proliferation in the *FGF19*

Table 1. Effects of FGF19 Signaling in Cancer

Effect	Reference
Proliferation	5,6,32,57-59
Survival	58,59
Chemotaxis/Motility	5
Adhesion	7

transgenic mice.[6] The nuclear accumulation of β-catenin observed in neoplastic cells from a subset of the liver tumors found in the *FGF19* transgenic mice potentially implicates activation of the Wingless/Wnt signaling pathway in this tumorigenic process.[6] In addition to this immunohistochemical evidence, cloning and sequencing of DNA from these tumor tissues revealed point mutations resulting in amino acid substitutions in and around the glycogen synthase-3β (GSK-3β) phosphorylation site of β-catenin that were suggestive of β-catenin activation.[6] Taken together, these studies demonstrated that FGF19 promotes hepatocellular carcinoma in mice.

To investigate the role of FGF19 in human cancer, we evaluated its expression in primary human tumors and analyzed the consequence of therapeutic FGF19 neutralization in relevant tumor models. *FGF19* and *FGFR4* were coexpressed in primary human hepatocellular carcinomas, lung squamous cell carcinomas, and colon adenocarcinomas.[5] Relative to the corresponding normal tissues, *FGF19* was found to be overexpressed in these cancers, as well as in a subset of human colon cancer cell lines.[5] *FGF19* was also strongly expressed in livers that had undergone cirrhosis, a preneoplastic condition that often leads to liver carcinoma.[5] To assess the importance of FGF19 in tumor growth and development, an anti-FGF19 blocking antibody (1A6) that selectively inhibits the interaction of FGF19 with FGFR4 was generated. 1A6 inhibited FGF19-modulated fibroblast growth factor receptor substrate 2 (FRS2) and mitogen-activated protein kinase (MAPK) signaling in Hep3B HCC cells and completely relieved repression of *CYP7A1* gene expression in HepG2 HCC cells.[5] To further determine whether FGF19 neutralization could inhibit tumor growth in vivo, HCT116 and Colo201 colon cancer xenograft mouse models were utilized. In comparison to xenograft mice treated with a control antibody, 1A6 treated mice showed a statistically significant growth inhibition of both HCT116 (60% inhibition, P = 0.01, n = 5 at 15 mg/kg 1A6, twice weekly treatment) and Colo201 (64% inhibition, P = 0.03, n = 5 at 30 mg/kg 1A6, twice weekly treatment) tumors.[5] In both tumor models, analysis of excised tumors showed that 1A6 significantly inhibited FGFR4, FRS2, MAPK, and β-catenin activation.[5] In order to evaluate 1A6 for in vivo activity, 1A6 was tested in the *FGF19* transgenic mice that had previously been demonstrated to develop liver tumors. In this study, *FGF19* transgenic mice were first treated with diethylnitrosamine (DEN), which accelerated tumor formation by 50%. All of the mice treated with control antibody had multifocal, large liver tumors throughout the liver lobes; with the exception of one 1A6 treated mouse that had a single small tumor, the 1A6 treated animals had no liver tumors.[5] Liver weights and tumor volumes from 1A6 treated transgenic mice were significantly lower than in control antibody treated mice.[5] These findings suggest that FGF19 may be involved in human tumor pathogenesis, notably in the liver, in early neoplastic progression.

While the aforementioned studies implicate FGF19 in tumorigenesis, the molecular mechanism by which this endocrine ligand functions in this manner is unclear. As previously mentioned, evidence of β-catenin activation was observed in neoplastic cells from *FGF19* transgenic mice. β-Catenin plays a key role in maintaining cell-cell adhesion and as a downstream effector in the Wnt signaling cascade. Interaction of E-cadherin with α- and β-catenin is essential for stable cell-cell adhesion and association of these proteins is regulated by tyrosine phosphorylation of β-catenin.[26,27] A complex of proteins containing adenomatous polyposis coli, axin, conductin, and GSK-3β regulates the stability of cytoplasmic β-catenin by targeting β-catenin for ubiquitination and proteasomal degradation. Activation of Wnt signaling results in GSK-3β inactivation and subsequent nuclear accumulation of β-catenin where β-catenin can then interact with TCF/LEF transcription factors to activate corresponding target gene expression.[28-30] Deregulation of Wnt signaling may be important for the initiation and/or progression of several malignancies (e.g., HCC, colorectal, ovarian, endometrial and prostate cancers, and melanoma).[28-31] Furthermore, the anti-FGF19 antibody 1A6 repressed β-catenin activation in colorectal cancer xenograft models. The effect of FGF19 on the β-catenin signaling pathway was recently assessed. The treatment of HCT116 colon cancer cells with FGF19 resulted in a dose-dependent increase in β-catenin phosphorylation, accompanied by a concomitant loss in E-cadherin binding.[7] FGF19 also increased GSK-3β phosphorylation and active β-catenin, and led to the activation of β-catenin/TCF4-regulated transcription.[7] Treatment of HCT116 cells with 1A6 reduced these effects; phospho-GSK-3β and active β-catenin levels were significantly reduced and FGF19-induced β-catenin/TCF4-regulated gene expression (i.e., *CCND1, CD44, JUN, COX2,* and *UPAR*) was consistently repressed.[7]

The Role of FGFR4 in Modulating FGF19 Activity

The normal activity of FGF19 is primarily modulated through the binding and activation of FGFR4. FGFR4 has recently been reported to uniquely mediate FGF19-induced hepatocyte proliferation.[32] Using a series of FGF19 truncation and chimeric constructs, Wu et al identified that amino acid residues 38 to 42 of FGF19 are sufficient to confer FGFR4 activation and increased hepatocyte proliferation, a process that is believed to be a prerequisite for HCC formation.

Contrary to the other FGFR family members, FGFR4 does not have alternatively spliced variants. Although FGFR4 is not alternatively spliced, its function appears to be altered by polymorphisms (summarized in Table 2). The contribution of FGFR4 polymorphisms to hepatocellular carcinoma was recently evaluated. A comprehensive sequence analysis of FGFR4 was conducted on 57 pairs of matched HCC and normal tissue samples.[33] Three known single nucleotide polymorphisms (i.e., Val10Ile, Leu136Pro, and Gly388Arg) and five previously unreported amino acid sequence alterations (i.e., Asp126Asn, Thr179Ala, Gly426Asp, Asp709Gly, and truncation at amino acid residue 450) were identified.[33] These polymorphisms appear to be germ line alterations as they were found in both the tumor and corresponding normal tissues of the respective samples. The Gly388Arg polymorphism was the most highly distributed variant.[33] Compared to the Gly/Gly genotype patients, homozygous Arg/Arg individuals had increased circulating levels of AFP, the embryonic tumor cell marker characteristic for liver cancer.[33] *FGFR4* mRNA expression was also found to be increased by at least 2-fold in 31.6% (18 of 57) of the HCC samples compared to the matched normal liver samples.[33]

Increased expression and genotype distribution of FGFR4 in the liver provides a potential link between FGFR4 and progression of hepatocellular carcinoma. To further investigate this potential functional role of FGFR4 in HCC, tumor responses were measured

Table 2. Effect of FGF19 on FGFR4 Polymorphisms in Cancer

Polymorphism	Indication	Effect of Polymorphism/Effect of FGF19	Reference
Arg388	Hepatocellular carcinoma	Increased circulating levels of AFP/FGF19 treatment increased AFP secretion and FRS2α phosphorylation in cultured cells	33
	Breast cancer	• Increased cell motility/Unknown	38
		• Promote elongated morphology in cell line/Unknown	38
		• Associated with resistance to therapy/Unknown	60
		• Decreased disease free survival time/Unknown	38
		Not indicative of disease free survival time/Unknown	51
	Colon cancer	Decreased disease free survival time/Unknown	38,40
		Not predictive of disease progression/Unknown	40
	Lung cancer	• Decreased age at onset and survival; Increased stage of disease/Unknown	40,44
		• Shorter survival in node positive patients/Unknown	45
		No association with disease prognosis/Unknown	54
	Prostate cancer	• Increased receptor stability and sustained activity/Unknown	43
		• Increased incidence and/or occurrence of aggressive disease/Unknown	41,43
		No association with survival differences/Unknown	33,53
	Soft tissue sarcomas	Decreased cumulative overall and metastasis-free survival/Unknown	50
	Bone sarcomas	No significant difference/Unknown	50
	Malignant gliomas	No association with risk of survival differences/Unknown	52
	Melanoma	No association with risk of survival differences/Unknown	55,60
	Squamous and basal cell carcinomas	No association with risk to disease development/Unknown	55
	Head/neck squamous cell carcinomas	Associated with reduced overall survival and with advanced tumor stage/Unknown	47,48
Gly388	Head/neck squamous cell carcinomas	Increased risk of developing disease/Unknown	46
Tyr367Cys	Breast Cancer	Enhanced ERK activation and cell proliferation/No effect	39

in HuH7 and HepG2 HCC cell lines, which are both homozygous for the Arg388 allele, stimulated with FGF19.[33] The stimulation of these cell lines with FGF19 increased AFP secretion in culture, as compared to untreated cells.[33] Phosphorylation of the FRS2α adaptor protein was also stimulated by FGF19 treatment of HuH7 and HepG2 cells, as compared to untreated cells, indicating activation of the FGFR4 signaling pathway.[33] Conversely, the suppression of FGFR4 expression using siRNA reduced the secretion of AFP.[33] The treatment of HCC cell lines with an FGFR inhibitor reduced the FGF19- and serum-stimulated AFP secretion, blocked proliferation and invasion, and induced apoptosis in vitro.[33] Based on these findings the authors associated the activity of the FGF19-FGFR4 axis with the progression of liver cancer. Furthermore, the potential role of FGFR4 mutants in FGF19 signaling is highlighted here, given that the HCC cell lines utilized in this study were homozygous for the Arg388 allele.

The role of FGFR4 in liver cancer progression was also studied in genetically engineered mice. It was shown that the repression of *Fgfr4* expression in mice increased liver injury and fibrosis induced by carbon tetrachloride.[34] Also, the induction of hepatocarcinogenesis by DEN was accelerated in *Fgfr4-null* mice compared to their wild-type littermates.[35] These findings suggested that Fgfr4 protects the hepatocytes against acute and chronic injury and plays a cancer suppressor function in liver.[34,35] It is unclear why Fgfr4 was found to have a protective effect in these reports. As already described, *FGF19* transgenic mice, as well as wild-type mice injected with recombinant FGF19, were shown to develop HCC.[6] Furthermore, inducible knockdown of *Fgfr4* expression in a xenograft model showed reduced tumor growth.[7] The studies in which Fgfr4 showed protective effects utilized endogenous mouse Fgf15. Thus, the potential differential effects of Fgf15 and FGF19 on the FGFR signaling pathway may in part explain this difference in observations.

FGFR4 is also associated with breast cancer. *FGFR4* was found to be amplified 2- to 4-fold in 10% of primary breast tumors.[36] A recent study suggests that resistance to chemotherapy, such as doxorubicin, is associated with *FGFR4* upregulation.[37] To address this potential link, the authors generated an anti-FGFR4 antibody (10F10) and tested it against various breast cancer cell lines that endogenously express FGFR4. Administration of 10F10 reduced phospho-ERK levels that were stimulated by FGF19 treatment in the breast cancer cell lines, as compared to untreated cells.[37] Treatment of the breast cancer cell lines with 10F10 resulted in increased rates of apoptosis following doxorubicin induction, as compared to control antibody treatment, correlating with inhibition of FGF19 signaling through FGFR4.[37]

FGFR4 polymorphisms have also been identified in breast carcinomas. As is the case for HCC, a Gly388Arg mutation was identified in the MDA-MB-453 breast cancer cell line.[38] In a panel of breast cancer cell lines the genotype frequency was 58% for homozygous Gly alleles, 31.5% for heterozygous Gly/Arg alleles and 10.5% for homozygousArg alleles.[38] Among 145 breast cancer patients 46% were found to be homozygous Gly/Gly, 43% heterozygous Gly/Arg, and 11% homozygous Arg/Arg.[38] An analysis of the genomic DNA from the circulating white blood cells from a subset of these patients demonstrated the same genotype as seen in the tumor tissue demonstrating that the Gly to Arg conversion was a germ line polymorphism.[38] The distribution of the allele frequency was also shown to be similar in healthy individuals.[38] The authors showed that the Gly/Arg388 genotype was significantly prevalent among patients with metastatic breast cancer that had recurrence within 62 months.[38] However, no Gly388 patients had suffered from metastatic disease at the time of the study.[38] The authors suggested that the Arg388 allele

FGF19 AND CANCER

represents a determinant that is innocuous in healthy individuals but predisposes cancer patients to significantly accelerated disease progression. In addition, the presence of the Arg allele in breast cancer patients was associated with resistance to adjuvant therapy.[38] Infection of breast cancer cells with a recombinant retrovirus encoding *FGFR4* with the Gly allele decreased cell migration compared to the parental cell line.[38] However, upon infection with a recombinant retrovirus encoding *FGFR4* with the Arg allele the cells adopted an elongated morphology associated with a mesenchymal phenotype and scattered distribution.[38] The Gly388Arg polymorphism was reported to modulate cancer cell migration in vitro and to be associated with breast cancer prognostic parameters.[38]

Although a potential link between the Gly388Arg mutant and breast cancer progression has been identified, there have not been any reports addressing the possible effects of FGF19 on this receptor polymorphism. However, given that FGF19 stimulation of HCC cell lines containing the Gly388Arg mutant leads to activation of the FGFR4 signaling pathway, FGF19 may have similar effects on this receptor polymorphism in breast cancers.

Interestingly, an FGFR4 Tyr367Cys mutant has also been identified in MDA-MB-453 breast cancer cells.[39] This mutant is reported to be constitutively active due to spontaneous dimerization of the mutant receptor and independent of ligand stimulation.[39] When MDA-MB-453 cells were treated with FGF19, enhanced ERK activation and cell proliferation were not observed.[39] In contrast, when MDA-MB-361 cells which express wild-type FGFR4 were administered FGF19, ERK phosphorylation and cell proliferation increased.[39] Inhibition of FGFR4 Tyr367Cys expression in MDA-MB-453 cells by small interfering RNAs (siRNAs) reduced MAPK signaling (i.e., ERK phosphorylation), reduced proliferation rate, and depleted cell viability, as compared with control siRNA.[39] The ligand independence of the FGFR4 Tyr367Cys mutant was confirmed by the inability of 10F10, the anti-FGFR4 antibody that was raised against the extracellular domain of FGFR4, to downregulate MAPK signaling.[39]

FGFR4 overexpression and polymorphisms (i.e., Gly388Arg) have been associated with a number of other cancers, including colorectal carcinoma,[38,40] prostate cancer,[41-43] lung cancer,[40,44,45] squamous cell carcinoma,[46-48] melanoma,[49] and soft tissue sarcomas.[50] The utility of this FGFR4 Gly388Arg polymorphism as a predictive marker for disease free survival time has been debated. After independent analysis of cancer patient populations, several groups have reported that the Gly388Arg status cannot be used to predict disease free survival time for patients with breast carcinoma,[40,51] colon carcinoma,[40] malignant gliomas,[52] prostate cancer,[33,53] lung cancer,[54] basal cell carcinoma, squamous cell carcinoma, and melanoma.[55] Despite these disparate findings, little to no role for FGF19 signaling has been reported in any of these additional cancer types thus far. KLB, a key component required for FGF19 signaling through FGFR4, is not expressed at significant levels in these tissues. However, KLB may be able to function in a similar manner to KL, in which, in addition to being cell-associated, the extracellular portion of KL is secreted,[56] thus facilitating FGF19 activity in tissues where it is not normally expressed. Overexpression of FGF19 by these tumors may also facilitate FGF19/FGFR4 signaling in an autocrine manner by acting to increase the local concentration of ligand. FGF19 is overexpressed in colon adenocarcinoma.[5,11] It may also be possible that FGF19 does not have a substantive role in the genetic background of these cancers. Further investigation is required to address these issues.

CONCLUSION

As previously described, the FGF-FGFR signaling pathway is thought to play important roles in a number of human cancers. The FGF19-FGFR4 signaling axis in particular promotes hepatocyte proliferation and the development and progression of HCC in vivo.[6] Thus, FGF19 is an attractive target for the treatment of liver cancers and other potential neoplasms. As such, a neutralizing antibody against FGF19 has been shown to inhibit FGF19-mediated signaling and tumor growth in mice.[5] The recent identification of KLB as a critical player in the FGF19/FGFR4 signaling complex adds another level of regulation to this endocrine FGF signaling pathway. The relative expression levels and tissue distribution of FGFR4 and KLB are key in determining the effects of FGF19. This may provide additional targets for therapeutic intervention in cancer, especially given the recent findings that FGF19 may affect multiple FGFR signaling pathways[8,17] and certain FGFR4 polymorphisms are constitutively active (i.e., signaling in a ligand-independent manner).[39] Targeting FGFR4 and/or KLB with antibody therapeutics and/or small molecule inhibitors may also prove to be beneficial in the treatment of appropriate human cancers.

REFERENCES

1. Ornitz DM, Itoh N. Fibroblast growth factors. Genome Biol 2001; 2:REVIEWS3005.
2. Eswarakumar VP, Lax I, Schlessinger J. Cellular signaling by fibroblast growth factor receptors. Cytokine Growth Factor Rev 2005; 16:139-149.
3. Sleeman M, Fraser J, McDonald M et al. Identification of a new fibroblast growth factor receptor, FGFR5. Gene 2001; 271:171-182.
4. Powers CJ, McLeskey SW, Wellstein A. Fibroblast growth factors, their receptors and signaling. Endocr Relat Cancer 2000; 7:165-197.
5. Desnoyers LR, Pai R, Ferrando RE et al. Targeting FGF19 inhibits tumor growth in colon cancer xenograft and FGF19 transgenic hepatocellular carcinoma models. Oncogene 2008; 27:85-97.
6. Nicholes K, Guillet S, Tomlinson E et al. A mouse model of hepatocellular carcinoma: ectopic expression of fibroblast growth factor 19 in skeletal muscle of transgenic mice. Am J Pathol 2002; 160:2295-2307.
7. Pai R, Dunlap D, Qing J et al. Inhibition of fibroblast growth factor 19 reduces tumor growth by modulating beta-catenin signaling. Cancer Res 2008; 68:5086-5095.
8. Kurosu H, Choi M, Ogawa Y et al. Tissue-specific expression of betaKlotho and fibroblast growth factor (FGF) receptor isoforms determines metabolic activity of FGF19 and FGF21. J Biol Chem 2007; 282:26687-26695.
9. Lin BC, Wang M, Blackmore C et al. Liver-specific activities of FGF19 require Klotho beta. J Biol Chem 2007; 282:27277-27284.
10. Nishimura T, Utsunomiya Y, Hoshikawa M et al. Structure and expression of a novel human FGF, FGF-19, expressed in the fetal brain. Biochim Biophys Acta 1999; 1444:148-151.
11. Xie MH, Holcomb I, Deuel B et al. FGF-19, a novel fibroblast growth factor with unique specificity for FGFR4. Cytokine 1999; 11:729-735.
12. Inagaki T, Choi M, Moschetta A et al. Fibroblast growth factor 15 functions as an enterohepatic signal to regulate bile acid homeostasis. Cell Metab 2005; 2:217-225.
13. Choi M, Moschetta A, Bookout AL et al. Identification of a hormonal basis for gallbladder filling. Nat Med 2006; 12:1253-1255.
14. Asada M, Shinomiya M, Suzuki M et al. Glycosaminoglycan affinity of the complete fibroblast growth factor family. Biochim Biophys Acta 2009; 1790:40-48.
15. Goetz R, Beenken A, Ibrahimi OA et al. Molecular insights into the klotho-dependent, endocrine mode of action of fibroblast growth factor 19 subfamily members. Mol Cell Biol 2007; 27:3417-3428.
16. Harmer NJ, Pellegrini L, Chirgadze D et al. The crystal structure of fibroblast growth factor (FGF) 19 reveals novel features of the FGF family and offers a structural basis for its unusual receptor affinity. Biochemistry 2004; 43:629-640.

FGF19 AND CANCER 193

17. Zhang X, Ibrahimi OA, Olsen SK et al. Receptor specificity of the fibroblast growth factor family. The complete mammalian FGF family. J Biol Chem 2006; 281:15694-15700.
18. Kuro-o M. Endocrine FGFs and Klothos: emerging concepts. Trends Endocrinol Metab 2008; 19:239-245.
19. Kurosu H, Ogawa Y, Miyoshi M et al. Regulation of fibroblast growth factor-23 signaling by klotho. J Biol Chem 2006; 281:6120-6123.
20. Wu X, Ge H, Lemon B et al. Selective activation of FGFR4 by an FGF19 variant does not improve glucose metabolism in ob/ob mice. Proc Natl Acad Sci USA 2009; 106:14379-14384.
21. Tomiyama K, Maeda R, Urakawa I et al. Relevant use of Klotho in FGF19 subfamily signaling system in vivo. Proc Natl Acad Sci USA 2010; 107:1666-1671.
22. Grose R, Dickson C. Fibroblast growth factor signaling in tumorigenesis. Cytokine Growth Factor Rev 2005; 16:179-186.
23. Korc M, Friesel RE. The role of fibroblast growth factors in tumor growth. Curr Cancer Drug Targets 2009; 9:639-651.
24. Fausto N. Mouse liver tumorigenesis: models, mechanisms and relevance to human disease. Semin Liver Dis 1999; 19:243-252.
25. Spangenberg HC, Thimme R, Blum HE. Serum markers of hepatocellular carcinoma. Semin Liver Dis 2006; 26:385-390.
26. Lilien J, Balsamo J. The regulation of cadherin-mediated adhesion by tyrosine phosphorylation/ dephosphorylation of beta-catenin. Curr Opin Cell Biol 2005; 17:459-465.
27. Nelson WJ, Nusse R. Convergence of Wnt, beta-catenin, and cadherin pathways. Science 2004; 303:1483-1487.
28. Clevers H. Wnt/beta-catenin signaling in development and disease. Cell 2006; 127:469-480.
29. Peifer M, Polakis P. Wnt signaling in oncogenesis and embryogenesis—a look outside the nucleus. Science 2000; 287:1606-1609.
30. Polakis P. Wnt signaling and cancer. Genes Dev 2000; 14:1837-1851.
31. Morin PJ. beta-catenin signaling and cancer. Bioessays 1999; 21:1021-1030.
32. Wu X, Ge H, Lemon B et al. FGF19-induced hepatocyte proliferation is mediated through FGFR4 activation. J Biol Chem 2010; 285:5165-5170.
33. Ho HK, Pok S, Streit S et al. Fibroblast growth factor receptor 4 regulates proliferation, anti-apoptosis and alpha-fetoprotein secretion during hepatocellular carcinoma progression and represents a potential target for therapeutic intervention. J Hepatol 2009; 50:118-127.
34. Yu C, Wang F, Jin C et al. Increased carbon tetrachloride-induced liver injury and fibrosis in FGFR4-deficient mice. Am J Pathol 2002; 161:2003-2010.
35. Huang X, Yang C, Jin C et al. Resident hepatocyte fibroblast growth factor receptor 4 limits hepatocarcinogenesis. Mol Carcinog 2009; 48:553-562.
36. Jaakkola S, Salmikangas P, Nylund S et al. Amplification of fgfr4 gene in human breast and gynecological cancers. Int J Cancer 1993; 54:378-382.
37. Roidl A, Berger HJ, Kumar S et al. Resistance to chemotherapy is associated with fibroblast growth factor receptor 4 up-regulation. Clin Cancer Res 2009; 15:2058-2066.
38. Bange J, Prechtl D, Cheburkin Y et al. Cancer progression and tumor cell motility are associated with the FGFR4 Arg(388) allele. Cancer Res 2002; 62:840-847.
39. Roidl A, Foo P, Wong W et al. The FGFR4 Y367C mutant is a dominant oncogene in MDA-MB453 breast cancer cells. Oncogene 2010; 29:1543-1552.
40. Spinola M, Leoni V, Pignatiello C et al. Functional FGFR4 Gly388Arg polymorphism predicts prognosis in lung adenocarcinoma patients. J Clin Oncol 2005; 23:7307-7311.
41. Ma Z, Tsuchiya N, Yuasa T et al. Polymorphisms of fibroblast growth factor receptor 4 have association with the development of prostate cancer and benign prostatic hyperplasia and the progression of prostate cancer in a Japanese population. Int J Cancer 2008; 123:2574-2579.
42. Wang J, Stockton DW, Ittmann M. The fibroblast growth factor receptor-4 Arg388 allele is associated with prostate cancer initiation and progression. Clin Cancer Res 2004; 10:6169-6178.
43. Wang J, Yu W, Cai Y et al. Altered fibroblast growth factor receptor 4 stability promotes prostate cancer progression. Neoplasia 2008; 10:847-856.
44. Falvella FS, Frullanti E, Galvan A et al. FGFR4 Gly388Arg polymorphism may affect the clinical stage of patients with lung cancer by modulating the transcriptional profile of normal lung. Int J Cancer 2009; 124:2880-2885.
45. Sasaki H, Okuda K, Kawano O et al. Fibroblast growth factor receptor 4 mutation and polymorphism in Japanese lung cancer. Oncol Rep 2008; 20:1125-1130.
46. Ansell A, Farnebo L, Grenman R et al. Polymorphism of FGFR4 in cancer development and sensitivity to cisplatin and radiation in head and neck cancer. Oral Oncol 2009; 45:23-29.
47. da Costa Andrade VC, Parise O, Jr., Hors CP et al. The fibroblast growth factor receptor 4 (FGFR4) Arg388 allele correlates with survival in head and neck squamous cell carcinoma. Exp Mol Pathol 2007; 82:53-57.
48. Streit S, Bange J, Fichtner A et al. Involvement of the FGFR4 Arg388 allele in head and neck squamous cell carcinoma. Int J Cancer 2004; 111:213-217.

49. Streit S, Mestel DS, Schmidt M et al. FGFR4 Arg388 allele correlates with tumour thickness and FGFR4 protein expression with survival of melanoma patients. Br J Cancer 2006; 94:1879-1886.
50. Morimoto Y, Ozaki T, Ouchida M et al. Single nucleotide polymorphism in fibroblast growth factor receptor 4 at codon 388 is associated with prognosis in high-grade soft tissue sarcoma. Cancer 2003; 98:2245-2250.
51. Jezequel P, Campion L, Joalland MP et al. G388R mutation of the FGFR4 gene is not relevant to breast cancer prognosis. Br J Cancer 2004; 90:189-193.
52. Mawrin C, Kirches E, Diete S et al. Analysis of a single nucleotide polymorphism in codon 388 of the FGFR4 gene in malignant gliomas. Cancer Lett 2006; 239:239-245.
53. Yang YC, Lu ML, Rao JY et al. Joint association of polymorphism of the FGFR4 gene and mutation TP53 gene with bladder cancer prognosis. Br J Cancer 2006; 95:1455-1458.
54. Matakidou A, El Galta R, Rudd MF et al. Further observations on the relationship between the FGFR4 Gly388Arg polymorphism and lung cancer prognosis. Br J Cancer 2007; 96:1904-1907.
55. Nan H, Qureshi AA, Hunter DJ et al. Genetic variants in FGFR2 and FGFR4 genes and skin cancer risk in the Nurses' Health Study. BMC Cancer 2009; 9:172.
56. Liu H, Fergusson MM, Castilho RM et al. Augmented Wnt signaling in a mammalian model of accelerated aging. Science 2007; 317:803-806.
57. Engstrom W, Granerus M. Effects of fibroblast growth factors 19 and 20 on cell multiplication and locomotion in a human embryonal carcinoma cell line (Tera-2) in vitro. Anticancer Res 2006; 26:3307-3310.
58. Miyake A, Nakayama Y, Konishi M et al. Fgf19 regulated by Hh signaling is required for zebrafish forebrain development. Dev Biol 2005; 288:259-275.
59. Siffroi-Fernandez S, Felder-Schmittbuhl MP, Khanna H et al. FGF19 exhibits neuroprotective effects on adult mammalian photoreceptors in vitro. Invest Ophthalmol Vis Sci 2008; 49:1696-1704.
60. Thussbas C, Nahrig J, Streit S et al. FGFR4 Arg388 allele is associated with resistance to adjuvant therapy in primary breast cancer. J Clin Oncol 2006; 24:3747-3755.

CHAPTER 13

UNDERSTANDING THE STRUCTURE-FUNCTION RELATIONSHIP BETWEEN FGF19 AND ITS MITOGENIC AND METABOLIC ACTIVITIES

Xinle Wu and Yang Li*

Amgen Inc, South San Francisco, CA, USA.
*Corresponding Author: Yang Li—Email: yangl@amgen.com

Abstract: FGF19 differs from the classical FGFs in that it has a much-reduced heparan sulfate proteoglycan binding affinity that allows it to act as endocrine hormone. Although FGF19 regulates several different metabolic activities, it still activates downstream signaling pathways through FGF receptors, in a similar manner to that seen in classical FGFs. Aberrant FGF signaling has been implicated in tumor development, and mouse models have confirmed that FGF19 has the potential to induce hepatocellular carcinoma. Treatment with anti-FGF19 antibody suppressed tumor progression in both FGF19 transgenic mice and colon cancer cell xenograft models. FGFR4, the predominant FGF receptor expressed in the liver, may play an important role in FGF19-mediated tumorigenesis. This review reports the current advances in understanding the structure-function relationship between FGF19 and its interactions with FGFRs, its physiological activities, and its differences from FGF21. The review also discusses strategies to separate the mitogenic and metabolic activities for the development of potential therapeutic molecules based on FGF19.

INTRODUCTION

FGF19 can act as a classic endocrine hormone to regulate bile acid homeostasis as well as glucose and lipid metabolism.[1-3] It is able to escape into circulation from its originating tissues because of its low affinity toward heparan sulfate of the extracellular matrix. Instead of relying on heparan sulfate, FGF19 requires βKlotho as a cofactor to activate FGF receptors (FGFRs).[4-6] βKlotho is mainly expressed in liver and adipose tissue. In the liver, FGF19 is able to suppress bile acid production by downregulating

---

Endocrine FGFs and Klothos, edited by Makoto Kuro-o.
©2012 Landes Bioscience and Springer Science+Business Media.

CYP7A1, the key enzyme in bile acid biosynthesis.[7] In adipose tissue, FGF19 may increase glucose uptake by upregulating *GLUT1*, a glucose transporter.[5,8]

There is compelling evidence for the involvement of deregulated FGF signaling in tumorigenesis.[9] FGF signaling can contribute to cell proliferation, survival, invasion, migration, and tumor angiogenesis.[9] FGF19 binds to multiple FGFRs in the presence of βKlotho and activates classical FGF downstream signaling.[4] FGF19 transgenic mice developed cancer in the liver, the target tissue for FGF19.[10] Hepatocellular carcinoma could be observed from as early as 10 month of age in those mice. 5-bromo-2-deoxyuridine (BrdU) labeling in the liver also confirmed enhanced proliferation from either transgenic mice or wild-type mice treated with recombinant FGF19 proteins.

FGF19 belongs to a subfamily of FGFs that also includes FGF21 and FGF23. There are significant overlapping properties between FGF19 and FGF21. For example, FGF21 also requires βKlotho as cofactor to activate FGFRs and can act as an endocrine hormone to regulate metabolism of glucose and lipids.[11] Both FGF19 and FGF21 transgenic mice have reduced fat mass and exhibit improved insulin sensitivity compared with wild-type littermates.[12,13] Injection of recombinant FGF19 or FGF21 proteins decreases serum glucose and insulin levels, improves glucose tolerance, improves liver steatosis, and reduces body weight in diabetic mice.[3] However, no evidence of hepatocellular carcinoma was observed in FGF21 transgenic mice,[13] and treatment with recombinant FGF21 protein did not increase proliferation in liver as determined by a BrdU assay.[14,15] There is also difference between FGF19 and FGF21 in their receptor specificities: Both FGF19 and FGF21 can activate FGFR 1c, 2c and 3c, while only FGF19 can efficiently activate FGFR4.[5,14,16]

This review reports the current advances in understanding the structure-function relationship between FGF19 and its interactions with FGFRs, its physiological activities, and its differences from FGF21. Strategies to separate the mitogenic and metabolic activities for the development of potential therapeutic molecules based on FGF19 will also be discussed.

FGF19 AND CANCER

FGFs and Cancer

Fibroblast growth factors (FGFs) are most common mitogens, they were first identified based on their activities in causing the proliferation of fibroblasts.[17] The connection between aberrant FGF/FGFR signaling and tumorigenesis has been demonstrated by various mouse genetic models as well as human genetic studies.[9,18-20] For example, more than 50% of bladder cancers have FGFR3 mutations and the majority of those mutations occur at a single position in the extracellular domain (S249C)[21] which resulted in auto-dimerization and constitutive receptor activation.[22,23] Another example is a common polymorphism of FGFR4, R388, which is associated with many cancer types, including prostate cancer and head and neck squamous cell carcinoma.[24,25] Biochemical study suggested that R388 mutation could result in increased receptor stability and causing sustained receptor activation.[26] Currently, several potential cancer therapies targeting FGF/FGFRs are been explored in different stages of clinical development.[9,19]

Activation of classical FGF signaling requires the formation of ternary complexes between FGF, FGFR, and heparan sulfate proteoglycans.[27,28] Upon ligand binding, FGFRs dimerize leading to conformational changes that activate intracellular kinase

FGF19 AND ITS MITOGENIC AND METABOLIC ACTIVITIES

domains.[29] Phosphorylation of the tyrosine residues on the receptor provides docking sites for downstream adaptor proteins, which can further activate different downstream intracellular signaling pathways.[30] One of the key adaptor proteins of FGFRs is FGF substrate 2 (FRS2) which after phosphorylation by FGFRs can recruit additional adaptor proteins such as son of sevenless (SOS) and growth factor receptor-bound 2 (GRB2) to activate RAS GTPase.[30] RAS GTPase promotes several downstream signaling, such as PLC/PKC, RAF/ERK and PI3K/PKB pathways.[19] These FGF downstream signaling pathways are involved in tumorigenesis over several aspects, including proliferation, survival, cell migration, invasion as well as angiogenesis. Therefore, FGF signaling has been implicated in a wide range of processes during tumor pathogenesis.

Effects of FGF19 on Cancer

FGF19 belongs to a unique FGF subfamily together with FGF21 and FGF23. Unlike other FGFs, this subfamily has much weakened affinity towards heparan sulfate proteoglycans.[4,31] The reduced affinity liberates FGF19 from its site of secretion and allows it to act as an endocrine hormone.[27] Another consequence of the decreased binding to heparan sulfate is FGF19 requires βKlotho as a protein cofactor to activate FGFRs. Therefore the target tissues of FGF19 are limited by the presence of βKlotho.[5]

Since FGF19 is still an FGF and will activated FGFRs and downstream signaling, its potential effects on tumorigenesis has been explored. It is known that FGF19 gene is located on human chromosome 11q13 and belongs to an evolutionary conserved cluster of genes that include FGF3, FGF4, and Cyclin D1.[32] This cluster is frequently amplified in breast cancer, head and neck squamous cell carcinoma, and bladder tumors.[33] Translocation of this cluster has also been observed in parathyroid tumors and B-cell lymphoma.[32] Furthermore, high expression levels of FGF19 were detected in several human tumors, including colon adenocarcinoma, lung squamous cell carcinoma, and hepatocellular carcinoma.[34]

The most compelling evidence supporting the role played by FGF19 in cancer development came from a transgenic FGF19 mouse model. In this model, human FGF19 was driven by myosin promoters and overexpressed by skeletal muscle. Elevated serum FGF19 protein levels were confirmed in those transgenic mice, and as early as 2 months of age, the hepatic mRNA levels for α-fetoprotein, an important biomarker for liver cancer, were upregulated.[10] In contrast to what is observed in wild-type control mice, H&E staining of the livers of 14-week-old FGF19 transgenic mice, clearly showed dense clusters of hepatocytes formed around a central vein with their nuclei polarized away from the endothelial basement membrane of the central vein.[10] Between 7 to 9 months of age, hepatocellular dysplasia became evident and dysplasia foci were also predominantly oriented around central veins.[10] Beginning at 10 months of age, FGF19 transgenic mice started to develop hepatocellular carcinomas. Tumors occurred in different liver lobes and were either solitary or multifocal. Histological staining showed the invasion of neoplastic cells and replacement of normal hepatocytes.[10] Expression analysis also demonstrated upregulation of TGF-α and c-*myc* in the transgenic livers.[10]

Hepatocytes are mitotically quiescent in normal liver. However, in cases of hepatocellular carcinoma, hepatocellular proliferation is a prerequisite for transformation. In vivo BrdU labeling was performed to assess the proliferation in FGF19 transgenic mice. BrdU-labeling index of hepatocytes was eight-fold higher in the transgenic mice than in age-matched wild-type mice at 2 to 4 months of age, demonstrating that hepatocelluar proliferation occurred before tumor development.[10] Like the cellular clusters and hepatocellular dysplasia observed from histology staining, BrdU-positive cells are also enriched around central veins.[10] Furthermore, 6 days after recombinant FGF19 protein was injected into normal mice with BrdU infusion, mice developed a significantly higher BrdU-labeling index, which is similar to the results seen intransgenic animals.[10]

Monoclonal anti-FGF19 antibody was generated and tested in tumor inhibition in FGF19 transgenic mice treated with diethylnitrosamine (DEN), which can accelerate tumor formation. Either anti-FGF19 antibody or control antibody was injected for six months in these animals. At the end of that period, all the animals treated with the control antibody developed multifocal, large hepatocellular carcinomas throughout the liver lobes, while almost none of the mice treated with FGF19 antibody had liver tumors.[34] The same anti-FGF19 antibody was also tested on mice receiving human colon cancer xenografts. Mice with xenografts of colon cancer cell lines HCT116 and Colo201 were injected with anti-FGF19 antibody twice weekly. Relative to the control groups, FGF19 antibody at day 35 suppressed tumor growths in HCT116 injected animals by 57% and in Colo201 injected animals by 64%.[34] It was therefore proposed that targeting FGF19 could be a valid strategy in certain cancer treatments.

FGF19 activates several signaling events in the liver that might contribute to the tumorigenesis. Besides ERK1/2, the Wnt pathway, which is known to cross-talk to FGF signaling, is also activated in FGF19 transgenic animals,[35] and 44% of neoplastic hepatocytes in FGF19 transgenic mice have nuclear staining for β-catenin.[10] β-catenin is downstream of the Wnt-signaling pathway and can enter the nucleus and interact with TCF/LEF family of transcription factors to regulate target gene expression. By using a TCF/LEF reporter assay, FGF19 has been shown to significantly increase TCF/LEF reporter activity in the colon cancer cell line HCT116.[35] Treating HCT116 cells with anti-FGF19 antibody reduced β-catenin target gene expression.[35] Another important function of β-catenin is to maintain cell-cell adhesion by linking cadherins to the actin cytoskeleton through α-catenin. Loss of cell-cell adhesion is associated with tumor invasion and metastasis. The binding of β-catenin to cadherins and α-catenin is regulated by phosphorylation of several critical tyrosine residues. FGF19 increases tyrosine phosphorylation of β-catenin and induces loss of E-cadherin binding to β-catenin in HCT116 cells, potentially contributing to tumor growth and invasion.[35]

Therefore, both in vivo and in vitro experimental results support the critical contribution from FGF19 to hepatocelluar carcinoma development in rodent models. The potential FGF receptor that may contribute to the tumorigenic activity of FGF19 in the liver will be discussed in the following sections.

STRUCTURAL BASIS FOR THE INTERACTIONS BETWEEN FGF19 SUBFAMILY MEMBERS AND THEIR RECEPTORS

Unlike other members of the FGF family, which function exclusively as paracrine molecules, the FGF19 subfamily members function as endocrine hormones. [2,3,18] FGF19 is secreted from the ileum and functions as an enterohepatic signal for the regulation of bile acid metabolism,[36] FGF21 is expressed predominantly in the liver and signals in adipose tissue,[11] and FGF23 originates from bone and regulates phosphate homeostasis in the kidney.[37] The main difference between the paracrine-acting FGFs and the FGF19 subfamily is the weak affinity of the subfamily members toward heparan sulfate (HS)

FGF19 AND ITS MITOGENIC AND METABOLIC ACTIVITIES

proteoglycans of the pericellular space. This weak affinity allows FGF19 subfamily members to escape from the extracellular compartment into circulation.[2,31] In addition, the presence of intramolecular disulfide bonds could increase the plasma stability of the FGF19 subfamily members, thereby allowing them to function as hormones and to act on distant tissues. While HS is required by paracrine-acting FGFs for high-affinity interactions and activation of FGFRs, the reduced affinity of FGF19 subfamily members to heparin and HS correlates with their inability to directly interact with FGFRs and their inefficient activation of FGFRs even at suprapharmacological doses.[38] The FGF19 subfamily members instead use single-transmembrane containing Klotho proteins to facilitate their interactions with FGFRs to compensate for the reduced affinity toward HS.[27] Two related Klotho proteins, αKlotho and βKlotho, both contain two homologous extracellular domains that share sequence homology to the β-glucosidase of bacteria and plants.[39,40] FGF21 and FGF23 selectively use βKlotho and αKlotho as coreceptors,[11,41,42] respectively, while FGF19 can function through either coreceptor in vitro (although the physiological significance of αKlotho in FGF19 function is unknown).[4-6] This requirement for either αKlotho or βKlotho provides another level of selectivity to FGFR signaling and restricts the target tissues for this endocrine FGF subfamily. FGF19 and FGF21 share extensive similarity in coreceptor βKlotho requirement and in their metabolic activities, but differ in their mitogenic effects. In the following sections we will review the structural basis for FGF19 function and will discuss what accounts for the differences between FGF19 and FGF21.

Receptor Specificity for FGF19 and FGF21

Since FGF19 and FGF21 require coreceptor βKlotho to interact and to activate FGFRs, the receptor specificity would first be determined by which FGFRs can interact with βKlotho. By using cotransfection and pull-down studies, it was demonstrated that βKlotho preferentially binds to FGFR1-3 c isoforms and to FGFR4, with the strongest binding to FGFR4, then to FGFR1c, FGFR2c, and the weakest binding to FGFR3c.[6,11,43]

To determine which FGFR isoforms can be activated by FGF19 and FGF21, various combinations of βKlotho and individual FGFRs were cotransfected into either the rat myoblast cell line L6 or the murine pro-B-cell line BaF3. Neither cell line normally responds to FGF treatment because of low or no expression of endogenous FGFRs, thus providing a relatively clean background for testing a specifically reconstituted receptor or a combination of receptors. Consistent with the preferential binding of βKlotho to the c isoforms of FGFR1-3 and FGFR4, in L6 cells, neither FGF19 nor FGF21 was able to activate FGFR1-3b isoform receptors, even in the presence of βKlotho (X.W. and Y.L. unpublished observations). Both FGF19 and FGF21 were able to activate FGFR1c, 2c, and 3c in the presence of βKlotho; however, they differed in their ability to activate FGFR4.[5,14] While FGF19 activated FGFR4, no significant signal was observed with FGF21 treatment on this receptor.[5,14] Similar observations were made in cotransfection studies of BaF3 cells. While FGF21 activated FGFR1c and FGFR3c in the presence of βKlotho, it had weaker activity on FGFR2c/βKlotho, and no activity was observed on FGFR4. In contrast, FGF19 activated all 4 receptors in the presence of βKlotho.[16]

In the absence of βKlotho, neither FGF19 nor FGF21 was able to activate signaling mediated through FGFRs 1-3c,[5,44] with the exception that FGF19 can still activate FGFR4 at high concentrations.[44] This differential coreceptor utilization was supported by binding studies. While βKlotho may recruit FGF19 and FGF21 to all four FGFR/βKlotho complexes,[5,11,44] heparin only promotes interaction between FGF19 and FGFR4.[44,45]

Therefore, while the interaction between FGF19 and heparin is much reduced compared with that seen in paracrine-acting FGFs, the weak binding is still able to promote interaction of FGF19 and FGFR4, leading to FGFR4 activation.

The receptor specificities of FGF19 and FGF21 observed in vitro are summarized in Figure 1. While both FGF19 and FGF21 can activate FGFR1c, 2c and 3c in a βKlotho dependent fashion, they differ in their ability to activate FGFR4. FGF19 can activate FGFR4 in a Klotho-dependent or heparin-dependent manner. In contrast, no significant signaling was observed on FGFR4 by FGF21. Given that the heparin-induced FGFR4 activation by FGF19 is relatively weak and only occurs at relatively high FGF19 concentrations, it is unknown whether these concentrations can be reached under physiological conditions.[44] In addition, given the weak interaction between βKlotho and FGFR2c or FGFR3c, the activity observed in vitro on these two receptors by FGF19 and FGF21 also require further study to understand their physiological significance.

FGF19 and FGF21 Domains Responsible for βKlotho Interaction

Amino acid sequence alignments of all FGFs demonstrate the presence of a much longer C-terminal domain in FGF19 subfamily members than in canonical FGFs such as FGF1 and FGF2.[46] The longer C-terminal domain may confer distinct functions to the FGF19 subfamily. On the basis of studies of a naturally occurring C-terminally truncated

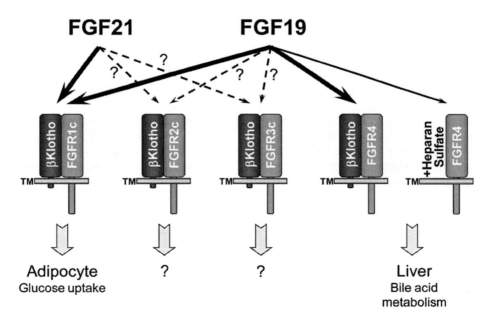

Figure 1. FGF19 and FGF21 receptor specificity and functions. Both FGF19 and FGF21 can activate FGFR1c, 2c, and 3c in the presence of βKlotho in vitro. In adipocytes, FGF19 and FGF21 increase glucose uptake mainly through FGFR1c activation. FGF19 can also activate FGFR4 either mediated through βKlotho or through heparin/heparan sulfate. In liver, FGF19 regulates bile acid metabolism through FGFR4 activation. "?" indicates unresolved research areas associated with the physiological significance of the observed FGF/receptor interactions.

FGF19 AND ITS MITOGENIC AND METABOLIC ACTIVITIES

FGF23 protein, it was suggested that the C-terminal tail of FGF23 may contribute to its interaction with αKlotho.[31] To test whether the C-terminal domains of this FGF subfamily confer the specificity toward the Klotho family members, chimeric FGF19 proteins with the C-terminal domains of FGF21 or FGF23 were constructed.[47] Since FGF21 and FGF23 are specific toward β- or α-Klotho, respectively, and since FGF19 may use either coreceptor in vitro,[47] if the C-terminal tail determines Klotho specificity, the chimeric FGF19 molecules would be expected to show an altered selectivity preference toward the two Klotho proteins. Indeed, in contrast to wild-type FGF19, the FGF19-21C molecule, which contains the FGF21 C-terminal 30 residues in place of the corresponding FGF19 sequences, resulted in specific activity toward βKlotho in both binding and receptor signaling assays; therefore, it becomes a more FGF21-like molecule.[47] In contrast, an FGF19-23C variant that contains the FGF23 C-terminal 81-residue substitution into FGF19 resulted in a change of specificity toward αKlotho only, a more FGF23-like feature.[47] In addition, the deletion of the FGF19 C-terminal 39 residues, only abolished the ability to activate FGFR1c, 2c, 3c, which completely depends on coreceptor βKlotho, while the ability to activate FGFR4 in a heparin-dependent manner was retained.[44] These results suggest that the C-terminal domains of FGF19 and FGF21 are important for interaction with coreceptor βKlotho.

To further understand the contribution of N- and C-termini to FGF21 function, sequential truncation mutants from the two ends were generated. While the N-terminal truncation had greater effects on the efficacy (or Emax) of signaling, the C-terminal truncations sequentially reduced the potency of the mutants in a functional reporter assay. The reduced potency of the mutants correlated with the reduced affinity toward βKlotho binding.[8,48] Deletions of as few as 2 residues from the C-terminal end substantially reduced the affinity toward βKlotho, and deletion of 14 residues completely abolished both binding and signaling activity. These truncation analyses further established the role of the C-terminal domain in βKlotho interaction. It is conceivable that the C-terminal tails in FGF19 subfamily members were acquired during evolution to compensate for their dramatically weakened affinity toward HS.

Heparin/Heparan Sulfate Interaction

The crystal structures of two subfamily members, FGF19 and FGF23, have been solved.[31,49] While both adopt the β-trefoil fold common to all FGF molecules, the conformation of the heparin-/HS-binding site (HBS) diverge completely from that of the paracrine-acting FGF molecules. The heparin-/HS-interaction site of paracrine-acting FGFs consists of a contiguous surface formed by nonlinear sequence segments including the β1-β2 loop 1and the β10-β12 region.[46] Although the primary sequences of HBS differ significantly among paracrine-acting FGFs, which may account for their differential affinity/specificity toward different heparin/HS, the HBSs are generally rich in basic amino acids to form positive electrostatic interactions with negatively charged Heparin/HS.

However, the HBS regions of FGF19 and FGF23 differ significantly both in primary sequence and in folding (Fig. 3C and 3D). The β1-β2 loops of FGF19 and FGF23 are much longer than those of the paracrine-acting FGFs and they form an extended structure protruding from the β-trefoil-like core domain.[31] In addition, the β10-β12 region has a unique alpha helical structure instead of the β11 strand present in the classical FGFs. This helix also protrudes from the β-trefoil core, thereby disrupting the continuous surface found in classical FGFs and creating a cleft between these two heparin-binding segments. These distinctive features of FGF19 and FGF23 were believed to be responsible for the poor affinity of these ligands toward heparin/HS.[46] In addition, these two extended segments also cause steric hindrance and push heparin/HS into a position unfavorable for interaction with FGFRs in a receptor complex.[31] Therefore, it was unclear from these earlier studies whether this surface may still be responsible for heparin/HS mediated FGF19 activation of FGFR4.

The structure of FGF21 has not yet been reported. Sequence alignment shows that the HBS of FGF21 is similar to that of FGF19 with a longer β1-β2 loop and a β10-β12 region; either of these characteristics may sterically hinder heparin/HS interaction.[49] Another feature of FGF21 that may further affect its interaction with heparin/HS is the surface charges. In contrast to FGF19, the surface charges along the HBS in the FGF21 model based on the FGF19 structure revealed incompatible interaction with heparin/HS.[14] Combination of both of these features may explain the fact that FGF21 has a even weaker affinity toward heparin than FGF19 has.[31]

FGF19 and FGF21 chimeric molecules were constructed to determine whether the β1-β2 loop and the β10-β12 region are still responsible for the heparin-/HS-induced FGF19/FGFR4 interaction and activation. The chimeras were designed to exploit the differences between FGF19 and FGF21 in heparin-induced activities. Since FGF21 has no interaction with heparin in direct binding assays[31] and cannot activate FGFRs in a heparin-/HS-dependent fashion, the β1-β2 loop and the β10-β12 region of FGF21 were substituted either alone or in combination into FGF19, replacing the corresponding FGF19 sequences (FGF19-1, FGF19-2, and FGF19-3; Fig. 2). These three chimeric FGF19 molecules carrying FGF21 sequences were still able to interact with both FGFR1c and FGFR4 in the presence of βKlotho. In addition, their ability to activate FGFR1c and FGFR4 receptor signaling in the presence of βKlotho was also preserved in receptor cotransfection studies in L6 cells.[15] These results suggest that the chimeric molecules are active and that the substitution of the β1-β2 loop and the β10-β12 region of FGF21 either individually or together did not significantly affect FGF19 folding.

In contrast to wild-type FGF19, however, these substitutions completely abolished heparin-induced direct interaction between FGF19 and FGFR4 in an ELISA binding assay. Furthermore, they also abolished HS-induced FGFR4 activation in an L6 signaling assay, suggesting that these two regions are essential to heparin-induced FGF19/FGFR4 interaction and activation.[15] In addition, FGF19 mutations K149A/R157A in the β10-β12 region were previously shown to significantly reduce the already weak affinity of FGF19 toward heparin.[31] The presence of these mutations also significantly reduced heparin-induced FGF19/FGFR4 interaction and HS-induced FGFR4 signaling by FGF19 (Wu and Li, unpublished observations). Together, these results strongly suggest that, as seen with the paracrine-acting FGFs, the β1-β2 loop and the β10-β12 region of FGF19 also participate directly in heparin/HS interaction and are responsible for heparin-/HS-induced interaction with and activation of FGFR4 by FGF19.

FGF19 Regions Responsible for FGFR Interaction

Several paracrine FGF/FGFR cocrystal structures have been solved, providing great insight into the interactions between classical FGFs and their receptors. Important interactions at the interfaces between FGF and the D2 domain, D2-D3 linker, and the D3 domain of FGFR have been identified. Additionally, it has been recognized that the βC'-βE loop in the D3 domain plays a critical role in receptor specificity determination.[46] However, in the absence of a cocrystal structure between FGF19 subfamily members and

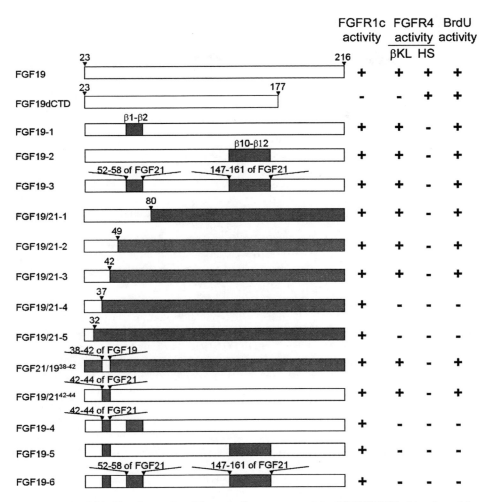

Figure 2. FGF19/21 chimeric proteins. Schematic diagram summarizing FGF19/FGF21 chimeric proteins and FGF19 variants that have been reported.[14,15,47,44] Their in vitro activities in L6 cells against FGFR1c, cotransfected with βKlotho, and FGFR4, with (indicated by βKL) or without (indicated by HS) βKlotho cotransfection, and their in vivo activities in inducing BrdU labeling in liver are also listed. "+" and "–" indicate qualitative assessment of whether activities were observed.

FGFRs, it is not known whether the structural determinants that govern the interactions between FGF19 and its receptors are similar to classical FGF/FGFR complexes, nor is it known how βKlotho might alter these specific interactions.

To gain insight into the specific interactions between FGF19 and its receptors, the differences between FGF19 and FGF21 were exploited to identify the region(s) of FGF19 responsible for FGFR signaling. FGF19 and FGF21 share significant sequence homology but differ in the ability to activate FGFR4 signaling (Fig. 1). The ability of the FGF19 variant FGF19dCTD, which lacks the C-terminal region to activate FGFR4, suggests that sequences important for FGFR4 interaction reside outside the C-terminal domain on the protein.[44] Thus, a series of FGF19/FGF21 chimeric molecules were generated that sequentially replaced the N-terminal region of FGF21 with the corresponding region of FGF19[14] (FGF19/21-1 to FGF19/21-5 as illustrated in Fig. 2) to identify the regions on FGF19 responsible for FGFR4 activation.

Receptor specificity was determined in vitro by signaling in L6 cells and by adipocyte glucose uptake assays. All the chimeras were able to activate FGFR1c and βKlotho receptor complexes in L6 cells to increase ERK phosphorylation and were able to induce glucose uptake in 3T3L1 adipocytes, where FGFR1c is the predominantly expressed receptor. These results indicate that the chimeras are properly folded and that their activities on FGFR1c/βKlotho were not compromised. However, the activities of these chimeras differ dramatically in their ability to activate FGFR4. In L6 cells transfected with FGFR4 and βKlotho, ERK-phosphorylation was only observed with FGF19/21 chimeras, FGF19/21-1, FGF19/21-2, and FGF19/21-3, which all contain FGF19 amino acid sequences 23-42 but not with chimeras, FGF19/21-4 and FGF19/21-5, that lack FGF19 sequences beyond residue 38 (Fig. 2). These analyses identified[38] WGDPI,[42] a segment on FGF19 between residues 38 and 42 that could play a critical role in FGFR4 activation.[14]

The ability of these 5 residues from FGF19 to confer FGFR4 activation was directly demonstrated by the chimera FGF21/19,[38-42] which substituted residues 38-42 from FGF19 replacing the corresponding sequences in FGF21 (Fig. 2).[14] As seen with FGF19 and FGF21, FGF21/19[38-42] induced ERK-phosphorylation in L6 cells transfected with FGFR1c and βKlotho and was active in adipocyte glucose uptake assays. However, in contrast to FGF21, FGF21/19[38-42] induced ERK-phosphorylation in L6 cells transfected with FGFR4 and βKlotho similar to FGF19 (Fig. 2).[14] These results suggest that these 5 amino acid residues on the N-terminal end of FGF19 are sufficient for FGFR4 activation. The reverse substitution was also constructed as chimera FGF19/21,[42-44] where these 5 residues in FGF19 were replaced by the corresponding sequences from FGF21 to determine whether mutations in these 5 amino acids are sufficient to abolish the FGFR4 activity of FGF19 (Fig. 2). As seen with FGF19 and FGF21, FGF19/21[42-44] induced ERK-phosphorylation in L6 cells transfected with FGFR1c and βKlotho and was active in adipocyte glucose uptake assays.[15] However, this substitution only partially affected FGFR4 activity, and FGF19/21[42-44] still retained the ability to induce ERK-phosphorylation in L6 cells transfected with FGFR4 and βKlotho.[15] These results suggest that removing the N-terminal region alone is not sufficient to abolish the FGFR4 activity of FGF19.

Therefore, it appears that the N-terminal 38-42 residue region and heparin-binding domains can make independent contributions to FGFR4 activation. Indeed, FGF19 variants with substitutions with the corresponding sequences from FGF21 in both the N-terminal residues 38-42 and one or both of the heparin-interaction regions completely lost the ability to activate FGFR4 signaling in L6 cells in either the absence or presence of βKlotho (FGF19-4, FGF19-5, and FGF19-6, Fig. 2). The combined changes in these two distinct regions had minimal impact on the ability of the mutant proteins to activate the FGFR1c/βKlotho receptor complex, suggesting that the FGF19 sequences in these two region determine its selective activation of FGFR4.[15]

Proposed Model for FGF19/FGFR4 Interaction

By using the apo-FGF19 structure (PDB code: 2P23)[31] overlaid with the FGF2 complex structure with FGFR1 (PDB code: 1FQ9)[15,50] a model for FGF19/FGFR interaction was built. The two regions identified by FGF19/FGF21 chimeric studies as being important for FGFR4 activation are mapped onto two opposite ends on the

FGF19 AND ITS MITOGENIC AND METABOLIC ACTIVITIES 205

FGF19 structure (Fig. 3D).[15] The FGF19/FGFR model suggests that the N-terminal 5 residues ([38]WPDPI[42]) of FGF19 may directly contact the βC'-βE loop of the receptor D3 domain (Fig. 3B). It has been postulated from the structural data that the specificity in paracrine FGF and FGFR interactions is achieved by the interplay between the flexible N-terminus of the FGFs and the flexible βC'-βE loop of the receptors.[46] It is

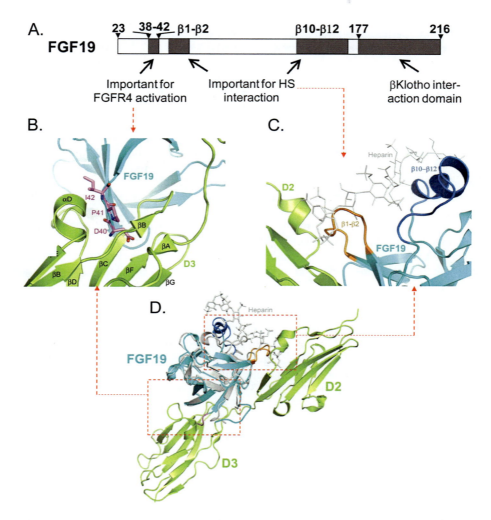

Figure 3. Structure of FGF19 domains. A) Schematic diagram showing FGFR4, heparan sulfate, and βKlotho interaction domains on FGF19 determined by functional studies from FGF19/21 chimeric proteins. B) Structural model of FGF19/FGFR D3 domain interface. FGF19 (PDB code: 2P23) is in cyan with the N-terminal residues [40]DPI[42] shown in magenta. C) Structural model of FGF19/FGFR D2 domain/heparin interface. β1-β2 loop of FGF19 is colored in orange, β10-β12 loop is colored in blue, and heparin is in gray. D) Ribbon representation of FGF19 structure superimposition on the complex structure of FGF2/FGFR1. FGF19 (PDB code: 2P23) is colored in cyan with the N-terminal residues [40]DPI[42] shown in magenta. In the ternary complex structure (PDB code: 1FQ9), FGF2 is colored in gray and FGFR1 in green. Heparin is shown in gray sticks.[15]

interesting to note that the unbiased FGF19/21 chimeric approach has identified this same region as being important for FGF19 specificity determination, suggesting that the endocrine FGFs may complex with the receptor in a manner similar to that seen with paracrine FGFs.

An unexpected finding came from FGF19/21,[42-44] which has mutations in the N-terminal 5 residues (38-42). In contrast to the expectation that this region may be important for βKlotho-dependent FGFR4 activation, this mutation did not abolish βKlotho-induced FGFR4 interaction/activation; rather, it did abolish heparin-induced FGFR4 interaction/activation (Fig. 2).[15] Given that the 38-42 region appears to be away from the putative HBS formed by the β1-β2 loop and the β10-β12 segments in this model (Fig. 3D), the effect of these 5 residues on heparin-induced FGFR4 interaction suggests an indirect role, perhaps through a shift in the position of FGF19 that affected HBS interaction with heparin. However, given the distinct features of the FGF19 HBS regions, such as the potential steric clash of heparin/HS along the surface formed by the β1-β2 loop and the β10-β12 segment, and the cleft formed between these two regions, whether heparin/HS interacts with FGF19 in a fashion that allows the formation of direct contact to the 38-42 region cannot be ruled out completely. However, the fact that FGF21/19[38-42] is not able to activate FGFR4 in a heparin-/HS-dependent manner at least suggests that the N-terminal 5 amino acid residues alone are not sufficient to induce a productive complex with FGFR4 mediated by heparin/HS.

Alignment of sequences in the βC'-βE loop of the four FGFRs that can be activated by FGF19 revealed that FGFR4 appears to have the shortest βC'-βE loop (Fig. 4A). The sequence alignment of the proposed N-terminal region interacting with the βC'-βE loop on FGF19 and FGF21 suggests that FGF19 may have a larger structure than FGF21 (Fig. 4B). Therefore, we propose that the ability of FGF19, but not FGF21, to efficiently activate FGFR4 may in part be determined by the interaction between the N-terminal regions of FGF19 and the βC'-βE loop of FGFR4. The smaller βC'-βE loop of FGFR4 may be better able to complement/accommodate the larger N-terminal region of FGF19 and may also allow FGF19 to bind more tightly to the D3 domain of FGFR4. Tighter binding in the D3 domain could allow more space in the interface with the D2 domain, thereby facilitating heparin/HS interaction. This could explain why only FGF19/FGFR4 interactions could be mediated by heparin/HS (Fig. 4C). In the case of FGF21, the smaller N-terminal region may not allow interaction with the smaller βC'-βE loop of FGFR4, therefore, it may activate FGFR4 significantly less than FGF19 would. In contrast, FGF21 will couple more efficiently to the larger βC'-βE loop of FGFR1c and will activate FGFR1c better than FGF19 would (Fig. 4C).[5,14]

The HBS of FGF19 also displayed unexpected activities. The additional mutations in the HBS of FGF19-4, -5, -6 compared with FGF19/21[42-44] abolished the βKlotho-dependent activation of FGFR4 still observed in FGF19/21[42-44] suggesting that HBS may also contribute to βKlotho-induced FGFR4 activation. The modeled structure suggests that the extended HBS may directly interact with the receptor (Fig. 3C). Therefore, it appears that, as seen with paracrine-acting FGFs, the N-terminal region of FGF19 (and by extension the N-terminal regions of the other subfamily members as well) may directly interact with the D3 region of FGFR to achieve receptor specificity and activation. But distinct contacts between its HBS and the receptor D2 region may also contribute to specific receptor interactions.[15]

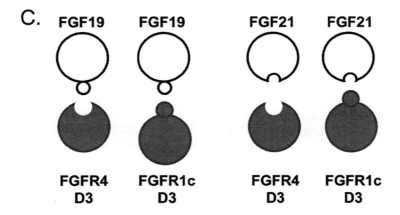

Figure 4. Proposed interaction model between βC'-βE loops from FGF receptors and FGF19 and FGF21. A) Alignment of sequences in the βC'-βE loops of FGFR1c, 2c, 3c, and 4. B) Sequence alignment of the proposed βC'-βE loop interacting regions of FGF19 and FGF21. C) Schematic diagrams of interactions between D3 domain of FGFR1c or FGFR4 and FGF19 or FGF21, respectively. The shorter βC'-βE loop of FGFR4 is represented as a concave surface, and the longer βC'-βE loop of FGFR1 is represented as a protrusion. The complementary FGF19 and FGF21 N-terminal regions are represented similarly.

ROLE OF FGFR4 IN GLUCOSE METABOLISM AND MITOGENESIS

Since its initial cloning,[51] several functions have been described for FGF19 and its mouse orthologue, FGF15.[1-3] The best understood function maybe its involvement in the regulation of bile acid metabolism, for which it acts as a target gene of FXR in the small intestine to serve as the enterohepatic signal for the regulation of *CYP7A1* expression in the liver through FGFR4 signaling.[7,36,52] Another physiological function of intestinal FGF19/FGF15 may be the regulation of gallbladder filling.[52,53] Under suprarphysiological conditions, additional functions of FGF19 have also been observed, such as its effects on glucose and energy homeostasis. When challenged with a high-fat diet, transgenic mice overexpressing FGF19 have lower serum glucose levels, improved glucose tolerance, and improved insulin sensitivity compared with their wild-type littermates.[12] Similar observations were made upon treatment of *ob/ob* mice with recombinant human FGF19. The treated mice exhibited lower serum glucose levels and improved insulin sensitivity,[54] suggesting that in addition to the ability to regulate bile acid metabolism, FGF19 also exerts effects on glucose and energy metabolism. In addition to these potentially beneficial effects of FGF19, the development of hepatocellular carcinoma (HCC) observed in transgenic FGF19 animals, indicates a mitogenic potential for FGF19.[10] In the next section, we will discuss studies that used FGF19 variants to elucidate the involvement of FGFR4 in the metabolic and mitogenic activities of FGF19.

FGFR4 Contribution to Metabolism

As discussed in previous sections, FGF19 is able to activate FGFRs1c, 2c, 3c, and 4, however, which one or combinations of these four receptors contribute to metabolic regulation has not yet been elucidated. Taking advantage of its structural features, an FGF19 variant that can only activate FGFR4 was constructed to determine the contribution of this receptor to the metabolic activity of FGF19.[44,55] The FGF19 feature that was exploited was that while the activation of receptors FGFR1c, 2c, and 3c completely depend on the presence of βKlotho, the ability of FGF19 to activate FGFR4 can be mediated through either heparin/HS or βKlotho.[44] Given that the βKlotho interaction domain resides in the C-terminal region of the FGF19 protein, an FGF19 variant, FGF19dCTD (Fig. 2),[44] with its C-terminal domain deleted is not expected to activate receptors that would depend on βKlotho for activity but can still activate FGFR4 by heparin/HS. This was demonstrated experimentally, first in a binding assay where the βKlotho-mediated interaction between FGF19dCTD and FGFR1c or FGFR4 was abolished while the heparin-induced binding to FGFR4 was unaffected. This correlated with receptor activities in L6 cells, in which FGF19dCTD was not able to activate ERK phosphorylation in cells cotransfected with βKlotho and FGFR1c, 2c, or 3c. In contrast, FGF19dCTD was able to activate cells transfected with FGFR4, presumably mediated by HS. Therefore, FGF19dCTD is an FGFR4-specific activator.

FGFR4 specificity was further demonstrated in animals injected with FGF19dCTD. ERK phosphorylation was observed in liver (where FGFR4 is predominantly expressed) but not in adipose tissue (where FGFR1c and 2c are the predominant receptors). Consistent with the signaling specificity, FGF19dCTD suppressed the expression of the target gene, *CYP7A1,* in the liver, presumably mediated through FGFR4 activation in a βKlotho-independent manner, suggesting that this FGFR4-specific activator may still be able to regulate bile acid metabolism in a manner similar to that done by wild-type FGF19. In contrast to wild-type FGF19, injection of FGF19dCTD into *ob/ob* diabetic mice did not lower plasma glucose levels in either an acute treatment or in a chronic 2-week study.[44] These results suggest that either FGFR4 is not important for glucose regulation or that activation of FGFR4 alone by FGF19 is not sufficient to recapitulate the ability to regulate glucose metabolism by wild-type FGF19. These findings may also suggest that activation of target tissues other than liver, including adipose, may provide significant contributions in the regulation of glucose homeostasis.

The inability of FGF19dCTD to regulate glucose metabolism is somewhat surprising, given that FGFR4 knock-out mice display glucose intolerance and insulin-resistant phenotypes[56] and that FGF19 has been shown to induce phosphorylation of FoxO1, a key

FGF19 AND ITS MITOGENIC AND METABOLIC ACTIVITIES 209

regulator of glucose metabolism, in primary mouse hepatocytes through FGFR4.[57] Whether this potential difference is caused by differences in the systems studied or represents different signaling pathways induced by FGF19 will need to be further defined in future studies.

FGFR4, Hepatocyte Mitogenesis, and Cancer

As a class, FGFs are known to induce mitogenesis and cell proliferation. Endocrine FGFs appear to have much reduced induction of cell proliferation in vitro, and tumor formations have not been reported for either FGF21 or FGF23 in vivo. However, the exception in this distinctive subfamily is FGF19, in which transgenic mice expressing FGF19 under the control of a muscle-specific promoter developed hepatocellular carcinomas by 10 months of age.[10] Despite the striking similarity in the ability of FGF19 and FGF21 to regulate metabolic homeostasis,[12,13,54,58,59] FGF21 transgenic mice (in contrast to FGF19 transgenic mice) neither developed liver tumors nor showed evidence of any other tissue hyperplasia up to 12 months of age.[13] In fact, forced overexpression of FGF21 may even delay hepatocarcinogenesis.[60]

The appearance of hepatocellular carcinomas in FGF19 transgenic mice was associated with increased mitotic activity in hepatocytes (as measured by incorporation of BrdU), which was detected in mice as young as 2 to 4 months of age. Additionally, wild-type FVB mice showed increased mitotic activity in hepatocytes in centrilobular regions of the liver after only 6 or 7 daily injections of FGF19.[10] However, such increased mitotic activity was not observed in the hepatocytes of animals injected with recombinant FGF21 under similar conditions, suggesting that FGF19 and FGF21 are distinct in this regard.[14] One possible explanation for the observed differential effects of FGF19 and FGF21 on hepatocyte proliferation may be the result of selective activation of a liver-expressed receptor by FGF19 but not by FGF21. As described in the previous section, the main difference in receptor specificity between these two endocrine FGFs is in their effects on FGFR4, where FGF19 is able to activate FGFR4, but not FGF21.

Given that FGFR4 is the predominant receptor expressed in the liver, it was reasonable to speculate that activation of this receptor may contribute to the hepatocyte proliferation observed with FGF19 treatment. The first evidence supporting this notion came from the analysis of FGFR4 expression patterns in the liver. In situ hybridization analysis revealed that the strongest signal for FGFR4 mRNA was present in hepatocytes adjacent to the central vein area and in a more dispersed staining away from central veins.[10] This coincided with the BrdU-labeled hepatocytes in which a pericentral distribution was also observed after FGF19 treatment.[10,14] More direct evidence supporting an FGFR4 role came from studies with FGF19 variants discussed in previous sections on structural activity studies of FGF19 (Fig. 2). The first direct evidence supporting this hypothesis was derived from experiments with FGF19dCTD, a C-terminally truncated variant of the FGF19 protein. The inability of FGF19dCTD to bind βKlotho explains its lack of activity toward FGFRs 1c, 2c, and 3c, and its specificity toward FGFR4 only. When tested in vivo for its effect on hepatocyte proliferation, FGF19dCTD increased BrdU incorporation in pericentral hepatocytes similar to FGF19, suggesting that activation of FGFR4 alone is sufficient to promote hepatocyte proliferation.[14]

Further evidence supporting this hypothesis came from analysis of FGF19/FGF21 chimeric proteins, which have helped the elucidation of structural determinants on FGF19 that are responsible for FGFR4 activation (Fig. 2). These analyses showed an absolute correlation between the ability to activate FGFR4, measured by ERK phosphorylation in vitro, and increased pericentral hepatocyte proliferation measured by BrdU labeling in liver in vivo (Fig. 2).[14,15] The correlation provides further evidence supporting the link between liver FGFR4 activation and hepatocyte proliferation. In addition, these studies identified a 5-amino acid region (residues 38-42) on FGF19 that is important and potentially sufficient for FGFR4 activation. When substituted into FGF21 in the form of FGF21/19[38-42] (Fig. 2), these 5 amino acids alone from FGF19 were sufficient to confer a gain-of-function phenotype on FGF21 with respect to both FGFR4 activation and increased hepatocyte proliferation (Fig. 2).[14,15]

Putting all this together, (1) the lack of increased hepatocyte proliferation by FGF21 treatment in direct comparison with FGF19; (2) the ability of FGF19dCTD, an FGFR4 specific activator, to increase pericentral hepatocyte BrdU incorporation; (3) the correlation among FGF19/FGF21 chimeric molecules in their ability to induce FGFR4 activation and to increase pericentral hepatocyte BrdU staining; and finally (4) the ability of just 5 residues from FGF19 to confer FGFR4 activation and increased BrdU labeling in pericentral hepatocytes, these observations provide compelling evidence that FGFR4 activation by FGF19 in hepatocytes may lead to increased proliferation.

There are many reports in the literature that implicate FGFR4 in tumorigenesis. Polymorphism in FGFR4 is associated with increased risks of several cancers.[24-26,61] For example, a common polymorphism of FGFR4 R388 is associated with many cancer types, including prostate cancer and head and neck squamous cell carcinoma.[24,25] Biochemical studies suggested that the R388 mutation results in increased receptor stability and sustained receptor activation.[26] FGFR4 is also frequently overexpressed in patients with hepatocellular carcinoma.[62] The anti-apoptotic effects of FGFR4 have been associated with resistance to chemotherapy, and siRNA against FGFR4 in liver cancer lines HuH7 is able to suppress α-fetoprotein production.[62,63] However, contradicting effects have also been reported. Genetic deletion of FGFR4 in mice has been reported to accelerate progression of DEN-induced hepatocellular carcinoma,[64] and overexpression of FGFR4 together with βKlotho has been reported to induce apoptosis and inhibit tumor cell proliferation in vitro,[65] suggesting a potential protective role of FGFR4 in suppressing hepatoma proliferation. Therefore, the contribution of FGFR4 in FGF19-induced tumor formation may require further clarification, and the use of FGFR4 knockout mice will provide a more definitive answer.

CONCLUSION

FGF19 and FGF21 are distinctive members of the fibroblast growth factor family that function as novel endocrine hormones. Their potent effects on normalizing glucose, lipid, and energy homeostasis in disease models have made them exciting new opportunities for combating the growing epidemics of diabetes and obesity. However, the observation of hepatocellular carcinoma formation in FGF19 transgenic mice has limited the consideration of this molecule to treat chronic metabolic diseases. Here, we have summarized some of the recent studies that provide a better understanding of the structure-function relationship of FGF19/FGF21 and the identification of the structural basis underpinning the distinct proliferative feature of FGF19 that is lacking in FGF21. The mapping of different domains responsible for interaction/activation with receptors and coreceptors, and the ability to interchange them between endocrine FGFs to alter

FGF19 AND ITS MITOGENIC AND METABOLIC ACTIVITIES

specific biological activities supports the concept of a modular evolution of these family members. The ability to separate and remove the undesired mitogenic effects from FGF19 also opens up a new avenue and provides a road map to engineer FGF19 as a potential therapeutic candidate for treating diabetes and obesity.

Future studies could be directed at gaining additional insight into FGF19/21/FGFR interactions as well as βKlotho/FGFR and FGF19/21/βKlotho interactions to give a more complete appreciation of the similarities and differences between endocrine and paracrine FGFs. The mitogenic activity of FGF19 should be further studied to determine whether it is caused by normal FGF19 biology. FGF15, the mouse orthologue of human FGF19, has not yet been tested for induction of hepatocyte proliferation at supraphysiological levels. Activation of FGFR4 by FGF19 in humans should also be studied to determine whether it also leads to mitogenesis. The answers to the above could increase understanding of the relevance of the observations in rodent to human biology.

ACKNOWLEDGEMENT

We thank Richard Lindberg for valuable discussions, and we thank Scott Silbiger for help in revising this manuscript.

REFERENCES

1. Jones S. Mini-review: endocrine actions of fibroblast growth factor 19. Mol Pharm 2008; 5:42-48.
2. Fukumoto S. Actions and mode of actions of FGF19 subfamily members. Endocr J 2008; 55:23-31.
3. Kharitonenkov A. FGFs and metabolism. Curr Opin Pharmacol 2009; 9:805-810.
4. Wu X, Ge H, Gupte J et al. Co-receptor requirements for fibroblast growth factor-19 signaling. J Biol Chem 2007; 282:29069-29072.
5. Kurosu H, Choi M, Ogawa Y et al. Tissue-specific expression of betaKlotho and fibroblast growth factor (FGF) receptor isoforms determines metabolic activity of FGF19 and FGF21. J Biol Chem 2007; 282:26687-26695.
6. Lin BC, Wang M, Blackmore C et al. Liver-specific activities of FGF19 require Klotho beta. J Biol Chem 2007; 282:27277-27284.
7. Holt JA, Luo G, Billin AN et al. Definition of a novel growth factor-dependent signal cascade for the suppression of bile acid biosynthesis. Genes Dev 2003; 17:1581-1591.
8. Micanovic R, Raches DW, Dunbar JD et al. Different roles of N- and C- termini in the functional activity of FGF21. J Cell Physiol 2009; 219:227-234.
9. Turner N, Grose R. Fibroblast growth factor signalling: from development to cancer. Nat Rev Cancer 10:116-129.
10. Nicholes K, Guillet S, Tomlinson E et al. A mouse model of hepatocellular carcinoma: ectopic expression of fibroblast growth factor 19 in skeletal muscle of transgenic mice. Am J Pathol 2002; 160:2295-2307.
11. Ogawa Y, Kurosu H, Yamamoto M et al. BetaKlotho is required for metabolic activity of fibroblast growth factor 21. Proc Natl Acad Sci U S A 2007; 104:7432-7437.
12. Tomlinson E, Fu L, John L et al. Transgenic mice expressing human fibroblast growth factor-19 display increased metabolic rate and decreased adiposity. Endocrinology 2002; 143:1741-1747.
13. Kharitonenkov A, Shiyanova TL, Koester A et al. FGF-21 as a novel metabolic regulator. J Clin Invest 2005; 115:1627-1635.
14. Wu X, Ge H, Lemon B et al. FGF19-induced hepatocyte proliferation is mediated through FGFR4 activation. J Biol Chem 2010; 285:5165-5170.
15. Wu X, Ge H, Lemon B et al. Separating mitogenic and metabolic activities of fibroblast growth factor 19 (FGF19). Proc Natl Acad Sci U S A 2010; 107:14158-14163.
16. Suzuki M, Uehara Y, Motomura-Matsuzaka K et al. BetaKlotho is required for fibroblast growth factor (FGF) 21 signaling through FGF receptor (FGFR) 1c and FGFR3c. Mol Endocrinol 2008; 22:1006-1014.
17. Gospodarowicz D. Localisation of a fibroblast growth factor and its effect alone and with hydrocortisone on 3T3 cell growth. Nature 1974; 249:123-127.
18. Beenken A, Mohammadi M. The FGF family: biology, pathophysiology and therapy. Nat Rev Drug Discov 2009; 8:235-253.
19. Knights V, Cook SJ. De-regulated FGF receptors as therapeutic targets in cancer. Pharmacol Ther 125:105-117.

20. Krejci P, Prochazkova J, Bryja V et al. Molecular pathology of the fibroblast growth factor family. Hum Mutat 2009; 30:1245-1255.
21. Cappellen D, De Oliveira C, Ricol D et al. Frequent activating mutations of FGFR3 in human bladder and cervix carcinomas. Nat Genet 1999; 23:18-20.
22. di Martino E, L'Hote CG, Kennedy W et al. Mutant fibroblast growth factor receptor 3 induces intracellular signaling and cellular transformation in a cell type- and mutation-specific manner. Oncogene 2009; 28:4306-4316.
23. Naski MC, Wang Q, Xu J et al. Graded activation of fibroblast growth factor receptor 3 by mutations causing achondroplasia and thanatophoric dysplasia. Nat Genet 1996; 13:233-237.
24. Streit S, Bange J, Fichtner A et al. Involvement of the FGFR4 Arg388 allele in head and neck squamous cell carcinoma. Int J Cancer 2004; 111:213-217.
25. Wang J, Stockton DW, Ittmann M. The fibroblast growth factor receptor-4 Arg388 allele is associated with prostate cancer initiation and progression. Clin Cancer Res 2004; 10:6169-6178.
26. Wang J, Yu W, Cai Y et al. Altered fibroblast growth factor receptor 4 stability promotes prostate cancer progression. Neoplasia 2008; 10:847-856.
27. Kurosu H, Kuro OM. The Klotho gene family as a regulator of endocrine fibroblast growth factors. Mol Cell Endocrinol 2009; 299:72-78.
28. Plotnikov AN, Schlessinger J, Hubbard SR et al. Structural basis for FGF receptor dimerization and activation. Cell 1999; 98:641-650.
29. Furdui CM, Lew ED, Schlessinger J et al. Autophosphorylation of FGFR1 kinase is mediated by a sequential and precisely ordered reaction. Mol Cell 2006; 21:711-717.
30. Eswarakumar VP, Lax I, Schlessinger J. Cellular signaling by fibroblast growth factor receptors. Cytokine Growth Factor Rev 2005; 16:139-149.
31. Goetz R, Beenken A, Ibrahimi OA et al. Molecular insights into the klotho-dependent, endocrine mode of action of fibroblast growth factor 19 subfamily members. Mol Cell Biol 2007; 27:3417-3428.
32. Katoh M, Katoh M. Evolutionary conservation of CCND1-ORAOV1-FGF19-FGF4 locus from zebrafish to human. Int J Mol Med 2003; 12:45-50.
33. Katoh M. WNT and FGF gene clusters (review). Int J Oncol 2002; 21:1269-1273.
34. Desnoyers LR, Pai R, Ferrando RE et al. Targeting FGF19 inhibits tumor growth in colon cancer xenograft and FGF19 transgenic hepatocellular carcinoma models. Oncogene 2008; 27:85-97.
35. Pai R, Dunlap D, Qing J et al. Inhibition of fibroblast growth factor 19 reduces tumor growth by modulating beta-catenin signaling. Cancer Res 2008; 68:5086-5095.
36. Inagaki T, Choi M, Moschetta A et al. Fibroblast growth factor 15 functions as an enterohepatic signal to regulate bile acid homeostasis. Cell Metab 2005; 2:217-225.
37. Fukumoto S, Yamashita T. FGF23 is a hormone-regulating phosphate metabolism—unique biological characteristics of FGF23. Bone 2007; 40:1190-1195.
38. Zhang X, Ibrahimi OA, Olsen SK et al. Receptor specificity of the fibroblast growth factor family. The complete mammalian FGF family. J Biol Chem 2006; 281:15694-15700.
39. Kuro-o M, Matsumura Y, Aizawa H et al. Mutation of the mouse klotho gene leads to a syndrome resembling ageing. Nature 1997; 390:45-51.
40. Ito S, Kinoshita S, Shiraishi N et al. Molecular cloning and expression analyses of mouse betaklotho, which encodes a novel Klotho family protein. Mech Dev 2000; 98:115-119.
41. Urakawa I, Yamazaki Y, Shimada T et al. Klotho converts canonical FGF receptor into a specific receptor for FGF23. Nature 2006; 444:770-774.
42. Kurosu H, Ogawa Y, Miyoshi M et al. Regulation of fibroblast growth factor-23 signaling by klotho. J Biol Chem 2006; 281:6120-6123.
43. Kharitonenkov A, Dunbar JD, Bina HA et al. FGF-21/FGF-21 receptor interaction and activation is determined by betaKlotho. J Cell Physiol 2008; 215:1-7.
44. Wu X, Ge H, Lemon B et al. Selective activation of FGFR4 by an FGF19 variant does not improve glucose metabolism in ob/ob mice. Proc Natl Acad Sci U S A 2009; 106:14379-14384.
45. Xie MH, Holcomb I, Deuel B et al. FGF-19, a novel fibroblast growth factor with unique specificity for FGFR4. Cytokine 1999; 11:729-735.
46. Mohammadi M, Olsen SK, Ibrahimi OA. Structural basis for fibroblast growth factor receptor activation. Cytokine Growth Factor Rev 2005; 16:107-137.
47. Wu X, Lemon B, Li X et al. C-terminal tail of FGF19 determines its specificity toward Klotho coreceptors. J Biol Chem 2008; 283:33304-33309.
48. Yie J, Hecht R, Patel J et al. FGF21 N- and C-termini play different roles in receptor interaction and activation. FEBS Lett 2009; 583:19-24.
49. Harmer NJ, Pellegrini L, Chirgadze D et al. The crystal structure of fibroblast growth factor (FGF) 19 reveals novel features of the FGF family and offers a structural basis for its unusual receptor affinity. Biochemistry 2004; 43:629-640.

FGF19 AND ITS MITOGENIC AND METABOLIC ACTIVITIES

50. Schlessinger J, Plotnikov AN, Ibrahimi OA et al. Crystal structure of a ternary FGF-FGFR-heparin complex reveals a dual role for heparin in FGFR binding and dimerization. Mol Cell 2000; 6:743-750.
51. Nishimura T, Utsunomiya Y, Hoshikawa M et al. Structure and expression of a novel human FGF, FGF-19, expressed in the fetal brain. Biochim Biophys Acta 1999; 1444:148-151.
52. Yu C, Wang F, Jin C et al. Independent repression of bile acid synthesis and activation of c-Jun N-terminal kinase (JNK) by activated hepatocyte fibroblast growth factor receptor 4 (FGFR4) and bile acids. J Biol Chem 2005; 280:17707-17714.
53. Choi M, Moschetta A, Bookout AL et al. Identification of a hormonal basis for gallbladder filling. Nat Med 2006; 12:1253-1255.
54. Fu L, John LM, Adams SH et al. Fibroblast growth factor 19 increases metabolic rate and reverses dietary and leptin-deficient diabetes. Endocrinology 2004; 145:2594-2603.
55. Wu X, Li Y. Role of FGF19 induced FGFR4 activation in the regulation of glucose homeostasis. Aging 2009; 1:1023.
56. Huang X, Yang C, Luo Y et al. FGFR4 prevents hyperlipidemia and insulin resistance but underlies high-fat diet induced fatty liver. Diabetes 2007; 56:2501-2510.
57. Shin DJ, Osborne TF. FGF15/FGFR4 integrates growth factor signaling with hepatic bile acid metabolism and insulin action. J Biol Chem 2009; 284:11110-11120.
58. Coskun T, Bina HA, Schneider MA et al. Fibroblast growth factor 21 corrects obesity in mice. Endocrinology 2008; 149:6018-6027.
59. Xu J, Lloyd DJ, Hale C et al. Fibroblast growth factor 21 reverses hepatic steatosis, increases energy expenditure, and improves insulin sensitivity in diet-induced obese mice. Diabetes 2009; 58:250-259.
60. Huang X, Yu C, Jin C et al. Forced expression of hepatocyte-specific fibroblast growth factor 21 delays initiation of chemically induced hepatocarcinogenesis. Mol Carcinog 2006; 45:934-942.
61. Sasaki H, Okuda K, Kawano O et al. Fibroblast growth factor receptor 4 mutation and polymorphism in Japanese lung cancer. Oncol Rep 2008; 20:1125-1130.
62. Ho HK, Pok S, Streit S et al. Fibroblast growth factor receptor 4 regulates proliferation, anti-apoptosis and alpha-fetoprotein secretion during hepatocellular carcinoma progression and represents a potential target for therapeutic intervention. J Hepatol 2009; 50:118-127.
63. Roidl A, Berger HJ, Kumar S et al. Resistance to chemotherapy is associated with fibroblast growth factor receptor 4 up-regulation. Clin Cancer Res 2009; 15:2058-2066.
64. Huang X, Yang C, Jin C et al. Resident hepatocyte fibroblast growth factor receptor 4 limits hepatocarcinogenesis. Mol Carcinog 2009; 48:553-562.
65. Luo Y, Yang C, Lu W et al. Metabolic regulator betaKlotho interacts with fibroblast growth factor receptor 4 (FGFR4) to induce apoptosis and inhibit tumor cell proliferation. J Biol Chem 2010; 285:30069-30078.

CHAPTER 14

FGF21 AS A THERAPEUTIC REAGENT

Yang Zhao, James D Dunbar and Alexei Kharitonenkov*

Eli Lilly and Company, Lilly Corporate Center, Indianapolis, Indiana, USA.
*Corresponding Author: Alexei Kharitonenkov—Email: a.kharch@lilly.com

Abstract: The prevalence of obesity and diabetes has been dramatically increasing during the last decade suggesting a greater patient need for more efficacious and safer drugs. Large molecule therapy has played an important role in diabetes since the discovery of insulin. This legacy was continued upon the introduction of Humulin (first recombinant insulin), Humalog (first engineered insulin) and Byetta (first incretin mimetic). Several other protein therapeutics, such as leptin, adiponectin, bone morphogenic protein-9 and others, are currently in or considered for therapeutic development. Among them, FGF21 is one of the most promising candidates given its outstanding pharmacologic benefits for nearly each and every abnormality of a metabolic disease and lack of apparent side effects in a variety of animal models. Thus, FGF21 represents a novel and appealing therapeutic reagent for Type 2 diabetes mellitus, obesity, dyslipidemia, cardiovascular and fatty liver diseases. The in vitro biology, genetic animal models and in vivo pharmacology of FGF21 will be discussed in this chapter.

INTRODUCTION

Since the discovery of the first fibroblast growth factor,[1] FGFs have been studied for almost 40 years. FGFs are present from nematodes to humans[2] and during embryonic development they play important functions in proliferation, migration and differentiation. In the adult organism, FGFs are critical for nervous system, wound healing, tissue repair and tumor angiogenesis.[3]

The currently identified 22 human FGFs mediate their cellular responses via activation of 4 known FGFRs (FGFR1-4).[3] Structurally, they have a single transmembrane domain, two or three IgG like domains (D1-D3), a heparin binding domain and a tyrosine kinase domain.[2] The 7-8 acidic residues between D1 and D2 (acid box) and a conserved positively

---

Endocrine FGFs and Klothos, edited by Makoto Kuro-o.
©2012 Landes Bioscience and Springer Science+Business Media.

FGF21 AS A THERAPEUTIC REAGENT

charged region within D2 are binding sites for Heparin.[4] The alternative splicing in D3 of FGFR1-3 results in b and c isoforms of FGFR with distinct ligand binding specificity and tissue expression profile with b-isoforms exclusively expressed in epithelial cells whereas c variants primarily present in mesenchymal cells.[5] In the classical FGF signaling, the binding of FGF to the extracellular domain of FGFR in complex with heparan sulfate proteoglycan (HSPG) that facilitates FGF/FGFR interaction leads to the receptor dimerization and tyrosine phosphorylation of its intracellular part.

The crystal structures of the ligand binding domain of FGFR1 and 2 in complex with FGF1 or 2 and of a ternary FGF/heparin/FGFR have been determined.[4,6,7] FGFRs have a longer juxtamembrane domain than other receptor tyrosine kinases do. The highly conserved region in the juxtamembrane domain provides a binding site for the phosphotyrosine binding (PTB) domains of FRS2α and β that play important and specific to FGFR signaling roles. Tyrosine phosphorylation of FRS2α and β stimulated by FGF recruits up to four Grb2/Sos complexes and leads to Ras/MAPK activation.[8] FRS2α can also bind Shp2 that in tyrosine phosphorylation form recruits additional Grb2 molecules. Biologically, FRS2α is essential for embryonic development.[9] At the cellular level, FRS2α is critical for FGF-stimulated MAPK and PI3K activation, cell migration and proliferation. FRS2α also recruits negative regulators such as Cbl which results in ubiquitination of both FGFR and FRS2α.[10] In addition, FGF-stimulated MAPK phosphorylates at least 8 threonine residues of FRS2α which attenuates its tyrosine phosphorylation. Mutagenesis of the threonine residues or treatment with a MEK inhibitor results in constitutive tyrosine phosphorylation in untreated cells.[11] The autophosphorylation on Tyr766 of FGFR1 recruits the SH2 domain of phospholipase Cγ (PLCγ) and therefore important for the downstream PI hydrolysis and Ca^{2+} release.[12]

The FGFR1 or 1c knockout mice show embryonic lethality due to defects in cell migration through the primitive streak, but no obvious phenotype was observed in FGFR1b knockout mice.[13-15] The FGFR2 knockout mice showed embryonic lethality due to placenta defects.[16] The FGFR2b knockout die immediately after birth due to developmental defects in the lung, limbs and other tissues.[17] Although the FGFR2c are viable, these mice have skull and bone developmental defects.[18] In contrast, FGFR3 knockout mice showed overgrowth in bone.[19,20] The FGFR4 knockout mice have no obvious developmental phenotype.[21]

FGFR1, 2 and 3 are important for human skeletal development. One gain of function mutation in FGFR3 is the cause of the most common form of human dwarfism which is consistent with the bone overgrowth phenotype in the knockout mice.[19,20] Somatic FGFR1 mutations can cause Kallmann syndrome (KAL2). This developmental disease is characterized by hypogonadism and development failure of olfactory bulb.[22] This is consistent with the FGFR1 knockout which has the same feature of KAL2 syndrome.[22] Translocation and fusion of FGFR1 with at least 5 genes have been found to cause myeloproliferative syndrome (MPS).[23] The resulting fusion kinases lack the juxtamembrane domain required for FRS2 binding. Tyr766, the binding site for PLC-γ, is required for the transforming activity.[24] Bcr-FGFR1 fusion kinase also recruits Bcr and results in myeloproliferative disorder related to chronic myeloid leukemia (CML).[25] Similarly, translocation and fusion of FGFR3 are associated with peripheral T-cell lymphoma and multiple myeloma.

The mammalian FGF family can be divided into several subfamilies: intracellular FGFs (iFGFs) including FGF11-14; the hormone-like FGFs (hFGFs) including FGF19 (FGF15)/21/23; and the canonical FGFs which function in a paracrine manner.[26] It has been proposed that FGF13-like gene is the ancestor of the iFGF subfamily and the evolutionary ancestor of the whole FGF family.[26] The ancestral FGFs for the subfamilies gained the functional diversity before vertebrates evolved. The subfamilies expanded during the evolution of early vertebrates. FGFs 1-23 were found in humans and mice but FGF15 is a mouse ortholog of human FGF19. Therefore the human FGF family has only 22 members.

The canonical FGFs have high affinity binding sites for heparin and heparan sulfate proteoglycan (HSPG).[26] The 2:2:2 FGF-FGFR-heparan sulfate dimers form when heparan sulfate is present.[27] The interaction with heparin and HSPG limits the diffusion of FGFs.[28,29] In contrast, the hFGFs only have low-affinity binding sites for heparin/HSPG and act in an endocrine manner.[30-32] hFGFs require coreceptors, Klotho or β-Klotho to activate FGFRs.[33,34] Most canonical FGFs serve essential functions in proliferation or differentiation during development. The hFGFs have distinct hormonal roles that regulate energy metabolism, phosphate and vitamin D metabolism and bile acid metabolism. However, some other FGFs also involve in metabolic process. FGF10 is involved in the differentiation of pancreas and white adipose tissue, whereas FGF16 is a specific factor for brown adipose tissue.[35-37] Some FGFRs were reported to be related to metabolism. Injection of a monoclonal FGFR1c antibody in rodents and monkeys caused strong hypophagia and weight loss.[38] Rapid weight loss was caused by decreased food intake and not by elevated energy expenditure or cachexia. The expression of a dominant negative FGFR1 in β cells led to diabetes in mice.[39] FGFR2 is important for the development of pancreas.[40-42] FGFR4 is involved in glucose and lipid metabolism, insulin sensitivity and bile acid synthesis.[43,44] The liver FGFR4 mRNA is subjected to regulations by fasting or insulin.[45]

The molecular weight of FGFs range from 17-34 kDa.[2] The FGFs share a core homology domain of nearly 120 amino acids that is flanked by highly variable sequences. The core region of FGF19 is most identical to FGF21 (38%) and FGF23 (36%). The hormone-like FGFs (FGF19, 21 and 23) act in an endocrine fashion to regulate bile acid, energy, glucose, lipid, phosphate and vitamin D homeostasis.[46] The heparin-binding region between the 10th and 12th β strand are dramatically different from other FGFs.[46] FGF19 has only 11 β-strands (missing the region between β10 and 12). In addition, a cleft between this region and the β1-β2 loop precludes the direct binding between the backbone of FGF19/23 and heparin/heparan sulfate.[46] There are two novel disulfide bonds found in FGF19.[47] One of them is conserved and stabilizes extended loops. These features suggest greater half life and a longer range of activity from the secreting cell. To function in their target tissues the hormone-like FGF19, 21 and 23 require Klotho/βKlotho that play a parallel to heparin/heparan sulfate role for classical FGFs. The extracellular domains of both Klotho proteins have two internal repeats which share 20-40% identity to the β-glucosidase and lactase glycosylceramidase.[48] αKlotho is primarily expressed in the distal convoluted tubules in the kidney[49] and βKlotho is expressed in liver, pancreas and adipose tissue.[33,48]

FGF15/19 is expressed by intestinal epithelium and is involved in hepatic bile acid feedback regulation.[32,50-52] βKlotho deficient mice have elevated bile acid synthesis, similar to the phenotypes of FGF15 or FGFR4 knockout mice.[43,51,53] The similar phenotypes suggest that βKlotho functionally interacts with the FGF19-FGFR4 signaling to regulate bile acid synthesis.

FGF21 regulates glucose and lipid metabolism.[30] FGF21 was first cloned by Itoh group from mouse embryos by homology based PCR.[54] FGF21 mRNA was first found to be most highly expressed in the liver and also in the thymus at lower levels.[54] The human FGF21 (209 aa) has 75% identity to mouse FGF21 (210 aa). FGF21 activity depends on βKlotho, a single transmembrane protein induced during adipocyte differentiation.[33] βKlotho interacts

FGF21 AS A THERAPEUTIC REAGENT

with several FGFRs to increase their ability to bind FGF21. Knockdown of βKlotho in adipocytes abolished FRS2α and ERK1/2 activation by FGF21 and FGF21 effect on glucose uptake.[33] FGF21 can activate FGFRs and downstream molecules including FRS2α and 44/42 MAPK (ERK1/2) in adipocytes.[30,55] 3T3-L1 preadipocytes overexpressing FGFRs do not respond to FGF21 suggesting a cofactor for FGF21 in adipocytes. Among all FGFRs idenitified, none of them showed differential expression between 3T3-L1 fibroblasts and adipocytes.[56] Using immunoprecipation and liquid chromatography/mass spectrometry/ mass spectrometry (LC/MS/MS), βKlotho was isolated and identified as coreceptor for FGF21.[56] Immunoprecipitation of FGFR1 or -2 from 3T3-L1 cells stably expressing βKlotho showed that βKlotho constitutively bound to FGFR, independent of FGF21.[56] In other words, βKlotho and FGFR exist as a preformed complex before FGF21 activates it. Forced expression of βKlotho rendered 293 cells the responsiveness to FGF21.[33] In BaF3 cells, which do not express endogenous FGFRs or heparan sulfate proteoglycan, βKlotho is required for FGF21 to activate FGFR1c and 3c.[57] Study of FGF21 in L6 cells revealed that L6 cells overexpressing βKlotho also activated FGFR3c in response to FGF21.[58] FGF21 also induces phophorylation of FRS2α in both FGFR2c/βKlotho expressing 3T3-L1 adipocytes and FGFR2c/βKlotho transfected L6 cells.[56,58] Based on these data, FGF21 can activate the c-isoforms of FGFRs in the presence of βKlotho.

FGF23 regulates phosphate and vitamin D homeostasis.[31] FGF23 derives from osteoblast lineage and regulates homeostasis in phosphate and vitamin D metabolism by regulating the sodium phosphate cotransporter and the vitamin D-metabolizing enzymes CYP27B1 and CYP24A1 in the kidney suggesting an FGF23 specific receptor in the kidney.[59-61] Klotho was identified as a receptor that binds to FGF23 in renal homogenate.[34] Klotho is mainly expressed in the kidney but not in renal cell lines. Overexpression of Klotho restored FGF23 responsiveness in renal cell lines.[34] The antiKlotho antibody administration in mice rendered FGF23 incompetence. The functional interaction between Klotho and FGF23 is also suggested by the similar phenotypes of Klotho-deficient mice and FGF23-null mice. Both mice have shorter lifespan, infertility, decreased bone mineral density and ectopic calcification, increase in renal phosphate reabsorption and Vitamin D synthesis.[62-64] Vitamin D restriction improved Klotho-deficient phenotypes suggesting abnormal mineral metabolism may be the cause of the ageing phenotype.[64] When various FGFR-Fc fusions were added to cells, only FGFR1(IIIc) significantly reduced Klotho-dependent FGF23 activity.[34] FGF23 injected mice only showed ERK1/2 phosphorylation in kidney, where Klotho is expressed.[33] These data suggest that Klotho and FGFR1 (IIIc) together function as receptor for FGF23.

The simultaneous interaction of FGF19, 21 and 23 to their cognate FGFRs and Klotho/βKlotho provide the affinity just enough to produce a metabolic but not mitogenic response. Furthermore, the restricted expression of Klotho/βKlotho limits the endocrine functions of FGF19, 21 and 23 to their specific tissues.

FGF21 IN VITRO FINDINGS

FGF21 activity was discovered in a high throughput screen using 3T3-L1 adipocyte glucose uptake as an assay system.[30] FGF21 was found to be a potent activator of glucose uptake in murine and human adipocytes. The effects of FGF21 on glucose uptake are insulin-independent since there is an additive effect when cells were treated with insulin and FGF21.[30] Unlike the rapid action of insulin, FGF21 needs hours to exert its effect on glucose uptake. This difference is explained by that insulin acts via GLUT4 translocation whereas FGF21 acts through GLUT1 mRNA upregulation.[30]

Under obesity and insulin resistant conditions, dysregulated lipolysis is a major contributing factor in this process. Lipolysis is regulated by perilipin (PLIN), a lipid droplet phosphoprotein phosphorylated by protein kinase A (PKA) under lipolytic stimulation. The phosphorylation of PLIN leads to the recruitment of lipases to the lipid droplet surface for subsequent TG hydrolysis. The role of FGF21 in adipocyte lipolysis has been controversial. While it has initially been reported that FGF21 stimulates basal lipolysis in murine adipocytes,[65] the attempt to reproduce these results failed with both murine and human FGF21. Instead, FGF21 was shown to attenuate catecholamine-and atrial natriuretic peptide (ANP) stimulated lipolysis in human and murine adipocytes after 3-day treatment.[66] The lipolysis inhibitory effect of FGF21 may contribute to its insulin sensitizing benefit in humans. FGF21 treatment does not change expression of lipolytic genes including hormone sensitive lipase (HSL), adipose-derived triglyceride lipase (ATGL), PLIN, protein kinase A regulatory protein 1 α (PKA R1α), posphodiesterase 3B (PDE3B) or peroxisome proliferators-activated receptor gamma (PPARγ).[66] But the treatment lowered PLIN protein level by 50%. The ability of FGF21 to inhibit lipolysis was further confirmed by increased nonesterified fatty acids (NEFAs) observed when FGF21 was knocked down following a ketogenic diet.[67] The antilipolytic effect of FGF21 is also consistent with reduced lipotoxicity on islet cells in FGF21 administered mice.[30]

FGF21 has been shown to promote adipocyte differentiation in a dose dependent manner.[30] The cotreatment with FGF21 and rosiglitazone synergistically increased 3T3-L1 adipocyte diffentiation.[68] FGF21 treatment in the presence of rosiglitazone also resulted in a dramatic increase in glucose uptake after 6 hours suggesting functional synergy between FGF21 and PPARγ pathways. Simultaneous treatment with FGF21 and rosiglitazone led to further elevation of Glut1 mRNA and protein levels compared to single treatment.[68] Likewise, pretreatment with rosiglitazone led to increased tyrosine phosphorylation of FGFR by FGF21 without changing FGFR protein levels.[68] Furthermore, rosiglitazone induced elevation of βKlotho,[56] which may partially explain the synergy between FGF21 and PPARγ. FGF21 itself was one of the top 5 genes encoding secreted proteins that are regulated by PPARγ agonists in the adipose tissue of *db/db* mice.[69] PPARγ regulates at least two adipogenic programs during 3T3-L1 cell differentiation.[70] Group 1 has canonical adipogenic genes such as FABP4/aP2, adiponectin and PLIN. Group 2 has a variety of genes including FGF21 and the oxidoreductase Ero1-Lα. PPARγ carrying mutation of F372 maintains the adipogenic program in response to thiazolidinediones (TZDs) but is unable to induce expression of FGF21 and Ero1-Lα.[70] PPARγ agonists target adipose tissue gene expression and improve whole-body insulin sensitivity. FGF21 may thus partially contribute to the antidiabetic activities of PPARγ agonists.

FGF21 has been detected in pancreas suggesting a broader expression profile for FGF21.[71] FGF21 protein levels are similar in skeletal muscle and liver in fasted mice.[72] In skeletal muscle specific Akt1 transgenic mice, FGF21 expression was upregulated in gastrocnemius muscle along with increase plasma concentration. The expression of FGF21 in skeletal muscle and its secretion is regulated by insulin via phosphatidylinositol 3-kinase (PI3K)/Akt pathway.[72] In obese mice, Akt1-mediated skeletal muscle growth led to a reduction in fat mass and improved metabolic phenotype.[73] It was proposed that Akt1 activation in skeletal muscle may result in secretion of a myokine that functions on liver, adipose or central nervous system to normalize metabolism. FGF21 mRNA and FGF21 protein in culture medium was also dramatically upregulated in C2C12 myocytes treated

FGF21 AS A THERAPEUTIC REAGENT

with adenovirus expressing myrAkt1.[72] FGF21 expression in the cell lysate and culture media was elevated by insulin treatment in C2C12 myocytes. Pretreatment with LY294002, a PI3K inhibitor, diminished both basal and insulin-stimulated FGF21 expression.[72] FGF21 may be one of the critical factors involved in the intertissue communication in skeletal muscle-specific Akt1 transgenic mice. Furthermore, FGF21 is significantly increased in hyperinsulinemic *ob/ob* mice.[74] In normal mice, insulin levels are low after fasting and high after feeding, but FGF21 expression is the opposite.[65,74]

FGF21 inhibits glucagon secretion from isolated rat islets without affecting insulin secretion.[30] In rat islets and INS1E cells, FGF21 activated ERK1/2 and Akt pathways.[75] The effect was transient and returned to basal level in 2 hours. Although FGF21 does not increase glucose-stimulated insulin secretion (GSIS) of islets isolated from normal rats, it elevates insulin secretion and insulin content from diabetic islets and increases β cell survival through ERK1/2 and Akt pathways.[75] These data are consistent with the view that insulin biosynthesis is not rate limiting for GSIS under healthy conditions but it becomes important when insulin demand is increased.[76,77] Islets exposed to both high glucose and palmitate showed a dramatic decrease in GSIS.[75] FGF21 treatment partially rescued GSIS and islet insulin content. Exposure of islets to FGF21 also reduced caspase 3/7 activity induced by a combination of high glucose and lipids or a combination of interleukin (IL) -1β and tumor necrosis factor (TNF) –α.[75] Akt-induced phosphorylation of Bcl-XL/Bcl-2-associated death promoter (BAD) may be the mechanism of the protective effect of FGF21 on β cells.

The ability of FGF21 to activate FGFR4 has been controversial. While FGF21 has not been shown to activate FGFR4 in H4IIE hepatoma cells[58] or primary hepatocytes,[78] soluble βKlotho binds to FGFR4 and FGF21 can form ternary complex with them.[33,56] More importantly, FGF21 has a lot of direct impacts on liver[79-81] and FGFR4 is the predominant FGFR in the liver.[82] FGF21 dose dependently lowered glucose production in H4IIE cells with an IC50 around 60 pM.[83] FGF19 also has direct effects on hepatocytes and suppress insulin stimulated fatty acid synthesis by inhibiting the expression of sterol regulatory element-binding protein-1c (SREBP-1c).[84] All these data is indicative that FGF21 can still signal via FGFR4.

A chimera of FGF19 with FGF21 C-terminus interacts with βKlotho and a chimera with FGF23 C-terminus only interacts with αKlotho.[85] These data suggest that the C-terminus is a region necessary for recognition of Klotho family proteins. FGF21 activity is sensitive to terminal deletions since removal of six amino acids from the N-terminus and four from the C-terminus each reduced FGF21 activity by 10-fold.[86] The 14 amino acids at the C-terminus of FGF21 is important for βKlotho interaction but the N-terminus interacts directly with FGFRs.[86,87] The FGF21 efficacy is dramatically reduced with N-terminal deletion of 6 aa and completely lost with 8 aa deletion. FGF21 with N-terminal deletions can inhibit full-length FGF21 activity, a feature of partial agonist.[87] The binding affinity between FGFR1c and βKlotho is low with Kd higher than 100 nM.[87] FGF21-βKlotho affinity is also weak with Kd around 15 nM. These data suggest that FGF21, βKlotho and FGFR1c may form a ternary complex. It has also been found that the N- and C-terminal domains of FGF23 are responsible for binding to FGFRs and Klotho, respectively.[88] The C-terminus of FGF19 is critical for binding to Klotho family proteins.[85] Both FGF19 and FGF21 can signal through FGFR1, 2 and 3 bound by βKlotho.[58] Both FGF19 and 21 bind to βKlotho-FGFR4, but only FGF19 signals through FGFR4 effectively. FGF19 but not FGF21 activates FGF signaling in hepatocytes that mainly express FGFR4 and reduce Cyp7A1 expression and bile acid synthesis.[58] In the gall bladder, all four FGFRs and βKlotho are expressed, with FGFR3 being most dominant.[89] FGF19 can increase gall bladder volume in FGFR4 deficient mice suggesting that FGF19 can signal through other FGFRs.[58]

FGF21 GENETIC MODELS

FGF21 transgenic mice were viable and resistant to diet induced obesity.[30] FGF21 transgenice mice showed reduced plasma glucose at 9 months. The plasma glucagon levels were lower in FGF21 transgenic mice because FGF21 inhibits glucagon secretion.[30] Under high fat high carbohydrate diet for 15 weeks, FGF21 transgenic mice were resistant to diet-induced obesity despite increased food intake normalized by body weight. While FGF21 transgenic mice resemble that of FGF19,[30] there is a fundamental in vivo difference. FGF19 transgenic mice developed histologically detectable liver tumors.[90] FGF19 administered mice had increased hepatocellular proliferation after 6 days of treatment. FGF21 transgenic mice did not form tumors after 10 months of age,[30] or even protected from cancer formation[82].

Overexpression study revealed that mouse FGF21 regulates lipolysis and ketogenesis.[65] FGF21 is a mediator of PPARα actions. Fasting, ketogenic diet (KD) or PPARα agonists induce liver FGF21 expression.[65,67,74] There was a striking absence of hepatic FGF21 in PPARα null mice. Mouse FGF21 overexpression also decreases physical activity and promotes torpor, a transient hibernation-like state in small animals, which is however unlikely to be a result of a direct FGF21 action but rather compensatory effect in these energy-deficient animals. As adaptation to starvation, mammals shift from carbohydrates to ketone bodies as major fuel source. During fasting, fatty acids are mobilized from fat to liver where they are oxidized to acetyl-CoA. Ketone bodies are produced from the acetyl-CoA and used as energy source in other tissues. Peroxisome proliferator-activated receptor α (PPARα) is a nuclear receptor required for normal starvation response. PPARα knockout mice accumulate liver triglyceride and become hypoketonemic and hypoglycemic during fasting.[91-93] FGF21 is required for maximal liver fatty acid oxidation and ketogenesis under ketogenic diet.[67] FGF21 did not induce ketogenesis by increasing mRNA but rather increase the protein levels of HMGCS2 and CPT1α in the liver.[65] FGF21 treatment caused a significant reduction in liver triglyceride in fasted PPARα knockout mice. Brain can sense starvation and lead to adaptations such as torpor to conserve energy. FGF21 showed slow but significant, nonsaturable and unidirectional influx across blood-brain barrier.[94] The small amount of FGF21 (0.5% per gram of brain tissue) that could enter central nervous system (CNS) may be physiologically relevant to fasting response.

Adenoviral knockdown of FGF21 led to fatty liver, increased blood NEFA, triglyceride and cholesterol and decreased serum ketones.[67] FGF21 knockdown was associated with reduction of β oxidation gene expression including very long (ACADVL), long (ACADL) and medium (ACADM) chain acetyl-CoA dehydrogenase and carnitine palmitoyltransferase 1 a (CPT1a) and acyl-CoA oxidase (ACOX1) without an effect on lipid synthesis genes. FGF21 knockdown in mice on KD suppressed gene expression of enzymes involved in ketogenesis and induced by KD including hydroxymethylglutaryl-CoA synthase 2 (HMGCS2) and β-hydroxybutyrate dehydrogenase (BDH).[67] FGF21 liver gene expression is negatively correlated with liver triglyceride in obese rats.[95] Hypocaloric high protein diet improves hypertriglyceridemia partially by stimulating hepatic lipolysis and lipid utilization via FGF21.[95] Both PPARα and FGF21 are increased after 8 hours of fasting[96] which is consistent with FGF21 being a direct target gene of PPARα.[65,67,74]

FGF21 knockout (KO) mice had mild weight gain and slightly disturbed glucose tolerance after 24 weeks.[97] Increased adiposity in FGF21 KO mice was due to increased food intake and decreased oxygen consumption. FGF21 KO mice have dramatic

FGF21 AS A THERAPEUTIC REAGENT

metabolic response to KD including weight gain rather than weight loss, hepatosteatosis and impaired ketogenesis and glucose homeostasis.[97] The higher leptin levels in FGF21 KO led to lower agouti-related peptide (AgRP) and higer neuropeptide Y (NPY) levels. Both liver triglyceride and liver glycogen stores were increased by FGF21 deficiency.[97] The hepatic expression of PGC1α and β were lower is FGF21 KO mice. The SREBP1c maturation was increased in the liver of FGF21 KO mice on KD.[97] The pheonotypes of FGF21 KO mice are consistent with those of FGF21 transgenic mice[30] and are most dramatic in response to KD.

Forced hepatocyte-specific overexpression of FGF21 had no effect on liver development and response to injury.[82] No abnormal hyperplasia or preneoplasia was observed in the transgenic mice for 1 year.[30,82] FGF21 overexpression delayed the appearance of liver tumors induced by diethylnitrosamin (DEN).[82] The number of surface tumors per liver was reduced by at least 50% in the FGF21 transgenic mice. However, hepatocellular carcinoma (HCC) was similar between transgenic and wild type mice suggesting that although FGF21 delays hepatoma onset, it may not slow down progression once hepatomas are formed.[82] In normal hepatocytes and early stage of tumorigenesis, resident FGFR4, a suppressor of hepatocarcinogenesis, is dominant. It has been proposed that FGF21 may delay adenoma development by activating resident hepatocyte FGFR4.[82]

The metabolic phenotypes of FGF21 transgenic mice resemble those of FGF19 transgenic mice, which had increased energy expenditure, reduced fat mass and improved insulin sensitivity.[52] Increased brown adipose tissue and decreased acetyl coenzyme A carboxylase 2 (ACC2) mRNA in the liver are two mechanisms for the increased energy expenditure. The decreased ACC2 expression in the liver also led to reduced liver triglyceride.[52]

FGF21 CHRONIC AND ACUTE ADMINISTRATION IN VIVO

FGF21 administration in *ob/ob*, *db/db* mice and obese ZDF rats significantly lowered blood glucose, triglyceride and fasting insulin and improved glucose clearance during oral glucose tolerance test (OGTT).[30] The benefincial effects of insulin require multiple dosing but FGF21 induced changes that can last for at least 24 hours. FGF21 has advantages over insulin because of the lack of weight gain and the lack of hypoglycemia in FGF21 treated animals or FGF21 transgenic mice under fasted or fed conditions. This data also rule out the possibility that FGF21 is insulin mimetic. There were no mitogenicity, hypoglycemia or weight gain observed in diabetic or wild type animals treated with FGF21.[30]

Macaca mulatta (rhesus monkey) has spontaneous development and progreesion of diabetes that is similar to human Type II diabetes (T2DM).[98-101] When diabetic rhesus monkeys were treated with FGF21 for 6 weeks, body weight and fasting plasma glucose, fructosamine, triglyceride, insulin and glucagon were reduced (Fig. 6).[102] FGF21 lowered LDL-cholesterol and increased HDL-cholesterol, improved cardiovascular risk factors. While PPARγ activators, or thiazolidinediones (TZDs), have many unwanted side effect including weight gain and edema,[103] FGF21 induced glucose lowering and weight loss without increased fluid retention, hypoglycemia or mitogenecity.[102]

FGF21 can also be used as anti-obesity treatment.[79] FGF21 treatment for 2 weeks in diet-induced obese (DIO) and *ob/ob* mice lowered body weight by 20%. No decrease in food intake, nutrient malabsorption, or effect on physical activity was observed. FGF21 transgenic mice are resistant to weight gain induced by high fat diet.[30] Small but significant weight reduction was shown in diabetic monkeys treated with FGF21.[102] FGF21 infusion in DIO mice increased oxygen consumption and total energy expenditure.[79] The effects of FGF21 on energy expenditure and oxygen consumption were similar during dark and light cycles suggesting that there are no diurnal variations in FGF21 mode of action. FGF21 treatment also resulted in increased core body temperature in obese animals under fed state.[79] But reduction of core body temperature was also observed when FGF21 was tested in lean mice under fasting state.[65] FGF21 also reduced hepatosteatosis and respiratory quotient (RQ) and increased fecal fat content suggesting that lipid oxidation is increased and fat is preferred over carbohydrate as fuel source.[79] In liver, SCD1 was most dramatically affected gene by FGF21 administration. SCD1-null mice are lean and have reduced liver triglyceride.[104] In white adipose tissue (WAT), FGF21 induced PGC1α and UCP1, both of which play important roles in energy metabolism.[105] FGF21 treated animals also showed modest hyperphagia which may be a feedback due to increased energy expenditure.[79]

The anti-obesity function of FGF21 was confirmed in DIO mice by Xu et al.[80] Dose dependent reduction of blood glucose, insulin, cholesterol, triglyceride, NEFA, β-hydroxybutyrate and leptin was observed with FGF21 in DIO mice. The reduced NEFA and ketone bodies argue against increase in ketogenesis and lipolysis. In human, only a weak correlation between FGF21 and ketone bodies was observed.[106] Liver section staining and biochemical analysis suggest that liver and muscle triglyceride along with plasma ALT/AST levels were reduced in DIO mice treated with FGF21.[80] FGF21 did not further increase fatty acid oxidative genes induced by high fat diet but it did reduce nuclear SREBP1c and its target gene expression. Oxygen consumption was increased on day 2 of FGF21 treatment.[80] CO_2 production, energy expenditure and physical activity were also increased by FGF21. The decreased physical activity can account for 50% of weight gain in DIO rodents.[107] Increased levels of UCP-1 and -2 in brown adipose tissue suggest that nonexercise activity thermogenesis may also contribute to the increased energy expenditure.[80] In contrast to earlier study,[79] respiratory quotient was not changed by FGF21 in DIO mice. In the study of *ob/ob* mice treated with FGF21, hepatic glucose output was reduced and cardiac glucose uptake was increased after 7 days.[81] In DIO mice treated with FGF21, multiple tissues including liver, adipose tissue, skeletal muscle and heart showed improved insulin sensitivity.[80] The effect on skeletal muscle and heart may be indirect since βKlotho expression is low in these tissues.

FGF21 partially reverses the hypoketonemia an hypertriglyceridemia in PPARα knockout mice.[65] Both fibrates and FGF21 lower LDL cholesterol, increase HDL cholesterol and improve insulin sensitivity in dyslipidemic rhesus monkeys.[102,108] Both PPARα agonists and FGF21 prevent diet-induced obesity and improve insulin sensitivity in rodents.[30,109] FGF21, as PPARα direct target gene in liver, may contribute to the therapeutic effects of fibrates.

Hperinsulinemic-euglycemic clamp studies clearly demonstrated the effects of FGF21 on insulin sensitivity and glucose fluxes upon chronic and acute administration in *ob/+* and *ob/ob* mice.[81] Acute FGF21 administration reduced liver glucose production, increased hepatic glucogen store and decreased glucagon and increased glucose clearance in *ob/+* mice but not in *ob/ob* mice.[81] Chronic FGF21 treatment reduced fasting glucose, insulin and NEFA levels in *ob/ob* mice.[81] The improvements in insulin sensitivity are largely due to FGF21 effects in liver since chronic/acute FGF21 administration did not change skeletal muscle or adipose glucose uptake in *ob/+* or *ob/ob* mice. Akt phosphorylation was increased in the liver during the clamp in FGF21 treated *ob/ob* mice.[81]

FGF21 AS A THERAPEUTIC REAGENT

Most recently, the effect of acute FGF21 administration was reexamined in *ob/ob* and DIO mice.[83] In *ob/ob* mice, FGF21 decreased blood glucose, insulin and amylin by 40-60% decrease within 1h of injection. In DIO mice, FGF21 lowered fasting glucose and insulin within 3 h of treatment. The acute glucose lowering and insulin-sensitizing effects of FGF21 are due to its actions in liver and adipose tissue.[83] The half life of intravenously (i.v.) injected FGF21 is 1-2 h.[102,83] Cmax was achieved at 0.1 h and 2 h after intraperitoneal (i.p.) and subcutaneous (s.c.) injection, respectively. The bioavailability of FGF21 after a single i.p. or s.c. dose of 1 mg/kg was about 50%.[83] The maximum glucose lowering of 40% was reached within 1 hour of injection and lasted for at least 6 hours. The glucose returned to baseline between 6 to 24 h. GTT in *ob/ob* or DIO mice 3 h after the injection showed improved glucose tolerance. FGF21 did not affect insulin or glucagon secretion in islets isolated from mice acutely treated with FGF21.[83] FGF21 increased ERK1/2 phosphorylation in liver and adipose tissue but not in heart and skeletal muscle after acute injection in vivo. FGF21 also had little effects on Akt phosphorylation in tissues.[83] Thus, unlike PPARγ agonists or metformin, which have a slow transcriptionally mediated mechanism to sensitize insulin action, FGF21 resulted in rapid insulin sensitization. FGF21 is known to increase glucose uptake in murine adipocytes by increasing Glut1 mRNA.[30] FGF21 also regulate lipogenic genes by reducing SREBP1c processing.[80] Therefore FGF21 may act through a variety of mechanisms.

Chronic infusion of FGF21 for 8 weeks in *db/db* mice almost brought fed glucose to normal levels.[75] Furthermore, the number of islets per pancreas and β cells per islet were both increased. FGF21 did not induce islet proliferation but improved β cell survival and function in diabetic animals.[75] At the whole animal level, there are two potential mechanisms for FGF21's protective effect on β cells. FGF21 reduces plasma glucose and triglyceride levels which will in turn reduce glucolipotoxicity on β cells; FGF21 directly reduces β cell apoptosis via Akt pathway. As Type 2 diabetes patients show progressive β cell loss due to an increased rate of apoptosis.[110] Increasing β cell survival can be an appealing therapeutic approach for Type 2 diabetes.

FGF21 IN HUMANS

In a study with 76 normal human subjects, circulating FGF21 levels vary dramatically and did not correlate to BMI, age, blood glucose, LDL or HDL cholesterol or triglyceride.[106] However, in studies in metabolically challenged subjects, serum FGF21 is shown to positively correlate with age, adiposity (BMI, waist circumference, waist-to-hip ratio and fat percentage), insulin resistance (fasting insulin and HOMA-IR) and triglyceride.[111-113] It is also negatively correlated with HDL cholesterol.[111,112] Fasting plasma FGF21 levels were also significantly increased in patients with T2DM compared with nondiabetic controls.[114] Plasma FGF21 levels were significantly higher in Cushing's syndrome patients and obese patients compared with lean control subjects.[115] All subjects' FGF21 levels positively correlated with BMI. Median serum FGF21 levels were at least 15-fold higher in chronic hemodialysis (CD) patients than in control subjects.[116]

Chronic malnutrition in humans is associated with reduced plasma FGF21. Plasma FGF21 was significantly reduced in anorexia nervosa (AN) compared with the control group.[117] The increased lipolysis in AN patients[118] may be due to decreased FGF21 since FGF21 attenuates lipolysis in human adipocytes in vitro.[66] Baseline FGF21 levels before treatment is positively correlated with body weight gain in AN patients after partial realimentation.[117] FGF21 mRNA is also higher in the liver and adipose tissue of *db/db* mice compared with the lean littermates.[111] The increase of serum FGF21 in obese individuals can be explained by a compensatory response of FGF21 or FGF21 resistance, similar to the well known insulin resistance and leptin resistance. FGF21 is strongly correlated with A-FABP, adipokine that is an independent risk factor for metabolic syndrome, Type 2 diabetes and cardiovascular diseases.[119] The adipose tissue is also an important source of FGF21 production. FGF21 may act in an autocrine/paracrine manner to regulate adipocyte biology.

Nondiabetic patients with hypertriglyceridemia had 2-fold increased FGF21[106] and fenofibrate further increased FGF21 levels by nearly 30% (Galman et al, 2008). Both 3-week low caloric diet and 3-month fenofibrate treatment elevated FGF21 levels.[120] FGF21 mRNA in visceral fat but not subcutaneous fat showed increase in obese subjects versus healthy subjects. Insulin infusion for 3 hours during isoglycaemic-hyperinsulinaemic clamp elevated FGF21 in T2DM group but not in healthy subjects.[120] Similar treatment increased muscle FGF21 expression in healthy males.[112]

CONCLUSION

FGF21 is a master metabolic regulator that improves glucose and lipid profiles in rodent and monkey models of metabolic disease. In addition, FGF21 reduces body weight and decreases cardiovascular risk factors. FGF21 is regulated by PPARα in liver and is physiologically linked to the biology of fasting and ketogenesis. FGF21 also has synergistic effect with PPARγ in adipose tissue explaining its insulin-sensitizing nature. The major sites of FGF21 actions include liver, adipose tissue and pancreas. As a potential therapeutic reagent, FGF21 has diverse metabolic benefits without obvious side effects such as body weight gain, hypoglycemia, fluid retention or mitogenecity thus making it as an attractive candidate for clinical development.

REFERENCES

1. Gospodarowicz D. Localisation of a fibroblast growth factor and its effect alone and with hydrocortisone on 3T3 cell growth. Nature 1974; 249(453):123-127.
2. Ornitz DM, Itoh N. Fibroblast growth factors. Genome Biol 2001; 2(3):REVIEWS3005.
3. Eswarakumar VP, Lax I, Schlessinger J. Cellular signaling by fibroblast growth factor receptors. Cytokine Growth Factor Rev 2005; 16(2):139-149.
4. Schlessinger J, Plotnikov AN, Ibrahimi OA et al. Crystal structure of a ternary FGF-FGFR-heparin complex reveals a dual role for heparin in FGFR binding and dimerization. Mol Cell 2000; 6(3):743-750.
5. Orr-Urtreger A, Bedford MT, Burakova T et al. Developmental localization of the splicing alternatives of fibroblast growth factor receptor-2 (FGFR2). Dev Biol 1993; 158(2):475-486.
6. Plotnikov AN, Hubbard SR, Schlessinger J et al. Crystal structures of two FGF-FGFR complexes reveal the determinants of ligand-receptor specificity. Cell 2000; 101(4):413-424.
7. Plotnikov AN, Schlessinger J, Hubbard SR et al. Structural basis for FGF receptor dimerization and activation. Cell 1999; 98(5):641-650.
8. Kouhara H, Hadari YR, Spivak-Kroizman T et al. A lipid-anchored Grb2-binding protein that links FGF-receptor activation to the Ras/MAPK signaling pathway. Cell 1997; 89(5):693-702.
9. Hadari YR, Gotoh N, Kouhara H et al. Critical role for the docking-protein FRS2 alpha in FGF receptor-mediated signal transduction pathways. Proc Natl Acad Sci USA 2001; 98(15):8578-8583.

FGF21 AS A THERAPEUTIC REAGENT

10. Wong A, Lamothe B, Lee A et al. FRS2 alpha attenuates FGF receptor signaling by Grb2-mediated recruitment of the ubiquitin ligase Cbl. Proc Natl Acad Sci USA 2002; 99(10):6684-6689.
11. Lax I, Wong A, Lamothe B et al. The docking protein FRS2alpha controls a MAP kinase-mediated negative feedback mechanism for signaling by FGF receptors. Mol Cell 2002; 10(4):709-719.
12. Mohammadi M, Honegger AM, Rotin D et al. A tyrosine-phosphorylated carboxy-terminal peptide of the fibroblast growth factor receptor (Flg) is a binding site for the SH2 domain of phospholipase C-gamma 1. Mol Cell Biol 1991; 11(10):5068-5078.
13. Deng CX, Wynshaw-Boris A, Shen MM et al. Murine FGFR-1 is required for early postimplantation growth and axial organization. Genes Dev 1994; 8(24):3045-3057.
14. Yamaguchi TP, Harpal K, Henkemeyer M et al. fgfr-1 is required for embryonic growth and mesodermal patterning during mouse gastrulation. Genes Dev 1994; 8(24):3032-3044.
15. Partanen J, Schwartz L, Rossant J. Opposite phenotypes of hypomorphic and Y766 phosphorylation site mutations reveal a function for Fgfr1 in anteroposterior patterning of mouse embryos. Genes Dev 1998; 12(15):2332-2344.
16. Xu X, Weinstein M, Li C et al. Fibroblast growth factor receptor 2 (FGFR2)-mediated reciprocal regulation loop between FGF8 and FGF10 is essential for limb induction. Development 1998; 125(4):753-765.
17. De Moerlooze L, Spencer-Dene B, Revest JM et al. An important role for the IIIb isoform of fibroblast growth factor receptor 2 (FGFR2) in mesenchymal-epithelial signalling during mouse organogenesis. Development 2000; 127(3):483-492.
18. Eswarakumar VP, Monsonego-Ornan E, Pines M et al. The IIIc alternative of Fgfr2 is a positive regulator of bone formation. Development 2002; 129(16):3783-3793.
19. Deng C, Wynshaw-Boris A, Zhou F et al. Fibroblast growth factor receptor 3 is a negative regulator of bone growth. Cell 1996; 84(6):911-921.
20. Colvin JS, Bohne BA, Harding GW et al. Skeletal overgrowth and deafness in mice lacking fibroblast growth factor receptor 3. Nat Genet 1996; 12(4):390-397.
21. Weinstein M, Xu X, Ohyama K et al. FGFR-3 and FGFR-4 function cooperatively to direct alveogenesis in the murine lung. Development 1998; 125(18):3615-3623.
22. Dode C, Hardelin JP. Kallmann syndrome: fibroblast growth factor signaling insufficiency? J Mol Med 2004; 82(11):725-734.
23. Macdonald D, Aguiar RC, Mason PJ et al. A new myeloproliferative disorder associated with chromosomal translocations involving 8p11: a review. Leukemia 1995; 9(10):1628-1630.
24. Roumiantsev S, Krause DS, Neumann CA et al. Distinct stem cell myeloproliferative/T lymphoma syndromes induced by ZNF198-FGFR1 and BCR-FGFR1 fusion genes from 8p11 translocations. Cancer Cell 2004; 5(3):287-298.
25. Demiroglu A, Steer EJ, Heath C et al. The t(8;22) in chronic myeloid leukemia fuses BCR to FGFR1: transforming activity and specific inhibition of FGFR1 fusion proteins. Blood 2001; 98(13):3778-3783.
26. Itoh N, Ornitz DM. Functional evolutionary history of the mouse Fgf gene family. Dev Dyn 2008; 237(1):18-27.
27. Mohammadi M, Olsen SK, Goetz R. A protein canyon in the FGF-FGF receptor dimer selects from an a la carte menu of heparan sulfate motifs. Curr Opin Struct Biol 2005; 15(5):506-516.
28. Moscatelli D. High and low affinity binding sites for basic fibroblast growth factor on cultured cells: absence of a role for low affinity binding in the stimulation of plasminogen activator production by bovine capillary endothelial cells. J Cell Physiol 1987; 131(1):123-130.
29. Flaumenhaft R, Moscatelli D, Rifkin DB. Heparin and heparan sulfate increase the radius of diffusion and action of basic fibroblast growth factor. J Cell Biol 1990; 111(4):1651-1659.
30. Kharitonenkov A, Shiyanova TL, Koester A et al. FGF-21 as a novel metabolic regulator. J Clin Invest 2005; 115(6):1627-1635.
31. Shimada T, Mizutani S, Muto T et al. Cloning and characterization of FGF23 as a causative factor of tumor-induced osteomalacia. Proc Natl Acad Sci USA 2001; 98(11):6500-6505.
32. Lundåsen T, Gälman C, Angelin B et al. Circulating intestinal fibroblast growth factor 19 has a pronounced diurnal variation and modulates hepatic bile acid synthesis in man. J Intern Med 2006; 260(6):530-536.
33. Ogawa Y, Kurosu H, Yamamoto M et al. BetaKlotho is required for metabolic activity of fibroblast growth factor 21. Proc Natl Acad Sci USA 2007; 104(18):7432-7437.
34. Urakawa I, Yamazaki Y, Shimada T et al. Klotho converts canonical FGF receptor into a specific receptor for FGF23. Nature 2006; 444(7120):770-774.
35. Bhushan A, Itoh N, Kato S et al. Fgf10 is essential for maintaining the proliferative capacity of epithelial progenitor cells during early pancreatic organogenesis. Development 2001; 128(24):5109-5117.
36. Ohuchi H, Hori Y, Yamasaki M et al. FGF10 acts as a major ligand for FGF receptor 2 IIIb in mouse multi-organ development. Biochem Biophys Res Commun 2000; 277(3):643-649.
37. Konishi M, Mikami T, Yamasaki M et al. Fibroblast growth factor-16 is a growth factor for embryonic brown adipocytes. J Biol Chem 2000; 275(16):12119-12122.

38. Sun HD, Malabunga M, Tonra JR et al. Monoclonal antibody antagonists of hypothalamic FGFR1 cause potent but reversible hypophagia and weight loss in rodents and monkeys. Am J Physiol Endocrinol Metab 2007; 292(3):E964-E976.

39. Hart AW, Baeza N, Apelqvist A et al. Attenuation of FGF signalling in mouse beta-cells leads to diabetes. Nature 2000; 408(6814):864-868.

40. Revest JM, Suniara RK, Kerr K et al. Development of the thymus requires signaling through the fibroblast growth factor receptor R2-IIIb. J Immunol 2001; 167(4):1954-1961.

41. Celli G, LaRochelle WJ, Mackem S et al. Soluble dominant-negative receptor uncovers essential roles for fibroblast growth factors in multi-organ induction and patterning. EMBO J 1998; 17(6):1642-1655.

42. Elghazi L, Cras-Méneur C, Czernichow P et al. Role for FGFR2IIIb-mediated signals in controlling pancreatic endocrine progenitor cell proliferation. Proc Natl Acad Sci USA 2002; 99(6):3884-3889.

43. Yu C, Wang F, Kan M et al. Elevated cholesterol metabolism and bile acid synthesis in mice lacking membrane tyrosine kinase receptor FGFR4. J Biol Chem 2000; 275(20):15482-15489.

44. Huang X, Yang C, Luo Y et al. FGFR4 prevents hyperlipidemia and insulin resistance but underlies high-fat diet induced fatty liver. Diabetes 2007; 56(10):2501-2510.

45. Shin DJ, Osborne TF. FGF15/FGFR4 integrates growth factor signaling with hepatic bile acid metabolism and insulin action. J Biol Chem 2009; 284(17):11110-11120.

46. Goetz R, Beenken A, Ibrahimi OA et al. Molecular insights into the klotho-dependent, endocrine mode of action of fibroblast growth factor 19 subfamily members. Mol Cell Biol 2007; 27(9):3417-3428.

47. Harmer NJ, Pellegrini L, Chirgadze D et al. The crystal structure of fibroblast growth factor (FGF) 19 reveals novel features of the FGF family and offers a structural basis for its unusual receptor affinity. Biochemistry 2004; 43(3):629-640.

48. Ito S, Kinoshita S, Shiraishi N et al. Molecular cloning and expression analyses of mouse betaklotho, which encodes a novel Klotho family protein. Mech Dev 2000; 98(1-2):115-119.

49. Tsujikawa H, Kurotaki Y, Fujimori T et al. Klotho, a gene related to a syndrome resembling human premature aging, functions in a negative regulatory circuit of vitamin D endocrine system. Mol Endocrinol 2003; 17(12):2393-2403.

50. Holt JA, Luo G, Billin AN et al. Definition of a novel growth factor-dependent signal cascade for the suppression of bile acid biosynthesis. Genes Dev 2003; 17(13):1581-1591.

51. Inagaki T, Choi M, Moschetta A et al. Fibroblast growth factor 15 functions as an enterohepatic signal to regulate bile acid homeostasis. Cell Metab 2005; 2(4):217-225.

52. Tomlinson E, Fu L, John L et al. Transgenic mice expressing human fibroblast growth factor-19 display increased metabolic rate and decreased adiposity. Endocrinology 2002; 143(5):1741-1747.

53. Ito S, Fujimori T, Furuya A et al. Impaired negative feedback suppression of bile acid synthesis in mice lacking betaKlotho. J Clin Invest 2005; 115(8):2202-2208.

54. Nishimura T, Nakatake Y, Konishi M et al. Identification of a novel FGF, FGF-21, preferentially expressed in the liver. Biochim Biophys Acta 2000; 1492(1):203-206.

55. Ibrahimi OA, Zhang F, Eliseenkova AV et al. Biochemical analysis of pathogenic ligand-dependent FGFR2 mutations suggests distinct pathophysiological mechanisms for craniofacial and limb abnormalities. Hum Mol Genet 2004; 13(19):2313-2324.

56. Kharitonenkov A, Dunbar JD, Bina HA et al. FGF-21/FGF-21 receptor interaction and activation is determined by betaKlotho. J Cell Physiol 2008; 215(1):1-7.

57. Suzuki M, Uehara Y, Motomura-Matsuzaka K et al. betaKlotho is required for fibroblast growth factor (FGF) 21 signaling through FGF receptor (FGFR) 1c and FGFR3c. Mol Endocrinol 2008; 22(4):1006-1014.

58. Kurosu H, Choi M, Ogawa Y et al. Tissue-specific expression of betaKlotho and fibroblast growth factor (FGF) receptor isoforms determines metabolic activity of FGF19 and FGF21. J Biol Chem 2007; 282(37):26687-26695.

59. Murer H, Hernando N, Forster I et al. Proximal tubular phosphate reabsorption: molecular mechanisms. Physiol Rev 2000; 80(4):1373-1409.

60. Riminucci M, Collins MT, Fedarko NS et al. FGF-23 in fibrous dysplasia of bone and its relationship to renal phosphate wasting. J Clin Invest 2003; 112(5):683-692.

61. Wikvall K. Cytochrome P450 enzymes in the bioactivation of vitamin D to its hormonal form (review). Int J Mol Med 2001; 7(2):201-209.

62. Shimada T, Kakitani M, Yamazaki Y et al. Targeted ablation of Fgf23 demonstrates an essential physiological role of FGF23 in phosphate and vitamin D metabolism. J Clin Invest 2004; 113(4):561-568.

63. Sitara D, Razzaque MS, Hesse M et al. Homozygous ablation of fibroblast growth factor-23 results in hyperphosphatemia and impaired skeletogenesis and reverses hypophosphatemia in Phex-deficient mice. Matrix Biol 2004; 23(7):421-432.

64. Yoshida T, Fujimori T, Nabeshima Y. Mediation of unusually high concentrations of 1,25-dihydroxyvitamin D in homozygous klotho mutant mice by increased expression of renal 1 alpha-hydroxylase gene. Endocrinology 2002; 143(2):683-689.

FGF21 AS A THERAPEUTIC REAGENT

65. Inagaki T, Dutchak P, Zhao G et al. Endocrine regulation of the fasting response by PPARalpha-mediated induction of fibroblast growth factor 21. Cell Metab 2007; 5(6):415-425.
66. Arner P, Pettersson A, Mitchell PJ et al. FGF21 attenuates lipolysis in human adipocytes—a possible link to improved insulin sensitivity. FEBS Lett 2008; 582(12):1725-1730.
67. Badman MK, Pissios P, Kennedy AR et al. Hepatic fibroblast growth factor 21 is regulated by PPARalpha and is a key mediator of hepatic lipid metabolism in ketotic states. Cell Metab 2007; 5(6):426-437.
68. Moyers JS, Shiyanova TL, Mehrbod F et al. Molecular determinants of FGF-21 activity-synergy and cross-talk with PPARgamma signaling. J Cell Physiol 2007; 210(1):1-6.
69. Muise ES, Azzolina B, Kuo DW et al. Adipose fibroblast growth factor 21 is up-regulated by peroxisome proliferator-activated receptor gamma and altered metabolic states. Mol Pharmacol 2008; 74(2):403-412.
70. Wang H, Qiang L, Farmer SR. Identification of a domain within peroxisome proliferator-activated receptor gamma regulating expression of a group of genes containing fibroblast growth factor 21 that are selectively repressed by SIRT1 in adipocytes. Mol Cell Biol 2008; 28(1):188-200.
71. Kharitonenkov A, Shanafelt AB. Fibroblast growth factor-21 as a therapeutic agent for metabolic diseases. BioDrugs 2008; 22(1):37-44.
72. Izumiya Y, Bina HA, Ouchi N et al. FGF21 is an Akt-regulated myokine. FEBS Lett 2008; 582(27):3805-3810.
73. Izumiya Y, Hopkins T, Morris C et al. Fast/Glycolytic muscle fiber growth reduces fat mass and improves metabolic parameters in obese mice. Cell Metab 2008; 7(2):159-172.
74. Lundåsen T, Hunt MC, Nilsson LM et al. PPARalpha is a key regulator of hepatic FGF21. Biochem Biophys Res Commun 2007; 360(2):437-440.
75. Wente W, Efanov AM, Brenner M et al. Fibroblast growth factor-21 improves pancreatic beta-cell function and survival by activation of extracellular signal-regulated kinase 1/2 and Akt signaling pathways. Diabetes 2006; 55(9):2470-2478.
76. Leibowitz G, Uçkaya G, Oprescu AI et al. Glucose-regulated proinsulin gene expression is required for adequate insulin production during chronic glucose exposure. Endocrinology 2002; 143(9):3214-3220.
77. Bollheimer LC, Skelly RH, Chester MW et al. Chronic exposure to free fatty acid reduces pancreatic beta cell insulin content by increasing basal insulin secretion that is not compensated for by a corresponding increase in proinsulin biosynthesis translation. J Clin Invest 1998; 101(5):1094-1101.
78. Inagaki T, Lin VY, Goetz R et al. Inhibition of growth hormone signaling by the fasting-induced hormone FGF21. Cell Metab 2008; 8(1):77-83.
79. Coskun T, Bina HA, Schneider MA et al. Fibroblast growth factor 21 corrects obesity in mice. Endocrinology 2008; 149(12):6018-6027.
80. Xu J, Lloyd DJ, Hale C et al. Fibroblast growth factor 21 reverses hepatic steatosis, increases energy expenditure and improves insulin sensitivity in diet-induced obese mice. Diabetes 2009; 58(1):250-259.
81. Berglund ED, Li CY, Bina HA et al. Fibroblast growth factor 21 controls glycemia via regulation of hepatic glucose flux and insulin sensitivity. Endocrinology 2009; 150(9):4084-4093.
82. Huang X, Yu C, Jin C et al. Forced expression of hepatocyte-specific fibroblast growth factor 21 delays initiation of chemically induced hepatocarcinogenesis. Mol Carcinog 2006; 45(12):934-942.
83. Xu J, Stanislaus S, Chinookoswong N et al. Acute glucose-lowering and insulin-sensitizing action of FGF21 in insulin resistant mouse models—Association with liver and adipose tissue effects. Am J Physiol Endocrinol Metab 2009.
84. Bhatnagar S, Damron HA, Hillgartner FB. Fibroblast growth factor-19, a novel factor that inhibits hepatic fatty acid synthesis. J Biol Chem 2009; 284(15):10023-10033.
85. Wu X, Lemon B, Li X et al. C-terminal tail of FGF19 determines its specificity toward Klotho coreceptors. J Biol Chem 2008; 283(48):33304-33309.
86. Micanovic R, Raches DW, Dunbar JD et al. Different roles of N- and C- termini in the functional activity of FGF21. J Cell Physiol 2009; 219(2):227-234.
87. Yie J, Hecht R, Patel J et al. FGF21 N- and C-termini play different roles in receptor interaction and activation. FEBS Lett 2009; 583(1):19-24.
88. Yamazaki Y, Tamada T, Kasai N et al. Anti-FGF23 neutralizing antibodies show the physiological role and structural features of FGF23. J Bone Miner Res 2008; 23(9):1509-1518.
89. Choi M, Moschetta A, Bookout AL et al. Identification of a hormonal basis for gallbladder filling. Nat Med 2006; 12(11):1253-1255.
90. Nicholes K, Guillet S, Tomlinson E et al. A mouse model of hepatocellular carcinoma: ectopic expression of fibroblast growth factor 19 in skeletal muscle of transgenic mice. Am J Pathol 2002; 160(6):2295-2307.
91. Hashimoto T, Cook WS, Qi C et al. Defect in peroxisome proliferator-activated receptor alpha-inducible fatty acid oxidation determines the severity of hepatic steatosis in response to fasting. J Biol Chem 2000; 275(37):28918-28928.
92. Kersten S, Seydoux J, Peters JM et al. Peroxisome proliferator-activated receptor alpha mediates the adaptive response to fasting. J Clin Invest 1999; 103(11):1489-1498.

93. Leone TC, Weinheimer CJ, Kelly DP. A critical role for the peroxisome proliferator-activated receptor alpha (PPARalpha) in the cellular fasting response: the PPARalpha-null mouse as a model of fatty acid oxidation disorders. Proc Natl Acad Sci USA 1999; 96(13):7473-7478.

94. Hsuchou H, Pan W, Kastin AJ. The fasting polypeptide FGF21 can enter brain from blood. Peptides 2007; 28(12):2382-2386.

95. Uebanso T, Taketani Y, Fukaya M et al. Hypocaloric high-protein diet improves fatty liver and hypertriglyceridemia in sucrose-fed obese rats via two pathways. Am J Physiol Endocrinol Metab 2009; 297(1):E76-E84.

96. Palou M, Priego T, Sánchez J et al. Sequential changes in the expression of genes involved in lipid metabolism in adipose tissue and liver in response to fasting. Pflugers Arch 2008; 456(5):825-836.

97. Badman MK, Koester A, Flier JS et al. Fibroblast Growth Factor 21-Deficient Mice Demonstrate Impaired Adaptation to Ketosis. Endocrinology 2009; 150(11):Epub as doi:10.1210/en.2009-0523.

98. Schäfer SA, Hansen BC, Völkl A et al. Biochemical and morphological effects of K-111, a peroxisome proliferator-activated receptor (PPAR)alpha activator, in nonhuman primates. Biochem Pharmacol 2004; 68(2):239-251.

99. Angeloni SV, Glynn N, Ambrosini G et al. Characterization of the rhesus monkey ghrelin gene and factors influencing ghrelin gene expression and fasting plasma levels. Endocrinology 2004; 145(5):2197-2205.

100. Oliver WR Jr, Shenk JL, Snaith MR et al. A selective peroxisome proliferator-activated receptor delta agonist promotes reverse cholesterol transport. Proc Natl Acad Sci USA 2001; 98(9):5306-5311.

101. Walston J, Lowe A, Silver K et al. The beta3-adrenergic receptor in the obesity and diabetes prone rhesus monkey is very similar to human and contains arginine at codon 64. Gene 1997; 188(2):207-213.

102. Kharitonenkov A, Wroblewski VJ, Koester A et al. The metabolic state of diabetic monkeys is regulated by fibroblast growth factor-21. Endocrinology 2007; 148(2):774-781.

103. Boden G, Zhang M. Recent findings concerning thiazolidinediones in the treatment of diabetes. Expert Opin Investig Drugs 2006; 15(3):243-250.

104. Cohen P, Miyazaki M, Socci ND et al. Role for stearoyl-CoA desaturase-1 in leptin-mediated weight loss. Science 2002; 297(5579):240-243.

105. Handschin C, Spiegelman BM. Peroxisome proliferator-activated receptor gamma coactivator 1 coactivators, energy homeostasis and metabolism. Endocr Rev 2006; 27(7):728-735.

106. Gälman C, Lundåsen T, Kharitonenkov A et al. The circulating metabolic regulator FGF21 is induced by prolonged fasting and PPARalpha activation in man. Cell Metab 2008; 8(2):169-74.

107. Bjursell M, Gerdin AK, Lelliott CJ et al. Acutely reduced locomotor activity is a major contributor to Western diet-induced obesity in mice. Am J Physiol Endocrinol Metab 2008; 294(2):E251-E260.

108. Winegar DA, Brown PJ, Wilkison WO et al. Effects of fenofibrate on lipid parameters in obese rhesus monkeys. J Lipid Res 2001; 42(10):1543-1551.

109. Guerre-Millo M, Gervois P, Raspé E et al. Peroxisome proliferator-activated receptor alpha activators improve insulin sensitivity and reduce adiposity. J Biol Chem 2000; 275(22):16638-16642.

110. Butler AE, Janson J, Bonner-Weir S et al. Beta-cell deficit and increased beta-cell apoptosis in humans with type 2 diabetes. Diabetes 2003; 52(1):102-110.

111. Zhang X, Yeung DC, Karpisek M et al. Serum FGF21 levels are increased in obesity and are independently associated with the metabolic syndrome in humans. Diabetes 2008; 57(5):1246-1253.

112. Hojman P, Pedersen M, Nielsen AR et al. Fibroblast growth factor-21 is induced in human skeletal muscles by hyperinsulinemia. Diabetes 2009; 58(12):2797-801.

113. Li L, Yang G, Ning H et al. Plasma FGF-21 levels in type 2 diabetic patients with ketosis. Diabetes Res Clin Pract 2008; 82(2):209-213.

114. Chen WW, Li L, Yang GY et al. Circulating FGF-21 levels in normal subjects and in newly diagnose patients with Type 2 diabetes mellitus. Exp Clin Endocrinol Diabetes 2008; 116(1):65-68.

115. Durovcová V, Marek J, Hána V et al. Plasma concentrations of fibroblast growth factors 21 and 19 in patients with Cushing's syndrome. Physiol Res 2010; 59(3):415-22.

116. Stein S, Bachmann A, Lössner U et al. Serum levels of the adipokine FGF21 depend on renal function. Diabetes Care 2009; 32(1):126-128.

117. Dostálová I, Kaválková P, Haluzíková D et al. Plasma concentrations of fibroblast growth factors 19 and 21 in patients with anorexia nervosa. J Clin Endocrinol Metab 2008; 93(9):3627-3632.

118. Bartak V, Vybiral S, Papezova H et al. Basal and exercise-induced sympathetic nervous activity and lipolysis in adipose tissue of patients with anorexia nervosa. Eur J Clin Invest 2004; 34(5):371-377.

119. Xu A, Tso AW, Cheung BM et al. Circulating adipocyte-fatty acid binding protein levels predict the development of the metabolic syndrome: a 5-year prospective study. Circulation 2007; 115(12):1537-1543.

120. Mraz M, Bartlova M, Lacinova Z et al. Serum concentrations and tissue expression of a novel endocrine regulator fibroblast growth factor-21 in patients with type 2 diabetes and obesity. Clin Endocrinol (Oxf) 2009; 71(3):369-375.

INDEX

Symbols

1α-hydroxylase (CYP27B1) 31, 35, 36, 43, 45, 54, 57, 66, 68, 78, 93, 95, 111, 159, 160, 163, 164, 166, 217

24-hydroxylase (CYP24) 35, 36, 65, 66, 68, 78, 111, 159, 160, 166

A

αklotho 1, 7, 10, 17-19, 42, 43

Adipocyte 11, 36, 200, 204, 216-218, 223, 224

A disintegrin and metalloprotease (ADAM) 86, 128, 129

Affinity 1-3, 5, 12, 14-19, 29, 32-34, 45, 86, 102, 104, 163, 171-173, 184-186, 195, 197-199, 201, 202, 216, 217, 219

Aging 10, 25-29, 31-33, 70, 100, 104, 105, 126, 127, 130, 132, 138, 140, 142, 143, 148, 163

Agonist 11, 48, 51, 54, 148, 174, 175, 177, 178, 218-220, 222, 223

Alternative splicing 2, 5, 215

Angiotensin converting enzyme inhibitor 147

Angiotensin II receptor blocker (Ang II) 130, 131, 133, 134, 137, 143, 147, 148

Antagonist 11, 137, 138, 143, 147, 148, 164

Antibody 18, 45, 47, 48, 51, 55, 93, 102, 159, 162-167, 184, 187, 188, 190-192, 195, 198, 216, 217

B

Autosomal hypophosphatemic rickets (ADHR) 7, 11, 29, 41-52, 57, 71, 85, 93, 94, 112, 158, 162

Autosomal recessive Fanconi syndrome 41, 53

Autosomal recessive hypophosphatemia (ARHP) 43, 46, 48, 51, 52

βklotho 1, 7, 10, 11, 18, 33-36, 173, 175-177, 184, 185, 195-197, 199-202, 204-206, 208-211, 216-219, 222

β-trefoil 5, 12, 163, 201

Bile 2, 10, 11, 16, 25, 33-37, 171, 174-180, 184, 185, 195, 196, 198, 200, 207, 208, 216, 219
acid biosynthesis 171, 174-179, 196

Binding 1-3, 5-7, 11, 12, 14-19, 34, 45, 71, 74, 76-78, 86, 94, 100, 102-104, 111, 143, 148, 162, 163, 167, 171-176, 179, 185, 186, 188, 195-202, 204, 206, 208, 214-216, 219

Biomarker 79, 107, 114, 117-121, 126, 136, 137, 143-145, 149, 197

Bone formation 27, 67, 78, 111

Bone-kidney axis 10, 65, 73

Bone mineralization 29, 65-67, 71, 73, 75, 76, 79, 160, 166

Bone resorption 27, 68, 77-79, 85, 110

C

Calcification   28, 29, 51, 54, 57, 71, 72, 85, 87, 88, 105, 108, 111, 116, 117, 126, 130, 132, 134, 136, 139, 140, 142, 144, 147, 149, 159, 164, 166, 217

Calcitriol   29-31, 34-36, 52, 78, 85, 107, 110, 111, 113, 114, 146

Calcium   29-32, 35, 43, 45, 47, 48, 50-52, 54, 57, 65, 67, 68, 72, 73, 76, 87-89, 93-97, 105, 108, 110-113, 115, 120, 121, 127, 129, 134, 139-142, 161, 172

   receptor   30, 48, 51, 57, 96, 112, 121, 129

Cancer   11, 183, 184, 186-192, 195-198, 209, 210, 220

Carcinoma   11, 49, 179, 183, 184, 186-191, 195-198, 208-210, 221

Cardiac hypertrophy   105, 126, 130, 140

Cardiovascular disease (CVD)   108, 110, 116, 117, 130, 137, 224

Channel   32, 43, 45, 89, 100, 101, 103-105, 127, 146

Cholesterol 7-hydroxylase (CYP7A1)   33, 35, 36

Chronic kidney disease (CKD)   11, 27, 28, 53, 57, 67, 68, 70, 72, 73, 78, 85, 86, 89, 92, 95-98, 107-121, 126, 127, 129-132, 134, 135-149, 161

Clathrin   103

Coreceptor   1, 7, 10, 11, 14, 17, 25, 29, 32, 33, 41, 42, 44, 45, 50, 57, 92, 104, 126, 129, 163, 164, 184-186, 199, 201, 210, 216, 217

Crystal structure   1, 2, 6, 7, 11, 12, 14, 16-18, 173, 174, 201, 215

D

Dentin matrix acidic phosphoprotein 1 (Dmp1)   43, 66, 67, 74-77, 166

Dentin matrix protein 1 (DMP1)   41, 43, 50, 51, 58, 71, 74-77, 162

Diabetes   7, 11, 28, 77, 107, 108, 131-133, 140, 148, 180, 210, 211, 214, 216, 221, 223, 224

Dialysis   57, 92, 95, 107, 108, 110, 112, 113, 116, 118, 121

E

Egr   45, 94, 95, 163, 164

Endocrine   1-3, 7, 10-12, 15-19, 25, 26, 29-31, 33-37, 54, 56, 67, 68, 73, 98, 105, 111, 127, 129, 131, 134, 136, 145, 146, 172-176, 184-187, 192, 195-199, 206, 209-211, 216, 217

Endocytosis   103

Enterocyte   85, 176, 177

ERK   189-191, 197, 204, 208, 210

Estimated glomerular filtration rate (eGFR)   109, 113-115, 117, 143, 144

F

Farnesoid X receptor (FXR)   35, 36, 174, 175, 177, 178, 207

Fat   11, 27, 68, 104, 176, 179, 196, 207, 218, 220-224

Fatty acid   11, 35-37, 179, 218-220, 222

Fibroblast growth factor (FGF)   1-3, 5-7, 9-13, 15-19, 25, 29, 32, 34-37, 41-46, 50, 53, 58, 67, 68, 70, 77, 84, 86, 92, 104, 107, 110, 111, 126, 127, 129, 158, 162-164, 171-174, 177, 179, 183-187, 192, 195-203, 205-207, 209-211, 214-216, 219

   FGF15   10, 16, 33-35, 68, 158, 171, 176-178, 185, 207, 211, 215, 216

   FGF19   1, 2, 7, 9-18, 25, 33-36, 68, 162, 163, 171-180, 183-192, 195-211, 215-217, 219-221

   FGF21   1, 7, 9-11, 15, 16, 18, 25, 33-36, 162, 172, 173, 180, 184, 195-204, 206, 207, 209, 210, 214, 216-224

   FGF23   1, 2, 7, 9-19, 25, 29-32, 34-36, 41-58, 65-79, 84-90, 92-98, 104, 105, 107, 110-112, 114, 115, 117-121, 126, 127, 129, 130, 136, 139-141, 144, 146, 158-167, 172, 173, 180, 184, 196-199, 201, 202, 209, 216, 217, 219

INDEX

Fibroblast growth factor receptor (FGFR) 1-3, 5, 6, 10, 11, 15-19, 25, 29, 32, 33, 43-46, 50, 53, 58, 66, 68, 70, 72, 75, 77, 86, 93, 94, 104, 105, 111, 129, 141, 162-164, 167, 171-174, 177, 183, 185-188, 190, 192, 195-199, 202-207, 209, 211, 214-219
    FGFR1c 1, 2, 6, 10, 11, 14-19, 29, 33, 34, 36, 44, 45, 86, 93, 96, 164, 173, 199-204, 206-208, 216, 217, 219
    FGFR2c 2, 16, 173, 199, 200, 215, 217
    FGFR3c 2, 29, 86, 173, 199, 200, 217
    FGFR4 2, 10, 11, 16, 18, 29, 33-36, 45, 70, 86, 171-177, 179, 183-192, 195, 196, 199-211, 215, 216, 219, 221
        FGFR4 Arg388 189, 191, 196, 210
        FGFR4 polymorphism 188-190, 192
Fibrosis 126, 133, 137-140, 146, 147, 190

G

Galectin 32, 100, 102-104
Gallbladder 11, 171, 172, 176-179, 185, 207, 219
GalNAc transferase 3 (GALNT3) 41, 43, 47, 55, 57, 71, 162, 166
Glucose 2, 33, 34, 53, 70, 103, 104, 133, 148, 179, 195, 196, 200, 204, 207-210, 216-224
Glucuronidase 42, 101, 102, 141, 142

H

Heparan sulfate (HS) 1-3, 5-7, 11, 12, 14-19, 171-173, 195-203, 205, 206, 208, 215-217
Heparin 6, 7, 12, 14, 15, 17, 19, 68, 76, 101, 171, 174, 185, 186, 199-202, 204-206, 208, 214-216
Hepatocellular carcinoma 11, 179, 183, 184, 186-189, 195-198, 208-210, 221
Hepatocyte 16, 36, 174, 175, 179, 186, 188, 190, 192, 197, 198, 209-211, 219, 221
Hepatocyte proliferation 16, 186, 188, 192, 209-211

Hereditary hypophosphatemic disorder 68, 71, 78
Hereditary hypophosphatemic rickets with hypercalciuria (HHRH) 41, 43, 47, 48, 52
HMG-CoA reductase inhibitor 143, 147-149
Hormone 18, 25-27, 29, 30, 33, 41, 48, 49, 51, 54, 55, 57, 67-69, 72, 73, 78, 85, 104, 105, 110, 129, 134, 136, 144, 158, 160, 166, 171, 172, 174, 176, 177, 179, 184, 195-210, 215, 216, 218
Hyp 45, 48, 49, 51, 70, 71, 73-77, 87, 165, 166
Hyperparathyroidism 42, 47-52, 68, 70-73, 78, 85, 86, 95-97, 111-115, 120, 121, 126, 127, 130, 140, 141, 161
Hyperphosphatemia 11, 18, 29, 31, 48, 51, 54, 55, 57, 68, 70-73, 77, 79, 85-88, 92, 104, 105, 111-118, 130, 132, 141, 142, 144, 159, 164, 166
Hyperphosphatemic familial tumoral calcinosis (HFTC) 43, 47, 54, 57, 71

I

Ileum 175-177, 185, 198
Insulin 33, 104, 129, 133, 148, 179, 196, 208, 214, 216-219, 221-224
Insulin-like growth factor-1 (IGF-1) 33

K

Kidney 7, 10, 11, 16, 25-31, 34-36, 41, 45, 53, 57, 58, 65, 66, 68-70, 72, 73, 78, 79, 84-87, 92-95, 100-102, 104, 105, 107-112, 114, 116, 126-149, 160, 161, 163, 172, 185, 198, 217
Klotho 1, 7, 10, 11, 14, 17, 18, 25-37, 41-43, 46, 50, 57, 68, 70-72, 75, 84, 86-90, 92-97, 100-105, 111, 120, 126-149, 162-167, 171, 173, 179, 185, 199-201, 216, 217, 219
KLPH 35

L

Leptin 68, 214, 221, 222, 224
Life span 25, 29, 32, 45, 100, 130, 132, 143, 146
Ligand 1-3, 5, 7, 9-13, 15-19, 32, 34-36, 45, 71, 100, 102, 103, 174, 183-187, 191, 192, 196, 202, 215
Lipid 2, 28, 132-134, 142, 143, 147, 149, 179, 195, 196, 210, 216, 218-220, 222, 224
Liver 10, 11, 25, 33-36, 47, 93, 101, 102, 110, 129, 171, 174-177, 179, 180, 185-188, 190, 192, 195-198, 200, 203, 207-210, 214, 216, 218-224
lpr/lpr 139

M

Matrix extracellular protein (MEPE) 43, 48, 49, 71, 74, 76, 77
Metabolic rate 179
Metabolism 2, 7, 10, 35, 37, 46, 67, 68, 71, 75, 76, 78, 79, 84, 85, 89, 95, 104, 105, 107, 108, 110, 111, 113-117, 119-121, 127, 129, 133, 134, 136, 144, 149, 158-161, 179, 184, 195, 196, 198, 200, 207-209, 216-218, 222
Mineral metabolism 37, 67, 78, 95, 107, 108, 110, 111, 113-117, 120, 121, 127, 129, 134, 136, 144, 149, 159, 217
Mitogen-activated protein kinase (MAPK) 3, 44, 45, 86, 92-94, 96, 97, 163, 184, 187, 191, 215, 217
Mitogenic 44, 172, 179, 195, 196, 199, 208, 211, 217
Mortality 32, 57, 72, 86, 95, 107, 108, 110, 116-121, 127, 130, 136, 140, 142

N

NaPi-2a 31, 32, 41, 43, 45, 52, 53, 57, 68, 85, 86, 88, 93, 141, 142, 160, 161, 163-166
NaPi-2c 41, 43, 45, 52, 68, 85, 86, 141, 160, 161

NaPi-3 139, 142
Nephrolithiasis 52
NHERF-1 52, 53

O

Obesity 11, 180, 210, 211, 214, 218, 220-222
ob/ob 208, 219, 221-223
Osteoporosis 27, 52, 53, 77, 78
Oxidative stress 28, 132-134, 137, 138, 140, 142, 143, 147-149

P

Paracrine 1-3, 5-7, 9-13, 15-18, 45, 67, 68, 77, 105, 128, 129, 171, 173, 175, 184, 198-202, 205, 206, 211, 215, 224
Parathyroid 26, 30, 31, 34, 41, 46, 48, 50, 51, 54, 57, 58, 67, 68, 72, 78, 85, 86, 92-98, 100, 110-112, 114, 116, 127, 128, 136, 141, 158-161, 163, 197
Parathyroid gland 26, 30, 31, 34, 41, 50, 67, 68, 72, 78, 86, 94, 96, 98, 100, 127, 128, 136, 141, 159, 161
Parathyroid hormone (PTH) 30, 31, 34, 41-43, 47, 48, 50, 51, 53, 54, 57, 58, 65-70, 72, 73, 77, 78, 85, 92-98, 110-115, 117, 120, 121, 127, 129, 140, 141, 144, 158-161, 165, 166
Peroxisome proliferator-activated receptor (PPAR) 35, 36, 148, 179, 220
PHEX 11, 41, 43, 49-51, 58, 66, 67, 71, 73-77, 87, 162, 165
Phosphate 1, 2, 7, 10, 11, 18, 25, 28-32, 34, 35, 41-54, 57, 58, 65-73, 75, 77-79, 84-89, 97, 104, 105, 107, 108, 110-121, 129, 132, 136, 140-142, 158-166, 172, 198, 216, 217
Phosphate metabolism 85, 104, 105, 160
Phosphaturia 29, 30, 48, 51, 57, 72, 93, 107, 110, 111, 114, 120, 136, 141, 142
Phosphorus 57, 94, 107, 108, 110, 111, 113-117, 119-121, 127, 141
Pit1 139, 142
Pit2 139, 142

INDEX

Proliferation 16, 27, 96, 104, 136, 184-192, 196-198, 209-211, 214-216, 220, 223
Proteoglycan 74, 101, 171, 172, 195-197, 199, 215-217
PTH/PTHrP receptor (PTHR1) 43, 54, 97

R

Receptor tyrosine kinase 2, 183, 215
Renal phosphate excretion 29, 32, 41, 42, 48, 49, 104, 105
Rickets 7, 29, 41-47, 50-54, 66, 70, 71, 75, 76, 85, 93, 112, 158, 159, 161, 162, 165-167
ROMK 45, 100, 103-105

S

Secondary hyperparathyroidism (SHPT) 47, 48, 50, 52, 78, 85, 86, 95, 96, 111-115, 120, 121, 126, 140, 141, 161
Secreted klotho 32, 33, 100-105, 126, 128, 138, 139, 141
Sialidase 32, 33, 100, 101, 104, 105, 127, 142
Signaling 2, 3, 5, 7, 10, 15-18, 32-35, 46, 54, 75, 84, 86, 95, 104, 126, 138, 139, 146, 159, 163, 164, 171-180, 183, 184, 186-188, 190-192, 195-204, 207-209, 215, 216, 219
Sodium-dependent phosphate transporter 70
Specificity 1, 2, 7, 16, 18, 149, 184-186, 196, 199-202, 204-206, 208, 209, 215
Stages of CKD 85, 108, 113, 114, 116, 144, 145
Sucrose octasulfate (SOS) 7, 12-15, 197

T

Therapeutic 18, 45, 47, 48, 51, 87, 89, 117, 120, 126, 130, 137, 143, 145, 147-149, 158, 159, 165-167, 180, 184, 187, 192, 195, 196, 211, 214, 222-224
Thiazolidinedione (TZD) 133, 148, 218, 221
Transforming growth factor-b1 (TGF-b1) 138, 139, 146
TRPC6 104, 105
TRPV5 32, 89, 100-105
Tumor induced osteomalacia (TIO) 7, 11, 48, 49, 57, 71, 73, 78, 93, 112, 158, 161, 162

V

Vascular calcification 28, 29, 54, 87, 88, 105, 117, 126, 130, 132, 134, 136, 139, 140, 147, 149
Vitamin D 2, 10, 25, 28-31, 34-36, 41, 43, 47, 50, 52, 53, 65-69, 71-73, 75, 78, 79, 84-89, 93-97, 107, 110-114, 116, 120, 121, 127, 136, 137, 140, 141, 144, 145, 158-161, 216, 217

W

Wnt 33, 126, 129, 138, 143, 146, 187, 188, 198

X

X-linked autosomal hypophosphatemia (XLH) 11, 18, 42, 43, 46-52, 67, 71, 73, 74, 87, 112, 162, 165